BRITISH ART IN THE NUCLEAR AGE

Rooted in the study of objects, *British Art in the Nuclear Age* addresses the role of art and visual culture in discourses surrounding nuclear science and technology, atomic power, and nuclear warfare in Cold War Britain.

Examining both the fears and hopes for the future that attended the advances of the nuclear age, nine original essays explore the contributions of British-born and émigré artists in the areas of sculpture, textile and applied design, painting, drawing, photo-journalism, and exhibition display. Artists discussed include: Francis Bacon, John Bratby, Lynn Chadwick, Prunella Clough, Naum Gabo, Barbara Hepworth, Peter Lanyon, Henry Moore, Eduardo Paolozzi, Peter Laszlo Peri, Isabel Rawsthorne, Alan Reynolds, Colin Self, Graham Sutherland, Feliks Topolski and John Tunnard. Also under discussion is new archival material from *Picture Post* magazine, and the Festival of Britain.

Far from insular in its concerns, this volume draws upon cross-cultural dialogues between British and European artists and the relationship between Britain and America to engage with an interdisciplinary art history that will also prove useful to students and researchers in a variety of fields including modern European history, political science, the history of design, anthropology, and media studies.

Catherine Jolivette is Associate Professor of Art & Design, Missouri State University, USA, and author of Landscape, Art and Identity in

BRITISH ART: *HISTORIES AND INTERPRETATIONS SINCE 1700*

Series Editors:

David Peters Corbett, University of East Anglia, UK

Sarah Monks, University of East Anglia, UK

This series exists to publish new and rigorous scholarship of the highest quality on British art after 1700. Proposals will offer new bodies of research or new interpretations, ideally both, and should demonstrate a clear awareness of the proposed volume's contribution to current and wider art-historical debates.

We define British art broadly to mean art made in the British Isles or by British artists, and particularly welcome proposals which address the topic from international or comparative cultural perspectives. We also welcome proposals for intellectually ambitious studies concerning more localized areas, issues and themes within British art during this period.

Above all, we encourage proposals for books on British art which transcend the descriptive in order to offer a broader methodological and/or historiographical contribution to the discipline of art history

British Art in the Nuclear Age

Edited by
Catherine Jolivette

Routledge
Taylor & Francis Group
LONDON AND NEW YORK

First published 2014 by Ashgate Publishing

Published 2016 by Routledge
2 Park Square, Milton Park, Abingdon, Oxon OX14 4RN
711 Third Avenue, New York, NY 10017, USA

First issued in paperback 2018

Routledge is an imprint of the Taylor & Francis Group, an informa business

British Library Cataloguing in Publication Data
A catalogue record for this book is available from the British Library

The Library of Congress has cataloged the printed edition as follows:
British art in the nuclear age / edited by Catherine Jolivette.
pages cm. -- (British art: histories and interpretations since 1700)
Includes bibliographical references and index.
ISBN 978-1-4724-1276-8 (hardcover : alk. paper)
1. Art and popular culture--Great Britain--History--20th century. 2. Art and technology--Great Britain--History--20th century. 3. Art, British--20th century--Themes, motives. 4. Technology--Social aspects--Great Britain--History--20th century. 5. Cold War--Social aspects--Great Britain. I. Jolivette, Catherine, 1972- editor of compilation. II. Jacobi, Carol, author. Kind of cold war feeling.

N72.S6B75 2014
701′.050941--dc23

2014015624

ISBN 13: 978-1-138-54888-6 (pbk)
ISBN 13: 978-1-4724-1276-8 (hbk)

For Product Safety Concerns and Information please contact our EU representative:
GPSR@taylorandfrancis.com
Taylor & Francis Verlag GmbH, Kaufingerstraße 24, 80331 München, Germany.

Contents

Illustrations

Cover Poster for the Science Exhibition at the Festival of Britain, 1951. Photo © Science Museum/Science & Society Picture Library. All rights reserved.

Colour Plates

1 Peter Rose Pulham, *Ville Entière*, c. 1949–1951, oil on canvas, 130 × 95 cm, private collection © Estate of Peter Rose Pulham.

2 Francis Bacon, *Untitled (Marching Figures)*, 1952, oil on canvas, 198 × 137 cm, by kind permission of the Estate of Francis Bacon.

3 Barbara Hepworth (1903–1975), *Forms (Brown, Grey and White)*, 1941, gouache and pencil/paper, 50.3 × 35.3 cm, coll. Hepworth Estate, photo © The Hepworth, Wakefield.

4 Barbara Hepworth (1903–1975), *Sculpture with Colour (Deep Blue and Red)*, 1940, painted plaster and string on plaster base, 10.5 × 14.9 × 10.5 cm, coll. Tate Gallery, photo © Bowness, Hepworth Estate.

5 Lynn Chadwick (1914–2003), *Balanced Sculpture*, 1952, welded iron, h. 49 cm, exhibited British Pavilion, Venice Biennale, 1952, private collection.

6 Graham Sutherland, *The Origins of the Land*, 1950–1951, oil on canvas, 425.5 × 327.7 cm, Tate Gallery, London © Tate Images, © Tate, London 2013/Estate of Graham Sutherland.

7 Peter Lanyon, *St Just*, 1953, oil on canvas, 243.8 × 121.9 cm, private collection © Estate of Peter Lanyon.

8 Peter Lanyon, *Thermal*, 1960, oil on canvas, 182.9 × 152.4 cm, Tate Gallery, London © Tate, London 2013/Estate of Peter Lanyon.

9 Alan Reynolds, *Hopfields*, 1952, ink and watercolour, 19.5 × 35 cm, private collection © Sotheby's 2000/ the artist.

10 John Bratby, *Jean at the Basin*, 1955, oil on board, 121.9 × 78.1 cm, private collection, photograph National Portrait Gallery © The Estate of John Bratby/ Bridgeman Art Library.

11 John Bratby, *Jean and Still Life in front of a Window*, 1954, 122 × 108 cm, Southampton City Art Gallery © The Estate of John Bratby/Bridgeman Art Library.

12 Prunella Clough, *Bypass I*, 1960, oil on canvas, 96.5 × 112 cm, The Courtauld Gallery, London © Estate of Prunella Clough 2013. All rights reserved DACS.

Black and White Figures

Contributors

KATE ASPINALL is a freelance art historian and a consultant researcher for the Roy Lichtenstein Foundation. She recently completed her PhD at the University of East Anglia, focused on the role of drawing as a concept in early twentieth-century British visual culture. She is in the process of turning her manuscript into a monograph, titled 'The Paradox of Medium Specificity: Drawing Practice and Twentieth Century Modernism in Britain'.

ROBERT BURSTOW is Reader in History and Theory of Art at the University of Derby. After receiving his PhD from the University of Leeds, he was awarded a Post-doctoral Fellowship by the Paul Mellon Centre for Studies in British Art. He has contributed chapters to several books, including *Herbert Read: A British Vision of World Art* (Leeds: Leeds City Art Galleries in association with The Henry Moore Foundation and Lund Humphries, 1993), *Henry Moore: Critical Essays* (Aldershot: Ashgate, 2003), *Sculpture in 20th-Century Britain* (Leeds: Henry Moore Institute, 2003), *Sculpture and the Garden* (Aldershot: Ashgate, 2006) and *The History of British Art, 1870–Now* (London: Tate Publishing in association with the Yale Center for British Art, 2008). He has published essays and exhibition and book reviews in academic journals, such as *Apollo*, *Art History*, *Oxford Art Journal*, *The Sculpture Journal* and *The Journal of the Twentieth Century Society*. As a curator, he has organized several exhibitions of post-war British sculpture, including *Sculpture in the Home: Re-staging a Post-war Initiative* (Henry Moore Institute, Leeds, 2008).

FIONA GASKIN is a PhD student at the Courtauld Institute of Art in London where she is currently working on her dissertation, 'British Landscape Painting in the Cold War, 1945–1963'. Her MA thesis, 'British Tachisme in the Post War Period, 1945–1957', was published in *Artists and Patrons in Post-War Britain*, ed. Margaret Garlake, Courtauld Research Papers No. 2 (Aldershot: Ashgate, 2001). Other publications include 'Gillian Ayres's Mural of 1957 for South Hampstead High School for Girls', *The Burlington Magazine*, vol. CXLII, no. 1165 (April 2000): 236–238.

Carol Jacobi is Curator of British Art 1850–1915 at Tate Britain. She has taught at Birkbeck College, the Courtauld Institute and Oxford Brookes, and was Leverhulme Fellow in the History of Portraiture at the National Portrait Gallery in 2008. Her research champions international and interdisciplinary views of nineteenth- and twentieth-century British art. Her books include *William Holman Hunt: Painter, Painting, Paint* (Manchester: Manchester University Press, 2006) and *Out of the Cage: The Art of Isabel Rawsthorne* (London and New Haven: Yale University Press, forthcoming).

Catherine Jolivette is Associate Professor in the Department of Art and Design at Missouri State University. She received her PhD in History of Art from the Pennsylvania State University and her MA in Fine Art from Edinburgh University. Her research interests include the intersection of twentieth-century British art with developments in science and technology, the construction of landscape and the renegotiation of British national identity following the Second World War. She is the author of *Landscape, Art and Identity in 1950s Britain* (Aldershot: Ashgate, 2009); her article 'London Pride: 1951 and Figurative Sculpture at the South Bank Exhibition' was recently featured in *Sculpture Journal*; and her essay 'Science, Art and Landscape in the Nuclear Age' was published in the special issue of *Art History* 'New Approaches to British Art' in April 2012. Archival research for her chapter 'Representations of Atomic Power at the Festival of Britain' was sponsored by a research support grant from the Paul Mellon Centre for British Art in London.

Christoph Laucht is Lecturer in Modern History at Swansea University. His research interests include the cultural and social history of the atomic age (especially in Britain and the United States); the transnational history of the Cold War (in particular in Britain, Germany and the United States); and film and history. He is the author of *Elemental Germans: Klaus Fuchs, Rudolf Peierls and the Making of British Nuclear Culture, 1939–59* (Basingstoke: Palgrave Macmillan, 2012) and co-editor of *Divided, But Not Disconnected: German Experiences of the Cold War* (New York and Oxford: Berghahn Books, 2010), as well as *British Nuclear Culture*, a special issue of the *British Journal for the History of Science*, vol. 45, no. 4 (2012).

Simon Martin is Artistic Director of Pallant House Gallery in Chichester, UK. He received his MA in the History of Art from the Courtauld Institute of Art and his BA in the History of Art from the University of Warwick. He is the author of several catalogues and books including *Poets in the Landscape: The Romantic Spirit in British Art* (Chichester: Pallant House Gallery, 2007), *Colin Self: Art in the Nuclear Age* (Lausanne and Worthing: AVA, 2008), *John Tunnard: Inner Space to Outer Space* (Chichester: Pallant House Gallery, 2010), *Edward Burra* (Farnham: Lund Humphries, 2011) and *Eduardo Paolozzi: Collaging Culture* (Chichester: Pallant House Gallery, 2013), and was a contributor to the catalogue of the V&A exhibition 'British Design 1948–2012'.

Gregory Salter is post-doctoral researcher for the Documenting Homes Collection at the Geffrye Museum, London. He completed a PhD, titled 'Domesticity and Masculinity in 1950s British Painting', at the University of East Anglia in 2013. His research interests lie in post-war British art and, increasingly, the work and experiences of migrant artists in Britain in this period, particularly in terms of sexuality and nationhood.

Catherine Spencer is Lecturer in the School of Art History at the University of St Andrews. She completed her PhD, 'Fieldwork: Performing Social Science in North America, 1961–75', at the University of York in 2014, and has also taught at the University of Edinburgh. Her research has been supported by the Terra Foundation for American Art, the Arts and Humanities Research Council and the Getty Research Institute, and her writing has been published in *Art History*.

For T.

Acknowledgements

During the writing of this book we have been supported, individually and collectively, by many people. The idea for this edited volume was generated during the research forum, 'New Approaches to British Art, 1939–1969'. We are grateful to the Courtauld Institute of Art in London for hosting the conference in June 2010, to David Peters Corbett and Lisa Tickner for organizing the event, and for the support of the Courtauld Research Forum, the University of York, the University of East Anglia, the Henry Moore Foundation and the Paul Mellon Centre for Studies in British Art which enabled scholars to come together to discuss under-explored areas of twentieth-century British art. From this opportunity, and from subsequent conversations between the contributors, a book on British art in the nuclear age was able to take shape.

We are, of course, indebted to our home institutions: the Courtauld Institute of Art, the University of Derby, the University of East Anglia, the University of St. Andrews, Missouri State University, Swansea University, Pallant House Gallery and Tate Britain; and to our friends and colleagues who have supported us in ways too numerous to mention individually. Special thanks are given to Margaret Michniewicz and the editorial staff of Ashgate Publishing for their professionalism and encouragement in bringing this book to publication, to the external readers for Ashgate who have been particularly generous with their time and careful attention to the manuscript, and to the many copyright holders for their permission to reproduce the illustrations that support this text.

Catherine Jolivette

Introduction

Catherine Jolivette

This book explores the role of art and visual culture in discourses surrounding nuclear science and technology, atomic power and nuclear warfare during the first three decades that followed the end of the Second World War. While this time-frame coincides with the early to middle years of the Cold War,[1] the essays in this collection consider the visual response to both the fears and anxieties and the hopes for the future that attended the advances of the nuclear age.

The picturing of nuclear war (real and imagined) and apocalyptic themes is explored in light of artistic responses to the atomic bombing of Hiroshima and Nagasaki in August 1945; the development of the hydrogen bomb; the series of American nuclear tests at Bikini Atoll in the Pacific Ocean between 1946 and 1958, including the explosion of the first H-bomb in November 1952; and British nuclear tests Operation Hurricane off the coast of Western Australia in October 1952, and Operations Grapple, Grapple X, Grapple Y and Grapple Z in the Central Pacific from 1956 until 1958. With the Cuban missile crisis in October 1962, the threat of nuclear annihilation gained terrifying intensity, as witnessed by heterogeneous works of art that addressed such themes as personal fear, isolation, domesticity and consumer culture. The cultural contexts for the production of art vary greatly in relation to these events. While the authors of the essays in this book make reference to the Cold War period, the term 'Cold War' should not be seen as a generalized umbrella designation. The individual chapters deal with quite specific discursive territories and time-frames that are distinguished by the authors from the multiplicity of synchronous and diachronic Cold War contexts of other periods and locations during the years encompassed by this volume.

Similarly, in a book on 'British Art in the Nuclear Age' the terms 'nuclear' and 'Cold War' need not be synonymous. While the political threat of military aggression with the potential to deploy nuclear weapons forms the backdrop to many of the essays, the chapters in this collection explore a multiplicity of nuclear cultures in post-war Britain across a number of decades. For example, the positive embrace of nuclear physics is discussed in terms of popular representations of science and the concomitant iconography of nuclear technology such as atoms, diagrams, circuit boards, radio transmitters and crystals. Of these, the science of crystallography (in which Britain led the field) offered a particularly fertile source of inspiration for artists, especially during the 1940s and 1950s, and is explored in several of the essays in this book.

Informed by knowledge of the properties of the atom, shifts in scale, from micro to macro, are a device that can be seen in the work of a number of artists and designers of this period, and in the representation of a variety of nuclear-inflected landscapes, whether industrial, rural or military. Cartography and the mapping of boundaries (including East/West, communism/capitalism, country/city), along with Cold War themes of scrutiny and surveillance, wastelands and ruins, are investigated in this book. In a groundbreaking new way, nature and nuclear are read not as polarized opposites but as part of a common, historically specific, culture in which the visual forms, rather than merely reflects, the discourses of the age.

Atomic power as a source of energy is a topic that is explored, with a particular focus on the years leading up to the operation of the world's very first nuclear power plant which went online at Calder Hall in Cumberland in 1956. A range of responses to this new form of power can be charted, from a cautious hope of meeting the energy needs of the future – as was articulated at the Festival of Britain in 1951 – to heightened public concern following the radioactive discharges from a catastrophic fire at the Windscale reactor in 1957, to the celebration of British achievement in the technology of nuclear power at the British engineering stand at the Brussels World Fair in 1958.

While nuclear themes continued to influence British artists throughout the Cold War period and its aftermath, with atomic power and nuclear weapons providing inspiration for art produced in Britain and elsewhere well into the twenty-first century,[2] the primary focus of this volume is art from the 1940s through the 1960s. Works made during this time correspond to nuclear developments during the period from the end of the Second World War, through the launch of the British Campaign for Nuclear Disarmament in 1958, up to the Limited Test Ban Treaty that was signed in Moscow by representatives of the United States, the Soviet Union and Great Britain on 5 August 1963.[3] It is clear that the concerns of the Cold War would continue to preoccupy artists such as Colin Self, who would go on to produce some of his most celebrated investigations of nuclear subjects, including his series of *Fall Out Shelter* drawings after a trip to the United States in 1965, and scientists like Jacob Bronowski, who would explore the ethics of the nuclear age in writings and broadcasts up until his death in 1974. It is hoped that this book will prompt other historians to further explore the intersection of British art with nuclear subject matter through the 1970s, 1980s and beyond.

Rooted in the study of objects, this volume examines the contributions of British-born and émigré artists in the areas of sculpture, textile and applied design, painting, drawing, photojournalism, poster design and exhibition display. Works encompass figuration and abstraction, pop art, the Kitchen Sink painters, constructivism, social realism, expressionism and surrealism, drawing upon cross-cultural dialogues between British and European artists and the relationship between Britain and the United States of America. Far from insular in its outlook, this book engages with an interdisciplinary art history that examines the visual in terms of the very real and ineluctable significance of nuclear concerns to every area of life.

The compilation comprises nine essays by scholars from Britain, Europe and the United States that present new research on artists including: John Armstrong, Francis Bacon, Derek Boshier, John Bratby, Lynn Chadwick, Geoffrey Clarke, Prunella Clough, William Crosbie, Naum Gabo, Barbara Hepworth, Peter Lanyon, Bernard Meadows, Gustav Metzger, Henry Moore, Eduardo Paolozzi, Peter Laszlo Peri, Peter Rose Pulham, Isabel Rawsthorne, Alan Reynolds, Colin Self, Graham Sutherland, Feliks Topolski and John Tunnard. Also under discussion is new archival material from *Picture Post* magazine, and the Festival of Britain.

Towards a History of British Art in the Nuclear Age

Despite a wealth of visual material and archival sources, no historian has produced an extensive critical study of artistic responses in Britain to developments in nuclear technology and the threat of atomic war, even though the material record shows that such issues were at the forefront of public consciousness during the period under discussion.[4]

While historians have begun to consider the embrace of atomic imagery in applied arts and design in Europe and America, corresponding investigations in British art remain under-explored and works voicing anxieties about scientific discoveries have been largely ignored. This project investigates British art that addresses both the positive and negative aspects of this new science and technology, in essays by scholars from visual culture studies, the history of art and design, political history and other disciplines working on British art in the atomic age. The idea for this edited volume was generated during the research forum 'New Approaches to British Art, 1939–1969' that was held at the Courtauld Institute of Art in London in June 2010,[5] and in subsequent conversations between a number of the contributors. Further authors were brought into the project due to their work on Cold War cultures, and at various junctures the writers gathered together to share ideas and resources, and to respond to each other's research. It became clear to us that the exploration of British art in the nuclear age was an exciting and neglected area of twentieth-century art history.

The influence of atomic culture on the world of design is a topic that has generated considerable interest in recent years, with exhibitions on both sides of the Atlantic. Two examples include the exhibition at the Wellcome Trust, London, 'From Atoms to Patterns' (April–August 2008), an exhibition of avant-garde design based on x-ray crystallography translated into 1950s designs for fabric, furniture, ceramics and house wares;[6] and the exhibition at the Victoria and Albert Museum, London, 'Cold War Modern: Design 1945–1975' (25 September 2008–11 January 2009) which considered artistic responses to the nuclear threat as seen through objects from both sides of the Iron Curtain, including graphics, fine art, film and imaginary schemes such as Buckminster Fuller's 1962 geodesic dome over Manhattan.[7]

While richly illustrated exhibition catalogues have been produced, such as Brooke Kamin Rapaport and Kevin Stayton's *Vital Forms: American Art and Design in the Atomic Age, 1940–1960*,[8] surprisingly little critical art historical attention has been paid to the responses of painters, sculptors and other fine artists (as opposed to designers of applied arts), or to the particular concerns of British artists regarding nuclear threat.

On the topic of art during the Cold War, Serge Guilbaut's *How New York Stole the Idea of Modern Art: Abstract Expressionism, Freedom, and the Cold War*[9] has become a seminal reference, together with Eva Cockcroft's essay for *Art Forum*, 'Abstract Expressionism: Weapon of the Cold War'.[10] Christine Lindey has explored the lesser known subject of modern Russian art in *Art in the Cold War: From Vladivostok to Kalamazoo 1945–1962*,[11] while David Caute has authored an ambitiously broad historical survey of cultural competition (in dance, music, art, theatre and film) in the artistic centers of London, Paris, New York, Moscow and Berlin in his book *The Dancer Defects: The Struggle for Cultural Supremacy During the Cold War*.[12] No scholar, however, has produced a full-length study that examines the direct effect of Cold War politics on British art during the same period, despite the publication of important essays such as Robert Burstow's 'The Limits of Modernist Art as a "Weapon of the Cold War": Reassessing the Unknown Patron of the Monument to the Unknown Political Prisoner'.[13]

What is British Nuclear Culture?

In June 2010 a conference, 'British Nuclear Culture: Themes, Approaches and Perspectives', was held at the School of History at University of Liverpool to explore the unique nature of nuclear culture in twentieth-century Britain. Co-edited by Christoph Laucht (whose essay on *Picture Post* magazine appears in this volume) and Jonathan Hogg, a special issue of the *British Journal for the History of Science* on British nuclear culture based on these papers subsequently appeared in December 2012. While the Liverpool conference addressed such cultural forms as fiction, cartoons, films and music, this book extends the discussion further to include the discourses surrounding fine art in mid-twentieth-century Britain.

As the organizers of the Liverpool conference have acknowledged, since the publication of Paul Boyer's *By the Bomb's Early Light: American Thought and Culture at the Dawn of the Atomic Age*[14] in 1985, academic discussions of nuclear culture have largely focused on the United States; thus a British context has, to this point, been less clearly theorized. In the December 2012 issue of the *British Journal for the History of Science* the historiography of so-called 'British nuclear culture' is critically examined in a think-piece by Jeff Hughes of Manchester University titled, 'What is British Nuclear Culture? Understanding *Uranium 235*',[15] an essay which problematizes the generality of this term while questioning the assumptions behind each of its three constituents. Hughes argues that writers on 'nuclear culture' have tended to essentialize the nuclear to the detriment of historical analysis, reifying it rather than explaining it.[16]

Rather than a monolithic treatment of a homogenous or isolated 'nuclear culture', the collection of essays in this volume explores a variety of nuclear cultures in post-war Britain. While several essays explore imagery from the early post-war period to the early 1970s, others are contingent upon a specifically narrow time-frame. Discussing art made during the years 1945–1950, Carol Jacobi examines the early Cold War period and the aftermath of the atomic attacks on Japan, while Christoph Laucht's account begins in the same year and follows the photojournalism of *Picture Post* magazine through the development of the H-bomb and atomic energy, up to 1957 when the magazine closed. Catherine Jolivette takes on the specific case of representations of nuclear science and atomic power at the Festival of Britain during the summer of 1951, a time when British developments of nuclear weapons were shrouded in secrecy yet the prospect of nuclear-generated electricity was promoted as a solution to the country's growing energy needs. On the cusp of the full international realization of thermonuclear capabilities, the Festival stands at a historically specific, politically charged moment: the same autumn would see the electoral defeat of the post-war Labour government, within six months the ailing king was dead, and within a year the nation would become the world's third nuclear power. Moving beyond the 1950s, Simon Martin's chapter includes a discussion of painter Colin Self's well-documented response to the Cuban missile crisis and his subsequent trip to the United States in 1965, while Kate Aspinall looks at questions that attended developments in nuclear physics from the perspective of polymath Jacob Bronowski and artist Feliks Topolski in their collaboration on the 'Knowledge or Certainty' segment of the 1973 BBC television series *The Ascent of Man*. This project, realised almost three decades after the attacks on Hiroshima and Nagasaki, is compared to their respective contributions to the Festival of Britain, in which each encouraged subjective judgement as an anathema to unreflective technocracy. Across these chapters, 'nuclear' is read as having many different meanings and clearly has different connotations depending on whether the context under discussion is 1945, 1952 or 1963.

Claiming that 'Historians have tended to be in thrall to the *aura* of the nuclear and its practitioners',[17] Jeff Hughes has also questioned the national frameworks that have become what he calls the 'natural unit of analysis' in nuclear history. While this book does indeed (as the title suggests) examine visual representations of various nuclear themes within a national context, neither the art nor the issues are treated as insular. The essays explore a range of exchanges between British and European artists, including British artists abroad and émigré artists working in Britain, and the significance of the cultural and political relationship between Britain and America (not to mention the mobility of British scientists engaged in international ventures such as the Manhattan Project). Carol Jacobi investigates the interchanges between French and British artists during the early post-war period, while Christoph Laucht acknowledges the aesthetic debt that the British publication *Picture Post* had to German illustrated papers, including the appointment of German-speaking émigré Stefan Lorant, former editor-in-chief of the *Münchner Illustrierte Presse*, as its first editor. *Picture Post* itself published photojournalistic representations that were international in scope, and Laucht discusses how, through this medium, nuclear developments in Japan, Britain and the United States were communicated to a British audience. The protagonists in Kate Aspinall's study, Jacob Bronowski and Feliks Topolski, were both born in Poland, while Robert Burstow explores the contributions of émigré artists to the British art scene in his discussion of the sculptures of Russian Naum Gabom and Hungarian Peter Laszlo Peri, and Simon Martin investigates the 'Auto-Destructive Art' of Gustav Metzger who was born in Germany to Polish Jewish parents. Through these essays we see that discourses of the 'nuclear' in Britain were part of a broader international conversation, and that concepts of 'Britishness', 'East' and 'West', 'self' and 'other' resist any singular definition.

Citing Raymond Williams, Hughes notes that 'culture is one of the two or three most complicated words in the English language'.[18] Williams himself reminds us not only that 'culture' refers to high art, which in visual art might include painting, sculpture, prints, drawings, fine art photography, and so on, but also that 'culture is ordinary'[19] in terms of both the experiences of daily life in the lived environment and the enjoyment of readily accessible mass-circulated forms of cultural signification such as magazines and television. A consideration of British art in light of nuclear technologies under development during this period would explore both what Williams calls the 'ordinary common meanings' and the 'finest individual meanings'.[20] The essays in this book consider both fine art and popular representations of nuclear themes in print and broadcast media, including *Picture Post* magazine, the BBC television series *The Ascent of Man*, and the very heterogeneous representations of the 'nuclear' in the murals and exhibition displays at various locations throughout the Festival of Britain. Contributors to this volume consider the creation, reception and consumption of these images, looking at visual representation from the perspectives of the different ideological groups engaged with its meaning: the civil servants charged with organizing the displays at the Festival of Britain; the publishers of *Picture Post* magazine; popular scientists such as Jacob Bronowski who felt a level of ethical responsibility to communicate both the discoveries and philosophy of science to the lay public; and individual artists such as John Bratby and Prunella Clough. The authors demonstrate how visual art does not represent or reflect but rather contributes to and constitutes discourses of the nuclear during the period under discussion.

The Essays

When asked about the exhibition *This is Tomorrow*, held at the Whitechapel Gallery in London in 1956, Eduardo Paolozzi refuted the idea that the show had launched the British Pop movement: 'there was a sort of basic art. And I think we had this shadow of, perhaps it might have been the atom bomb and things like that; there was a kind of cold war feeling about art … it was more Art Povera than Pop art.' In her essay, '"A Kind of Cold War Feeling" in British Art, 1945–1952', Carol Jacobi retraces this proto-Povera art, exploring the dialogue between British and French outsider painters and sculptors whose maverick personae have obscured a moment of crisis/community in the formative years between 1944 and 1950. Jacobi's investigation of new primary sources brings into view a confluence of Francophile British artists forced back to London by the Occupation, including Peter Rose Pulham and Isabel Rawsthorne, and up-and-coming painters such as Francis Bacon and Eduardo Paolozzi. Their association in Paris after the Liberation, a moment when London and Paris were close as political allies, coincided with the testing, implementation and aftermath of the atomic bomb. The work that resulted was a clearly conceived alternative to British Neo-Romanticism that did not idealize nature. Its iconography and its means of communication implemented the violent, 'convulsive' quality being simultaneously explored by French artists and writers. The painters and sculptors of this chapter were neither established nor official war artists; they had spent the conflict in other roles, military service, fire fighting, intelligence, agriculture, crime, and so on. This inflected and expanded their post-war vision and made them sensitive to a 'cold war feeling'. They were alert to the visual possibilities of popular screen and print media and everyday objects, as well as specialist literary, philosophical, scientific, anthropological, political and even military material. This chapter shows how heterogeneous Cold War thinking might inform new readings of an anti-ideal strand neglected in British post-war art, originating in the mid 1940s and continuing up to the 1970s.

Robert Burstow's chapter, 'Geometries of Hope and Fear', investigates the ways in which nuclear science and technology figured in a variety of sculptural forms in early Cold War Britain. Burstow shows how, from the 1930s, the constructivist sculptors Barbara Hepworth and Naum Gabo embraced atomic science, encouraged by contact with the crystallographer J.D. Bernal. Through works alluding to the geometry of crystal structures, they signified optimistic hopes for increased human understanding in a Socialist society. The chapter goes on to examine how the creation and use of atomic weapons led surrealist, social-realist and Pop sculptors to make critical works about nuclear science and technology. From the ambivalent to the satirical, works by Henry Moore, Eduardo Paolozzi, Peter Laszlo Peri and Colin Self exemplify a range of sculptural representations of nuclear arms and the disarmament campaign (in which several sculptors and critics were active). Lastly, the chapter considers the extent to which the expressionist sculpture of Reg Butler, Lynn Chadwick, Geoffrey Clarke, Bernard Meadows and others has also been understood to reflect fears of nuclear warfare, despite an absence of explicit 'nuclear' signification. As this now familiar interpretation of their imagery has invariably been supported by Read's famous characterization of it as 'the geometry of fear', the chapter particularly interrogates the intended meaning of his epithet and how and when it became associated with the nuclear threat. Throughout the chapter, sculptors' and contemporary critics' explanations of these disparate formal and iconographical engagements with nuclear science and technology are scrutinized, alongside analysis of how they related to the aesthetic and ideological oppositions of the Cold War.

Christoph Laucht's essay, 'An Imagined Cataclysm Becomes Fact', examines how *Picture Post*, Britain's leading illustrated magazine, depicted real and imagined nuclear war through its combination of images and text. *Picture Post* has often been compared to the American *Life* magazine and heralded for its social commentary, especially its photography. Through its combination of images and texts, it helped to shape many Britons' views of atomic energy before television became a mass medium. This chapter focuses on two areas: first, it investigates the magazine's coverage of the atomic bombings of Hiroshima and Nagasaki. Here it examines visual and textual representations of real nuclear war. Second, the chapter looks at how *Picture Post* imagined a future nuclear war. In this part, particular attention is paid to the hydrogen bomb that moved the nuclear age in a much more destructive direction. The period under investigation runs from the magazine's 'Atom Age' issue in August 1945 – which informed readers for the first time about the dangers and prospects of atomic energy – until June 1957 when *Picture Post* folded.

Catherine Jolivette explores some of the different ways that visual displays were used to tell the story of British nuclear physics at the Festival of Britain in 1951. Exhibits and works of art from various Festival sites are examined in terms of the discourses that they engaged concerning both the potential (for better or for worse) of nuclear energy, and the impact of the atom on both the artistic output and popular imagination of the era. Within the nationwide programme of Festival events there were multiple exhibitions that showcased nuclear science: in London these included the *Exhibition of Science* at the Science Museum in South Kensington, a display on nuclear physics as part of the *Physical World* section of the *Dome of Discovery* at the *South Bank Exhibition*, and the *Growth and Form* exhibition at the Institute of Contemporary Arts (ICA); while in Scotland, Glasgow's Kelvin Hall played host to the *Exhibition of Industrial Power*. Works considered include designs by the Festival Pattern Group; John Tunnard's crystallographic mural for the South Bank's *Regatta Restaurant*; Richard Hamilton's installation at the ICA; Laurence Scarfe's *Atomic Energy* mural in the *Dome of Discovery*; and William Crosbie's atomic mural for Glasgow's *Hall of the Future*.

Fiona Gaskin's essay, 'The *Genius Loci* of Cold War Britain', describes how, during the early Cold War, the proliferation of nuclear capability between the superpowers created widespread feelings of vulnerability, anxiety and despair that found expression in a number of art forms from the late 1940s through the 1960s. While the hybrid chimeras and 'iron waifs' of the sculpture that was categorized by Herbert Read as belonging to the 'geometry of fear' produced some of the most celebrated examples of early Cold War imagery in British art, there existed a less well-known and less frequently analysed parallel in landscape painting. Within this genre, the mutant forms of sculpture found echoes in disquieting 'personages' constructed from organic elements. These troubling landscapes were often given disturbing and sometimes violent anthropomorphic qualities that read as metaphors for the nuclear threat. Gaskin's chapter looks at how the concept of *genius loci* – an idea already familiar in British art through surrealism and Neo-Romanticism – was adapted to the purpose of expressing the fears and anxieties of the Cold War, concentrating on the work of three artists that she identifies as representative of this tendency: Graham Sutherland, Peter Lanyon and Alan Reynolds.

Existing scholarship on the British artist John Bratby focuses on his position as one of the so-called 'Kitchen Sink Painters' in the 1950s, and particularly the group's connections to contemporary art criticism and politics. Gregory Salter's chapter, 'Cold War at Home', approaches Bratby's art differently, investigating instead its connections to discourses of the family in post-war Britain and Cold War anxieties towards the nuclear bomb. Connections between the family and the Cold War in Bratby's art are explored through the lens of contemporary psychoanalysis – in particular, the theories of Melanie Klein and D.W. Winnicott which were particularly influential at the time. In this context, the post-war family becomes a site of intense focus from the outside while also appearing to be constantly on the brink of conflict on the inside. Bratby's painted world of enclosed spaces and tense, often explosive relationships with his family lends itself to an interpretation built on contemporary psychoanalysis. With its half-obscured views from windows, cramped spaces and piles of food, it also presents the home as a shelter from the threats of outside – Bratby even imagines the effects of radiation on himself in a self-portrait. Drawing parallels with contemporary literature, this chapter traces the connections between anxieties about the family and nuclear war to present a fresh consideration of Bratby's art.

Prunella Clough's paintings from the 1950s and 1960s have received little critical attention: what exists focuses on Clough's shift from figuration to abstraction. Catherine Spencer's essay, 'Covert Resistance', offers a thematic re-assessment of Clough's work, arguing that her experience as a cartographer for the Military Railways division of ETOUSA (European Theater of Operations, United States Army) during the Second World War resurfaces during the 1950s and 1960s as a means of expressing anxieties about the boundaries – between East and West, communist and capitalist – invoked by the Cold War, the ambivalent ends of scientific development, and the shadow cast by nuclear threat. The map-like arrangement of paintings such as *Electrical Landscape* (1960) and *Electrical Installation* (1959) traverse what Clough called the 'urbscape'. These zones between the countryside and city proper, encompassing the manufacturing estate, factory and warehouse, disclose an artistic sensibility rooted in the intersecting industrial and military landscapes of post-1945 Britain. Clough's urbscapes deploy map forms to concatenate multiple references – circuit boards, ciphers, wires, blueprints, diagrams, cells, crystals and atoms – reconfiguring the technocratic 'white heat' rhetoric of Harold Wilson into a nuanced mediation on scientific advance. While Clough's shifts between macro and micro denote the impact of microscopy, her use of the bird's-eye view expresses paranoid scrutiny, and the pulsing iridescence of images like *Untitled* (1965) speaks menacingly of radiation and mutation. Linking Clough's Cold War 'urbscapes' with the British response to Abstract Expressionism, this chapter argues that the neglect of these images is symptomatic of what Barry Curtis calls the 'subdued and repressed' signs of Cold War conflict in Britain.

Kate Aspinall's chapter, 'The Aesthetic of Scientific Authority in a Nuclear Age', discusses how celebrity historian of science Jacob Bronowski interpreted Polish émigré artist Feliks Topolski's use of line as an analogy for science's use of subjective judgement in the 'Knowledge or Certainty' segment of Bronowski's 1973 BBC television series *The Ascent of Man*. The culmination of Bronowski's views on the relationship of scientific authority and epistemological method, the episode examines the ethics of interpreting scientific judgement through Topolski's particular dynamic aesthetic. As a visual embodiment of Bronowski's ideal epistemological approach, Topolski's drawings become instructive rather than merely aesthetic – helping a public removed from scientific realities to conceptualize the language and significance of subjective (and thus fallible) judgement as a possible alternative to the primacy of technology in decision making. Aspinall demonstrates how the seeds of Bronowski and Topolski's collaboration, which was formulated around the hotly debated 25-year anniversary of the atomic bombings in Japan, can be traced back to the developing paradoxes for nuclear scientists as authorities in the 1950s, and to Bronowski's decades-long engagement with these discussions.

Finally, in his essay, 'Painting the End', Simon Martin explores the response of artists including Derek Boshier, Richard Hamilton, Gustav Metzger, Colin Self and John Tunnard to the Cuban missile crisis and to the threat of nuclear annihilation in the 1950s and 1960s. The work of artists such as John Tunnard (who explored the nuclear threat from the more abstract point of view of science and technology, rather than its impact on the individual) is contrasted with that of the more figurative response of the pop artists, in particular Colin Self whose work was to do with personal fear and reflected consumer culture. These British-based artists are considered in relation to the response of artists in the United States and Europe to the atom bomb.

Nuclear Doubt

The essays in this book examine a complex and varied range of attitudes towards nuclear weaponry and atomic power. Responding to Spencer Weart's expression, *Nuclear Fear*,[21] this collection might perhaps be viewed in terms of what Catherine Spencer calls 'nuclear doubt', a phrase that suggests not just the suspicions and misgivings of the Cold War period but, more broadly, a lack of certainty.

In her essay for this volume, Spencer describes the paintings of Prunella Clough as employing 'strategies of covert resistance', in which they refuse to 'dictate what is seen'. Works like *Bypass I* (1960) resemble maps and yet, Spencer argues, they 'deploy cartographic language only to destabilize it'. Unlike military maps, which define boundaries and locate targets, Clough's paintings refuse to classify their subjects or grant surety about what is seen. This is what art can do: to the discourses of the nuclear age the contribution of the artist brings a moment of hesitation, it troubles the picture and destabilizes the viewing perspective. Paintings present an understanding of the world as intertwined with subjective, variable experience, rather than a set of markers that can be easily followed. Whereas maps are constructed as symbols of authority,[22] art allows for ambiguity, heterogeneity and a questioning of received knowledge.

In terms of art history, too, aporia might provide a useful entry into the readings of British art that responds to developments in nuclear technology. Across and within the essays in this collection, there is no singular interpretation or definition of British nuclear culture. Like Spencer's discussion of Clough, Fiona Gaskin's essay on the metamorphic landscape paintings of Graham Sutherland, Peter Lanyon and Alan Reynolds offers dislocations and a shifting rather than a fixed sense of terrain. Changes in scale, from the macrocosmic to the microcosmic, are explored, as they are also examined in Catherine Jolivette's consideration of the Festival of Britain's *Exhibition of Science* and the *Growth and Form* show held at London's ICA in 1951. This telescopic view is pinpointed by Spencer in her discussion of Hannah Arendt's account of the Sputnik launch of 1957, detailed in the book, *The Human Condition*.[23] As Spencer explains, Arendt saw the new perspectives offered by technology as providing only an illusory enhancement of human understanding promised by science; advances that were paradoxically accompanied by a reduced ability to control the consequences of our actions.

The idea of perspective, or viewpoint, is a theme that recurs in a number of chapters. Spencer demonstrates how Clough's paintings explore the aerial views of the decimated cities of Hiroshima and Nagasaki. These images are also discussed by Christoph Laucht in his essay on the photojournalism of *Picture Post* magazine, which, in addition, often featured close-up views of the destruction and concentric circles mapping the effects of blast, heat and radiation. Such photographs provide documentary accounts while, as Spencer notes, situating the viewer in the place of the bomber. It could also be argued that, beyond the role of the disconnected spectator, or detached consumer, the viewer is placed in an ethical position vis-à-vis the potential or actual devastation represented. Such a perspective demands what role, what responsibility, does the viewer have concerning recent atomic (and potential future thermonuclear) attacks?

This is the question asked in Kate Aspinall's chapter that explores the BBC television collaboration between scientist Jacob Bronowski and artist Feliks Topolski. Bronowski's episode, 'Knowledge or Certainty', sought to consider the 'viewpoint' of the informed, ethical citizen. Aspinall shows how, on a visit to Japan, Bronowski – a scientist – realized his humbling inability to visualize the ramifications of atomic science;[24] from then on he asserted that 'the individual needed not only to understand mathematical and scientific language, but also to be able to visualize this understanding'.[25] Art, for Bronowski, served as an aid for conceptualizing abstract ideas and brought a human face to the nuclear age (quite literally in the case of his partnership with Topolski.) Similarly, in *Picture Post*, contemporary photographs of individual survivors from the United States' Castle Bravo thermonuclear test in 1954 in the South Pacific brought home the human cost of the development of the H-bomb as images of the crew of the Japanese fishing trawler, the *Lucky Dragon*, communicated the misery of radiation sickness. Elsewhere, the visual was also able to bring together the realms of both imagined and actual threats, as in civil defence 'imagineering'[26] of the impact of projected future nuclear strikes.

Bronowski's television series, *The Ascent of Man*, stands at the apogee of the 'authored documentary' genre that included Sir Kenneth Clark's *Civilisation* series (BBC, 1969) and John Kenneth Galbraith's *The Age of Uncertainty* (BBC, 1977). In this format, a series of programmes, led by a charismatic presenter, gave a personalized view of history that married scholarly content and accessible, informative entertainment. Bronowski, who had also been an advisor for the *Exhibition of Science* at the Festival of Britain,[27] maintained a lifelong commitment to the popular presentation of science and firmly believed that the public had a duty to inform themselves about developments in nuclear technology. By the period under examination in this book, the scientist had come to be seen as an authority, especially in the face of what seemed to many as the inaccessible, and even frightening, world of nuclear physics. Bronowski, as educator and translator, encouraged wider levels of public literacy in scientific terminology and saw art as a mediator of this information.

Even more importantly, as Aspinall demonstrates, Bronowski saw artistic process as a metaphor for the cultivation of judgement that was the responsibility of both the scientist and the public at large. Artists and scientists, in Bronowski's view, shared 'not only values but a creative process' in which judgement was arrived at by witnessing the world, by recording subjective encounters and by synthesizing those experiences.[28] Art – as in Bronowski's use of Topolski – is 'instructive rather than merely aesthetic', with the artistic process itself (like that of the pursuit of scientific knowledge in the nuclear age) 'human, fallible, and yet insightful'.[29] Through art, 'authority' and 'uncertain knowledge' can go together, just as doubt – as a strategy of covert resistance – might mitigate nuclear fear.

Nuclear Art as Critical Matter

The cultural importance of the 'nuclear age' for Britain (as opposed to the well-documented analysis of nuclear cultures in the United States) is a topic that is under-explored. In their introduction to the December 2012 special issue on 'British Nuclear Culture' for the *British Journal for the History of Science*,[30] Jonathan Hogg and Christoph Laucht point to an expansion in British nuclear scholarship over the past decade that has coincided with a renewed interest in the cultural studies of the Cold War era, facilitated by the continued release of information under the 2005 Freedom of Information Act and the ever-improving accessibility of the digital archive.[31] As Hogg and Laucht have observed, it was Kirk Willis who, almost 20 years ago, invented the term 'British nuclear culture' to explore the relationships between atomic scientists and the public in the early twentieth century.[32] Since then, the scope of this 'nuclear culture' available for study has extended to encompass cultural documents relating to medicine, horticulture, food production, military technology and atomic energy, to name but a few of the multiple fields transformed by the application of nuclear science.[33] What emerges, in Britain and elsewhere, is a diversity of nuclear cultures engaged not only with the subject of nuclear weapons but with a variety of questions addressing not just military implications but also civilian concerns. As the essays in this volume demonstrate, the consequences of living in the nuclear age touched every area of life and raised issues that intersect with topics of domestic and foreign policy, economics, civil defence, gender, ethnicity, race and class.[34]

Art historians have much to contribute to the historical discourse by unpacking the visual document.[35] In addition to painting and sculpture, visual narratives stand alongside written texts in civil defence newsreels, posters and pamphlets; in print journalism and in the popular transmission of scientific information through authored television documentaries; and through exhibition displays such as those at the Festival of Britain. While the essays in this book offer a tightly chronological thematic account of British art during the middle years of the twentieth century, the authors also expose the visual material to scrutiny that lifts the works out of a rigidly time-based context through the employment of critical methodologies drawn from disciplines including psychology, anthropology and sociology.

In considering the role of the artist in discussions of nuclear culture, Simon Martin opens the concluding chapter of this volume with a question asked by *Time* magazine in 1952: 'How should a modern artist react to the atomic age?'[36] This query, central to the discussion of the visual material examined in this book, was posed in a review of contemporary Italian art on display at the Venice Biennale but the moral attention that it demanded could be extended to all visual artists working between 1945 and 1975. As Martin has observed elsewhere, while there had been official war artists for both World Wars, during the Cold War, which was more a 'war of ideologies', the psychological strain of living under nuclear threat was not something that could be easily represented, nor was it an aspect of British life that official channels wished to endorse.[37] While photojournalism recorded the devastation of the atomic attacks on Hiroshima and Nagasaki, and the iconic mushroom clouds of the nuclear tests of the 1950s, there was relatively little in the way of visual art as overt political propaganda. The role of the artist in the case studies explored within these pages is to navigate the nuclear age and to construct meaning through images. Art makes tangible and visible the 'nuclear'; from atomic formations on a microscopic scale to the macroscopic potential of atomic war. It communicates scientific ideas, and brings to them human concerns. It is able to give shape to an iconography of hope for the future, just as it has the ability to represent amorphous fears. As Kate Aspinall reminds us in her essay on artist Feliks Topolski and polymath scientist Jacob Bronowski, the 'two cultures' debate over the supposed rift between the arts and the sciences,[38] launched in 1959 by the Cambridge Rede lectures of physicist and novelist Charles Percy Snow and later countered by literary critic and academic Frank Raymond Leavis, did not necessarily present an insuperable breach. For thinkers like Bronowski, she asserts, there were not 'two cultures', but only one. It is in the spirit of his creative collaboration across disciplines that this volume is written.

Notes

1 The Cold War refers to the period of military tensions and international political hostilities (rather than direct military action which, due to the possession of nuclear weapons, could lead to their Mutual Assured Destruction) between the USSR and the United States of America and its allies in Western Europe. Although the dates for the beginning of the Cold War are often contested, most historians trace its origin to the period immediately following the end of the Second World War, with a denouement following the collapse of the Berlin Wall in 1989 and the dissolution of the Soviet Union in 1991.

2 For example, events and installations staged in London by Arts Catalyst and artists Chris Oakley, and Simon Hollington and Kypros Kyprianou with BANG,

the British Atomic Nuclear Group: see <http://www.artscatalyst.org/images/uploads/files/Nuclear_Press_Release.pdf>. See also the nuclear art forum hosted in 2008 by the Royal Society of Arts. Outside of Britain, Japanese artist Isao Hashimoto has produced a series of works on nuclear weapons, including a video which he began in 2003 titled '1945–1998' that charts the worldwide history of nuclear tests and explosions.

3 The text of the *Treaty Banning Nuclear Weapon Tests in the Atmosphere, in Outer Space and Under Water* can be found on the US Department of State website at <http://www.state.gov/t/isn/4797.htm>.

4 Two significant scholarly publications to consider the broad sweep of British art produced during the 1950s and 1960s are: James Hyman, *The Battle for Realism: Figurative Art in Britain During the Cold War, 1945–1956* (New Haven and London: Yale University Press, 2001); and Margaret Garlake, *New Art, New World: British Art in Postwar Society* (New Haven and London: Yale University Press, 1998). While Hyman's work does discuss the threat of nuclear war, his overwhelming concern is with figuration, thus scientifically informed imagery and landscapes play only a subsidiary role. Garlake's text provides a dense and detailed overview of the period through a collection of themed essays rather than a focused study on a single iconological issue. The specific influence of technological developments and atomic age fears on representations of a new and unstable environment remains an area of study for the most part unexplored.

5 Many thanks to David Peters Corbett and Lisa Tickner for organizing this conference.

6 See Lesley Jackson, *From Atoms to Patterns: Crystal Structure Designs from the 1951 Festival of Britain* (Yeovil: Richard Dennis, 2008).

7 David Crowley and Jane Pavitt, eds, *Cold War Modern: Design 1945–1970* (London: Victoria & Albert Museum, 2008).

8 Brooke Kamin Rapaport and Kevin Stayton, eds, *Vital Forms: American Art and Design in the Atomic Age, 1940–1960* (Brooklyn: Harry N. Abrams, 2001).

9 Serge Guilbaut, *How New York Stole the Idea of Modern Art: Abstract Expressionism, Freedom, and the Cold War* (Chicago: University of Chicago Press, 1983).

10 Eva Cockcroft, 'Abstract Expressionism: Weapon of the Cold War', *Art Forum*, vol. xii, no. 10 (1974): 38–41.

11 Christine Lindey, *Art in the Cold War: From Vladivostok to Kalamazoo 1945–1962* (London: New Amsterdam Books, 1990).

12 David Caute, *The Dancer Defects: The Struggle for Cultural Supremacy During the Cold War* (Oxford: Oxford University Press, 2003).

13 Robert Burstow, 'The Limits of Modernist Art as a "Weapon of the Cold War": Reassessing the Unknown Patron of the Monument to the Unknown Political Prisoner', *Oxford Art Journal*, vol. 20, no. 1 (Spring 1997): 68–80.

14 Paul Boyer, *By the Bomb's Early Light: American Thought and Culture at the Dawn of the Atomic Age*, 2nd edition (Chapel Hill: University of North Carolina Press, 1994). First edition published in 1985.

15 Jeff Hughes, 'What is British Nuclear Culture?' *British Journal for the History of Science*, vol. 45, special issue (December 2012): 495–518.

16 Ibid., 496.

17 Ibid.

18 Ibid., 495.

19 The title of his well-known essay, written in 1958. See Raymond Williams, *Resources of Hope: Culture, Democracy, Socialism* (London: Verso, 1989), 3–14.

20 Ibid.

21 See Spencer Weart, *The Rise of Nuclear Fear* (Cambridge, MA: Harvard University Press, 2012); Spencer Weart, *Nuclear Fear: A History of Images* (Cambridge, MA: Harvard University Press, 1988).

22 See Denis Wood and John Fels, *The Natures of Maps: Cartographic Constructions of the Natural World* (Chicago and London: University of Chicago Press, 2008), 7.

23 Hannah Arendt, *The Human Condition* (Chicago: University of Chicago Press, 1958).

24 Jacob Bronowski, *Science and Human Values* (London: Hutchinson & Co., 1961), 13–14.

25 See Kate Aspinall, Chapter 8: 'The Aesthetic of Scientific Authority in a Nuclear Age: Jacob Bronowski and Feliks Topolski'.

26 See David Monteyne, *Fallout Shelter: Designing for Civil Defense in the Cold War* (Minneapolis: University of Minnesota Press, 2011), 21–33.

27 For an account of popular science at the Festival of Britain, see Sophie Forgan, 'Festivals of Science', *The British Journal for the History of Science*, vol. 31, no. 2 (June 1998): 117–240.

28 See Aspinall, 'The Aesthetic of Scientific Authority in a Nuclear Age'.

29 Ibid.

30 Jonathan Hogg and Christoph Laucht, 'Introduction: British Nuclear Culture', *British Journal for the History of Science*, vol. 45, special issue (December 2012): 479–493, 479.

31 A good example of a publication made possible by greater access to British Cold War sites and to previously classified information is Wayne D. Cocroft and Roger J.C. Thomas, *Cold War: Building for Nuclear Confrontation 1946–1989* (London: English Heritage, 2003).

32 Kirk Willis, 'The Origins of British Nuclear Culture, 1895–1939', *Journal of British Studies*, vol. 34 (1995): 59–89, cited in Hogg and Laucht, 'Introduction: British Nuclear Culture', 481.

33 See, for example, Paige Johnson's account of 'atomic gardening', 'Safeguarding the Atom: The Nuclear Enthusiasm of Muriel Howorth', *British Journal for the History of Science*, vol. 45, special issue (December 2012): 551–571.

34 In this volume, Gregory Salter looks at the effects of nuclear anxiety on the family and domestic life in a study of masculinity that parallels commentary in Jonathan Hogg's essay for the *British Journal for the History of Science* which describes the effects of 'nuclearity' on women's perceptions of motherhood and the family (see Jonathan Hogg, '"The Family that Feared Tomorrow": British Nuclear Culture and Individual Experience in the Late 1950s', *British Journal for the History of Science*, vol. 45, special issue (December 2012): 535–549.

35 In *The Look of the Past: Visual and Material Evidence in Historical Practice* (Cambridge: Cambridge University Press, 2012), Ludmilla Jordanova has persuasively argued for the importance of the visual in the understanding of the past. From a shared conviction of this viewpoint, *British Art in the Nuclear Age* brings together the work of both historians and art historians in their readings of British nuclear visual culture.

36 'Outside is Everything', review of the 1952 Venice Biennale, *Time*, 26 May 1952.

37 Simon Martin and Marco Livingstone, *Colin Self: Art in the Nuclear Age* (Chichester: Pallant House and AVA Publishing, 2008), 19–20.

38 C.P. Snow, *The Two Cultures* (Cambridge: Cambridge University Press, 1959.) For a discussion of the ensuing controversy, see Guy Ortolano, *The Two Cultures Controversy: Science, Literature and Cultural Politics in Postwar Britain* (Cambridge: Cambridge University Press, 2009).

1

'A Kind of Cold War Feeling' in British Art, 1945–1952

Carol Jacobi

When asked about the Independent Group's exhibition *This is Tomorrow* held at the Institute of Contemporary Arts (ICA) in 1956, Eduardo Paolozzi recalled it as 'a sort of basic art'. The event is often seen as the launch of Pop, but he went on: 'I think we had this shadow of, perhaps it might have been the atom bomb and things like that; there was a kind of cold war feeling about art ... it was more Art Povera than Pop art.'[1] This chapter rewinds from the 'high' Cold War that coincided with defined art movements such as Pop, Tachisme, Abstract Expressionism and Povera to focus on the Cold War imagery of the transitional period 1945–52 that preceded *This is Tomorrow*. It will show that a distinct, 'basic' Cold War art emerged directly from wartime concerns and in dialogue with France at the moment of crisis/community of the post-war years.

The later 1940s saw an Anglo-French cultural network that overlapped artistic, philosophical and military spheres, open-eyed to the implications of the nuclear age: Francis Bacon (1909–1992), Peter Rose Pulham (1910–1956) and Isabel Rawsthorne (1912–1992) were at the centre of David Sylvester's so-called 'School of London' between 1948 and 1952;[2] Nigel Henderson (1917–1985), William Turnbull (1922–2012) and Eduardo Paolozzi (1924–2005) would, with Richard Hamilton, form the core of the Independent Group in 1952.[3] The interruption of the war brought these two generations together and coincided their first group and solo shows in the late 1940s at the new ICA (founded 1946) and Hanover Gallery (founded 1947), as well as the Lefevre and London Galleries. In 1993 Francis Morris' exhibition *Paris Post War: Art and Existentialism 1945–55*, held at the Tate Gallery, presented an inter-connected perspective of this period that David Mellor's catalogue essay argued might equally be applied to this British avant-garde.[4] Further work – Martin Harrison's *Transition: London Art World in the Fifties* at the Barbican Art Gallery, for instance – has followed but, as Catherine Jolivette has pointed out, Cold War aspects remain relatively under-researched.[5]

Between the German occupation of France in June 1940 and the Americans entering the war in December 1941, British and Europeans opposed to Nazi rule were unusually close in political and military terms. A correlation evolved between the Soho area of London and Left Bank Paris with a military flavour. Soho was a centre for European refugees and for French exiles orientated around Charles de Gaulle and expatriate Free French Forces, as well as for artists and writers brought together by the removal of Paris as their focus. Rawsthorne's autobiography likened The George pub near the BBC – 'noisy, animated/intellectual' – to Les Deux Magots and Café Flore, and the Prada Restaurant to the Petit St Benois.[6] In 1940, despite paper shortages, the critic Cyril Connolly and the wealthy collector Peter Watson (sponsor of the London Gallery) started the journal *Horizon*, which defiantly perpetuated French perspectives. Bacon, Rawsthorne and Rose Pulham had lived in France before the war and she remained in touch with the sculptor Alberto Giacometti, one of the close-knit group who worked on in German-occupied Paris.[7] These, Pablo Picasso, Francis Gruber, Jean Dubuffet and others, saw their artistic activity as resistance. Some, such as Gruber and the sculptor and designer Diego Giacometti (who stayed after his brother returned to their native Switzerland), collaborated directly with the British through underground Resistance cells. Rawsthorne worked with the Resistance through her black propaganda role in the Special Operations Executive (SOE).[8] Bacon and Rose Pulham served in Air Raid Precautions, and Henderson, Turnbull and Paolozzi in the forces.[9]

The emergence from these conditions of a post-war British avant-garde coincided exactly with the advent of the atomic era. Paris was liberated on 19 August 1944, Victory in Europe declared on 8 May 1945 and the first atomic plutonium fission device, the Trinity Test, detonated in New Mexico on 16 July 1945. In August news arrived of an atomic bombing of Hiroshima and Nagasaki on an incomprehensible scale.

The social and cultural transformations that this event brought about are conventionally presented as abrupt, like the blast itself. However, for artists born in the first two decades of the century, inheritors of the First World War, the political, class and ethnic strife of the 1920s and 1930s, the Spanish Civil War and then the Second World War, it was experienced not as change, but as continuity.[10] The Victory in Japan which the bombs secured did not, moreover, end fighting in the East; the Indochina Wars (September 1945–1954) were followed by the Korean War (1950–1953). The cooperation maintained between the Allied West and the USSR while they had a common enemy in Nazi Germany broke down and within two months of the bombings, in October 1945, their stand-off was described for the first time as 'Cold War' by George Orwell in his article 'You and the Atomic Bomb'.[11] US plans to test devices on Bikini Atoll were underway by Christmas. For this generation, the Cold War confirmed the modern age as fundamentally military. Early Cold War art was correspondingly different from the war art sanctioned during the Second World War; it did not have its discrete or official character and was interwoven with wider modern circumstance.

Rawsthorne's memory of watching the British Prime Minister, Winston Churchill, deliver the victory speech from a balcony in Whitehall makes an interesting contrast to the popular idea of VE Day jubilance: 'The crowd was quite subdued; the ceremony was quite touching in its simplicity. For one thing, the war in the Far East was still being fought with ferocity. I walked slowly through the parks where there were bonfires burning away and good-humoured crowds circling round. I felt totally alone, desolate.'[12] The first eye-witness reports of European devastation and the atrocities of concentration camps were beginning to be broadcast daily.[13] Shock, worry, exhaustion, bereavement and displacement had changed most people's lives. A few days later Rawsthorne set off for France. The sense of dislocation is expressed in an unusual image: 'The future was a large hole ... As the ship drew slowly away from the quay I looked down upon a large eel curving its way beside us. Amazing creature born in the Sargasso Sea, now here.' She was the first civilian in the circle to reach Paris that May, and Rose Pulham and Alberto Giacometti, her closest colleagues, were back by September. The city was empty apart from soldiers, and years of privation had altered the depleted Left Bank community. Those with Jewish connections had been in hiding, imprisoned or killed, others had died and Gruber was terminally ill with tuberculosis. Food and heating were as scarce as they were in London. Rawsthorne lived hand to mouth with the Giacomettis, and Rose Pulham returned to his bare attic overlooking the Seine.[14]

Despite the shared austerity, the fact that Paris had been occupied and London had not distinguished their artistic atmospheres in a way that would have important implications. During the war, British artists had aligned themselves with the enduring 'Englishness' of the poetic nude and landscape traditions of Paul Nash, Henry Moore and Graham Sutherland. The 'Neo-Romantic' tendency, as it was named in 1946, thrived under the patronage of the British War Advisory Scheme led by Kenneth Clark.[15] These were his 'loyal men and dreamy boys', as Lawrence Alloway, an Independent Group member, later expressed it.[16] The value of the wayward or iconoclastic artistic persona, by contrast, diminished. As early as 1938, masters like Vincent Van Gogh and Chaim Soutine were spoken of as 'no longer in practical connection to his community'.[17] Paolozzi recalled that by the 1940s Rose Pulham and Bacon seemed the only two British artists free of 'English values'.[18] They were, with Rawsthorne, outsiders, working without recognized war artist roles or financial or institutional support. Once the war was over, their anti-establishment approach found a context in France. There the ruthless public épuration ensured that any hint of a recent official role or recognition under German rule destroyed post-war careers, while solitary workers were fêted.[19]

Two Left Bank survivors who shaped this alternative were the writer-philosophers Jean-Paul Sartre and Simone de Beauvoir. Their existentialist ideas, with those of Albert Camus, valorized the outsider and crossed into mainstream culture, catalysed by Sartre's public lecture *L'existentialisme est un humanisme* on 29 October 1945.[20] Its success was a symptom of a need to redefine lives after 'History in its most terrible form' – as de Beauvoir put it – had shown established structures to fail.[21] The existential philosophy shifted authority away from god or the state to the individual, whose responsibility was to recreate their society through 'engaging' with life and living 'authentically'. 'Man is condemned to be free', wrote Sartre, 'there is no reality except in action.'[22] These ideas grew out of wartime experience: 'We were never so free as under the German occupation', Sartre claimed, 'each word became as precious as a declaration of principle.'[23] Camus' vision of man alone in the 'absurd', indifferent and deadly world had, similarly, been distilled in *Le Mythe de Sisyphe*, published in 1942.[24] In 1945, the ultimate manifestation of this condition became, for Sartre and others, the atomic threat.

Giacometti and Rawsthorne were part of Sartre's circle. They were squeezed into the audience at the lecture at Club Maintenant and had many friends in common – for instance, the writers and anthropologists Georges Bataille and Michel Leiris.[25] The crowd often dined at Café Flore and danced to jazz and bebop at nearby clubs such as the Tabou.[26] 'Basic' working apparel, such as Giacometti's loose jacket and scarf and Rawsthorne's black masculine knits, hair worn naturally loose and long, became the existential 'look'. The outsider artist, musician or poet '*maudit*', toiling in their modest studio, became the emblem of 'authenticity'. New books celebrated Soutine who had died in hiding from the Nazis in 1944, and Antonin Artaud, imprisoned in an asylum for most of the war, composed his famous *Van Gogh Le Suicidé de la Société*.[27] In 1948 Sartre's essay on Giacometti, 'La Recherche de l'Absolu', published in English in New York as well as in French in Paris, established the sculptor as existentialism's living paradigm.[28]

The lifting of the embargo of the Occupation brought a new cultural transnationalism beginning with the movements of American and Allied forces. More artists from Britain, North America and elsewhere returned to Europe. Bacon moved from London to Monte Carlo in the summer of 1946. He and Rose Pulham had met during the war and in France he began a friendship with Rawsthorne.[29] She encountered Paolozzi in May 1947, his last term at the Slade School of Art, and he moved to Paris that summer, where she introduced him to Rose Pulham, Giacometti and others. A *Horizon* article brought him to the attention of Jean Dubuffet.[30] David Sylvester arrived in the autumn and met Giacometti through Michel Leiris.[31] Nigel Henderson and Peter Watson made many trips between 1947 and 1948, and Artaud and other French writers appeared in *Horizon*. The famous café conviviality circulated ideas and the older hands bridged barriers of language.[32] The atmosphere was summed up in a popular Paris paper in 1947: 'The existentialists … meet from ten to midnight at the Bar Vert, rue Jacob, where they carve existentialist graffiti in the waterclosets and telephone booths … no obscenities, no hearts and arrows, but grave thoughts which ponder about the void, the tomb, suicide and Bikini Atoll …'[33] All except Rose Pulham were living in London by 1949 but they often travelled back and forth and continued to socialize in Soho.[34]

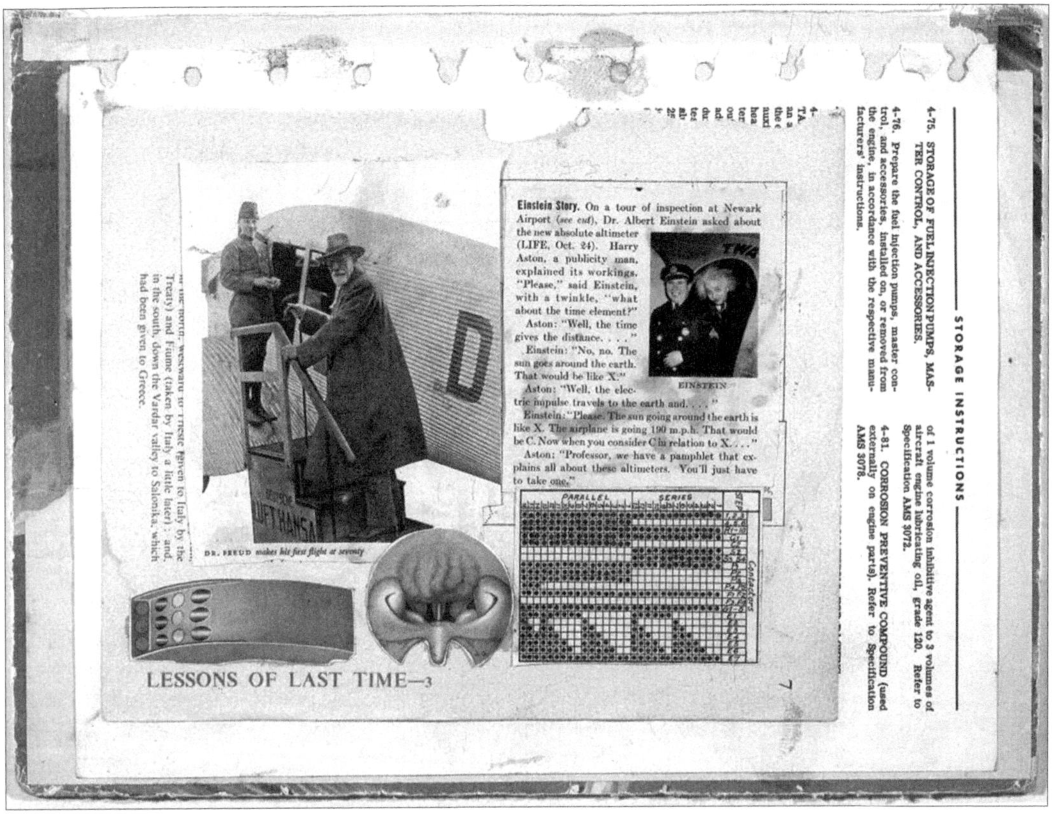

1.1 Eduardo Paolozzi, *Lessons of Last Time*, 1947, 22.9 × 31.1 cm, Tate © Estate of Eduardo Paolozzi.

These British artists were, if not politically *'engagé'* in Sartre's sense, then practically so. Even those who did not sympathize with Left Bank, left-wing ideologies (Bacon, for example) could not avoid engagement with their combative times.[35] Correspondingly, British avant-garde art moved on from pre-war interest in the abstract and surreal to alertness to the absurd in the real. This is conspicuous in Paolozzi's *Bunk* collages between 1941 and 1952, created from packaging, papers or journals scavenged and patched after the manner of his childhood scrapbooks and wartime 'make-do-and-mend'.[36] In Paris in 1947 he obtained a pile of magazines from a group of Americans.[37] The collages differed from those of Surrealists such as Max Ernst in that they and their materials addressed – as the title of the series from the famous paraphrase of Henry Ford, 'History is bunk',[38] suggests – contingent events rather than dreams.

Bunk illustrates how this group replaced conventional historical narrative with historical analogy, a strategy similar to Walter Benjamin's 'allegory'.[39] Several theorists were beginning to write about history in this way, as provisional and relational rather than narrative and sequential.[40] Paolozzi juxtaposed scraps from different time-frames in the manner of Amédée Ozenfant's influential book *L'Art* (1928, published in English in 1931) or photo and film montage, both enthusiasms of the circle.[41] *Lessons of Last Time* (Figure 1.1), for example, made in Paris in 1947, places an image of the psychoanalyst

Sigmund Freud at an aircraft door with a uniformed man, taken in 1926 (source unknown), next to a similar shot of Albert Einstein taken in March 1939, cut from the American magazine *Life*.[42] Both Jewish men had fled from the Nazis in the 1930s, but the Freud picture shows an attitude of departure, while Einstein arrives in New York. Einstein's presence in the USA, along with other refugee scientists, famously initiated the Manhattan Project which led to the atomic bomb. Below the pair of pictures is pasted a diagram which appears to show the components of an atomic explosion (shock wave and cloud) alongside another that looks like a console of buttons. The diagram could equally be an anatomical one and, together with the console, offers a 'Freudian' suggestion of female and male sexual parts. The two ways of viewing syncopate like the two men, two perspectives and two time-frames above.

Paolozzi clipped the red heading of the collage, 'Lessons of Last Time', from an editorial on the problems for post-war peace published by George Soule in 1942 in the American liberal journal, *The New Republic*.[43] The page discusses the frontiers of Greece and Italy. It is itself mounted over a sheet of storage instructions for aircraft components (source unknown). The juxtaposition recalls a statement in Orwell's 'You and the Atomic Bomb': 'We were once told that the aeroplane had "abolished frontiers"; actually it is only since the aeroplane became a serious weapon that frontiers have become definitely impassable ... The atomic bomb may complete the process by robbing the exploited classes and peoples of all power to revolt.' Unseen on the reverse of the picture of Einstein cut from *Life* is just such a 'revolt', a photograph of a woman spitting at German troops in a feature on Hitler's invasion of Czechoslovakia which also occurred in March 1939. The event had personal significance for Paolozzi; it coincided with Benito Mussolini's decision to sacrifice Italy's relationship with Britain for one with Germany, illustrated in a cartoon on the same page, which led in 1940 to the internment of British Italian men. Paolozzi was released but his father died during deportation to Canada.[44]

Allegory connected pre-war and wartime events with the present and future.

Cold War Beginnings

Because the Cold War imagery of this early period is interwoven with war imagery of other kinds, there is a danger of over-generalized readings. Before considering further examples, it is important to establish what images were associated specifically with nuclear war and if and when they were available to British artists. Atomic imagery was very limited in the West before 1952. Furthermore, the British avant-garde, with its close mixing of artists, philosophers, writers, journalists and military intelligence, viewed events from a different perspective than other occidental artists, notably those in the United States who have been the focus of writings and exhibitions on this subject up to now.[45]

The race between the Allies and the Nazi regime to create an atomic weapon was well known, public knowledge since 1939.[46] The first explosion, Trinity, in July 1945, was a covert test, so civilians did not hear of its realization until the second bomb was detonated over Hiroshima a month later. The first, defining images arrived over the radio. The BBC Home Service announced: 'Scientists, British and American, have made the atomic bomb at last. The first one was dropped on a Japanese city this morning.' A longer report added detail:

> ... the device was more than 2,000 times more powerful than the largest bomb used to date. An accurate assessment of the damage caused has so far been impossible due to a huge cloud of impenetrable dust covering the target. Hiroshima is one of the chief supply depots for the Japanese army ... The President [Harry S Truman] said the atomic bomb heralded the 'harnessing of the basic power of the universe'.[47]

British and other European newspapers carried reports from Japanese sources and by 8 August headlines were proclaiming two thirds of Hiroshima flattened and deaths in the hundreds of thousands.[48] On 9 August a similar announcement was made regarding the third bomb exploded over the city of Nagasaki.

Western artists saw no pictures of Hiroshima or Nagasaki. Many were made, of course, by Japanese witnesses, the photographer Satsuo Nakata, for example.[49] On 10 August 1945, the day following the catastrophe, the Japanese military photographer, Yōsuke Yamahata, took over 100 shots of Nagasaki and some were published in the national newspaper *Mainichi Shimbun* (Daily News).[50] The United States military and the Allied media censored and sifted images in the West, however. Only the mushroom clouds were shown and these were held back until the announcement of Japanese surrender on 20 August to introduce the means in the context of the ends.[51] In September Japanese film makers made a set of newsreels, *Effects of the Atomic Bomb*, and in October the Japan Science Council commissioned Shigeo Hayashi to undertake a photographic survey, but all films and photographs were confiscated by US forces. Their own air and ground photographic records were classified.[52] The cloud thus became the sole image in the West, representing the event while quite literally obscuring it.[53]

For nearly a year there was no other image to associate with the fate of the two cities and descriptions were defined by US sources. The few pictures of people with localized burns and a wrecked cathedral that appeared in *Life* magazine gave the impression of being so similar to those from European theatres of war, Coventry or Dresden, for example, that they discouraged understanding of the scale or difference of the devastation.[54] At the same time general accounts of the 'atomic era' or 'age' became commonplace in books and articles.[55] The dominant narratives were calls to control the new weapons, accompanied by awe and anticipation about the power and 'bright', energy-rich future that they might promise, the 'harnessing of the basic power of the universe'. This was represented after 1946 by the American Baruch Plan proposing international nuclear agreement.[56]

Nuclear anxiety crystallized around fears for the failure of this plan, summarized in the blunt phrase 'one world or none', coined by Freda Kirchwey in her US leftist journal, *The Nation*.[57] A book called *One World or None*, published in March 1946 by the Federation of Atomic Scientists newly formed for this purpose (including Albert Einstein, Robert Oppenheimer and Neils Bohr), became an instant best-seller.[58] One of the text's key visual tropes was the analogous blasted city, the potential for everyone's familiar skyline to become a Hiroshima or Nagasaki, and in the same year a nine-minute film summary of *One World or None* juxtaposed rare glimpses of the Hiroshima ruins with an artist's impressions of New York after a bomb.[59] In August the American Pulitzer Prize-winning writer John Hersey published a groundbreaking alternative view of individual victims; 'Hiroshima', an account of several months of research in Japan, appeared in a special edition of the *New Yorker* and as a hugely successful book.[60]

The British avant-garde circle spearheaded a parallel discourse. *Horizon* featured several articles. Its 1945 Christmas editorial anticipated the city imagery of *One World or None*: 'The race is now on between the Atomic Bomb and its desire to explode (and since London is the ideal target it will make for us as inevitably as eels for the Sargasso Sea) and between the efforts of human good will and intelligence to avert it'[61] (an interesting echo of Rawsthorne's eel image). The avant-garde artists' circle on both sides of the Channel was particularly informed about nuclear science. Surrealists took an interest in its world-changing potential and Georges Bataille was employed by President De Gaulle's new Commissariat à L'energie Atomique in 1945.[62] Those, such as Rawthorne, who had held intelligence roles were also knowledgeable. Left-wingers, especially Communists orientated towards the USSR, followed Cold War tensions very closely, moreover. One of these was, intriguingly, Donald Maclean who had been socializing with Rose Pulham, Rawsthorne, Giacometti and others since joining the Paris Foreign Office in 1939.[63] None were aware that he was secretly spying for Russia. He had passed cabinet minutes on the progress of the uranium bomb as early as September 1941.[64] By 1947 he was secretary of the Anglo-American Combined Policy Committee which guided atomic policy.

French and British philosophers associated with these groups spearheaded anti-atomic comment. When the Hiroshima explosion occurred, *Combat*, the underground newspaper of the French Resistance, followed its announcement of 'La Premiere Bombe Atomique' with a critical editorial from Camus.[65] Days later, Sartre's essay, 'Libération de Paris: une semaine d'apocalypse', noted the coincidence of the anniversary of the Liberation and the advent of the A-bomb, contrasting Paris' 'apocalypse' and 'explosion' of human will, as he termed it, with 'the blind powers of the young and … inhuman forces that have just won the victory'. He concluded: 'What the bomb represents is the negation of man, not only because it risks destroying the whole of mankind, but because it makes the most human qualities vain and ineffectual.'[66] Orwell's 'You and the Atomic Bomb' also saw the Bomb as a threat to individual human freedom as well as human existence. The main British voice was the veteran Bertrand Russell. He was admired by the circle, and his young collaborator, A.J. Ayer, was attached to the British Embassy in Paris with SOE in the summer of 1945, where he began lasting friendships with them.[67] Immediately after the Hiroshima bombing, Russell published 'The Bomb and Civilization', beginning a long and vocal anti-nuclear battle, eventually leading to the founding of the Campaign for Nuclear Disarmament, of which he was the first president.[68]

Some of the British press put these concerns in pictorial form. The edition of *Picture Post* carrying news of the Hiroshima bomb had on its cover an image of a child semi-silhouetted against a sea and stormy sky, a low sun barely breaking the clouds. The headline read 'Man enters the atomic age', but picked out in small letters against the landscape was the question 'Dawn – or dusk?'[69] One report was particularly significant. A left-wing British journalist, Wilfred Burchett, managed to reach Hiroshima 'ground zero' a month after the bomb detonated. He wrote for the London newspaper, the *Daily Express*. Burchett's eye-witness account, 'The Atomic Plague: "I Write This as a Warning to the World"', dominated the front page of the newspaper on 5 September 1945 (Figure 1.2).[70] The article began: 'Express Staff Reporter Peter Burchett [sic] was the first Allied staff reporter to enter the atom-bomb city. He travelled 400

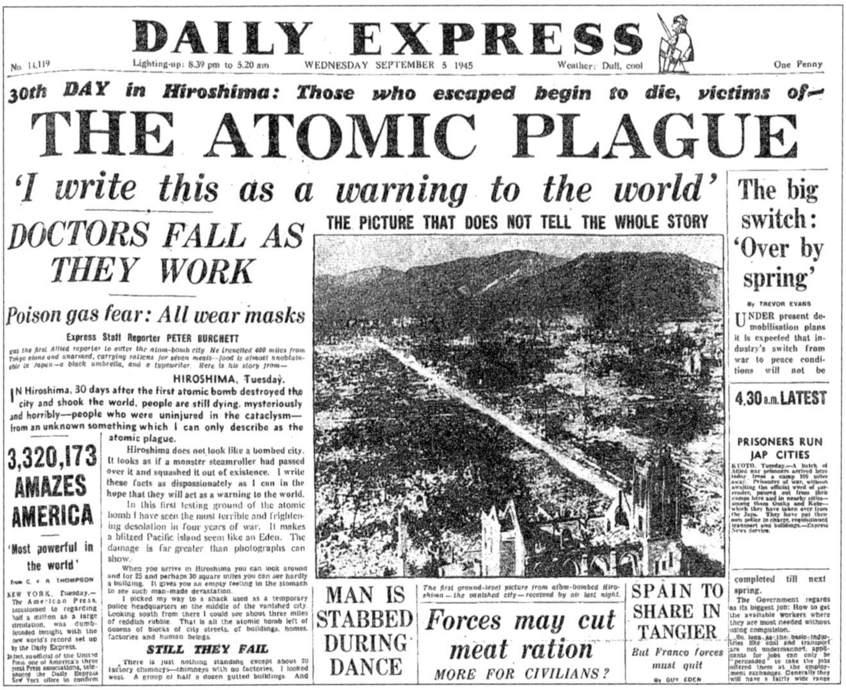

DAILY EXPRESS

No. 14,119 Lighting-up: 8.39 pm to 5.20 am WEDNESDAY SEPTEMBER 5 1945 Weather: Dull, cool One Penny

30th DAY in Hiroshima: Those who escaped begin to die, victims of—

THE ATOMIC PLAGUE

'I write this as a warning to the world'

DOCTORS FALL AS THEY WORK

Poison gas fear: All wear masks

Express Staff Reporter PETER BURCHETT

was the first Allied reporter to enter the atom-bomb city. He travelled 400 miles from Tokyo alone and unarmed, carrying rations for seven meals—food is almost unobtainable in Japan—a black umbrella, and a typewriter. Here is his story from—

HIROSHIMA, Tuesday.

IN Hiroshima, 30 days after the first atomic bomb destroyed the city and shook the world, people are still dying, mysteriously and horribly—people who were uninjured in the cataclysm—from an unknown something which I can only describe as the atomic plague.

Hiroshima does not look like a bombed city. It looks as if a monster steamroller had passed over it and squashed it out of existence. I write these facts as dispassionately as I can in the hope that they will act as a warning to the world.

In this first testing ground of the atomic bomb I have seen the most terrible and frightening desolation in four years of war. It makes a blitzed Pacific island seem like an Eden. The damage is far greater than photographs can show.

When you arrive in Hiroshima you can look around and for 25 and perhaps 30 square miles you can see hardly a building. It gives you an empty feeling in the stomach to see such man-made devastation.

I picked my way to a shack used as a temporary police headquarters in the middle of the vanished city. Looking south from there I could see about three miles of reddish rubble. That is all the atomic bomb left of dozens of blocks of city streets, of buildings, homes, factories and human beings.

STILL THEY FAIL

There is just nothing standing except about 20 factory chimneys—chimneys with no factories. I looked west. A group of half a dozen gutted buildings. And

3,320,173 AMAZES AMERICA

'Most powerful in the world'

From C. V. R. THOMPSON

NEW YORK, Tuesday.—The American Press, accustomed to regarding half a million as a large circulation, was dumbfounded tonight with the new world's record set up by the Daily Express.

In fact, an official of the United Press, one of America's three great Press associations, telephoned the Daily Express New York office to confirm

THE PICTURE THAT DOES NOT TELL THE WHOLE STORY

The first ground-level picture from atom-bombed Hiroshima—the vanished city—received by air last night.

MAN IS STABBED DURING DANCE

Forces may cut meat ration

MORE FOR CIVILIANS?

SPAIN TO SHARE IN TANGIER

But Franco forces must quit

By GUY EDEN

The big switch: 'Over by spring'

By TREVOR EVANS

UNDER present demobilisation plans it is expected that industry's switch from war to peace conditions will not be completed till next spring.

The Government regards as its biggest job: How to get the available workers where they are most needed without using compulsion.

So long as the basic industries like coal and transport are not undermanned, applicants for jobs can only be "persuaded" to take the jobs offered them at the employment exchanges. Generally they will have a fairly wide range

4.30 a.m. LATEST

PRISONERS RUN JAP CITIES

KYOTO, Tuesday.—A batch of Allied war prisoners arrived here today from a camp 100 miles away. Prisoners of war, without awaiting the official word of surrender, poured out from their camps here and in nearby cities—among them Osaka and Kobe—which they have taken over from the Japs. They have put their own police in charge, requisitioned transport and buildings.—Express News Service.

1.2 *Daily Express* (5 September 1945): 1, by kind permission of Daily Express/ Express Syndication.

miles from Tokyo alone and unarmed carrying rations for seven meals – food is almost unobtainable in Japan – a black umbrella, and a typewriter.' Almost a year before Hersey's *Hiroshima*, Burchett provided word pictures for the scale of the destruction, the poison rain washing radioactivity from the air into the soil below and the effects of radiation sickness striking down inhabitants and doctors and scientists who had rushed to the scene from all over Japan. Perhaps the most important element was a photograph; a rare appearance of one of those taken by Satsuo Nakata presented a devastated part of the city called Kami-Nagarekawa-cho. The unusually panoramic perspective might appear to distance the atrocity but the desert effect, the absence of humanity over such a huge area, dramatized the appalling, unprecedented uniformity described in Burchett's copy. The exposé prompted tighter censorship in the longer term, but cannot have failed to make an impact. Many left-wing *Daily Express* journalists – Tom Driberg, for instance – were Soho regulars and Rawsthorne had lived and worked with the newspaper's chief foreign correspondent, Sefton Delmer (latterly a senior SOE officer), between 1935 and 1944.[71]

Visual representations caught up with textual ones in 1952 when US troops left Japan. Hidden prints of *Effects of the Atomic Bomb* were immediately screened in Japanese cinemas, and in September photographs appeared in the Western mainstream press. *Life* magazine headed its seven-page spread: 'When Atom Bomb Struck – uncensored'.[72] The images were quickly joined by the spectacle of the first so-called hydrogen bomb explosion in November 1952, 450 times more powerful than the atom bomb, and news and photographs of the first Russian and British tests. At the same time, fallout from the Bikini Atoll test raised awareness of the potential of radioactive contamination to spread well beyond the target. The arms race and its consequences became conspicuous.[73] By 1955, a decade after the disaster, Yōsuke Yamahata's photographs went on exhibition in *The Family of Man* at the Museum of Modern Art, New York, curated by Edward Steichen. This toured the world. In the same year, a permanent Peace Memorial Museum was founded in Hiroshima.[74] The newly accessible eye-witness photographs and film of past and ongoing atomic activity fed the fresh visual vocabulary of the high Cold War period.

Early Cold War Aesthetic

In the West, early artistic representations of nuclear war manifested the mix of celebration and concern that was common in the first writings.[75] Artists working in North America responded most obviously; Salvador Dali's *Melancholy, Atomic Uranic Idyll* (1945, Museo Nacional Centro de Arte Reina Sofía) and Isamu Noguchi's *Humpty Dumpty* (1946, Whitney Museum of American Art) were prompt examples. Gavin Parkinson argues that the Surrealist circle in Paris were discouraged from making art that confronted nuclear arms because the majority of dissenting voices were religious ones from which they

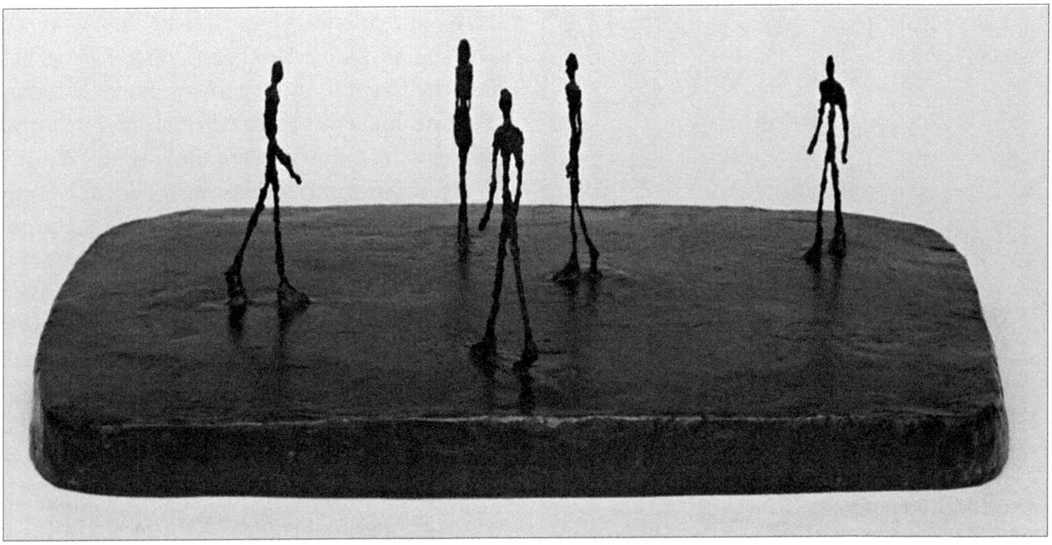

1.3 Alberto Giacometti, *La Place (City Square)*, 1948, bronze, 21.6 × 64.5 × 43.8 cm, MOMA © 2013 Artists Rights Society (ARS), New York/ ADAGP, Paris.

wished to distance themselves.[76] In Britain, war art was associated with the establishment, both religious and political. These considerations, combined with the lack of nuclear pictures, may account for the absence of an obvious body of British nuclear art before 1952. The 'shadow' of the Bomb can be observed, however, as Paolozzi said, as both cold war feeling and content.

The lean figures that Giacometti had been developing since 1937, 'skeletons in space', became the 'logo of existentialism' and their battered, X-ray appearance was soon associated with the fragility of man in a post-nuclear world.[77] Leiris described Giacometti's walking figures adrift in urban space, such as *La Place* (*City Square*) (Figure 1.3), using Shakespeare's image of 'unaccommodated man' in *King Lear*, a 'forked animal' cast out in the storm.[78] An American commentator summed up in 1952: 'Giacometti's space-devouring, ghostly figures, moving as in a dream from nowhere to nowhere, imply the overwhelming loneliness of modern man in a space that has overnight taken on a fearful atomic dimension.'[79] French and British artists linked to the Left Bank pursued a parallel 'basic', austere language in *grisaille* tones, gestural marks and distressed material surfaces. Gruber was a hit at the 1944 Salon de Liberation with an emaciated, modern, male nude which he called *Job* (Tate). Like Lear, the abject Biblical prophet's defiance of God in the whirlwind, evocative of the nuclear maelstrom, gave him a new secular currency. Rose Pulham and Rawsthorne read *The Book of Job* in 1946.[80] Between 1945 and 1947, Artaud and Dubuffet exhibited even more uncompromisingly damaged, abbreviated figures, and Jean Fautrier showed a series of harrowing, glutinous *Otages* (Hostages).[81]

Zoological motifs were pursued in a similar style.[82] Rawsthorne had begun studying bird and animal skeletons and desiccated insects and carcasses in 1939. Gruber took up this practice in the 1940s, as did the Swiss sculptor, Germaine Richier, when she returned to her nearby Paris studio in 1946. Her

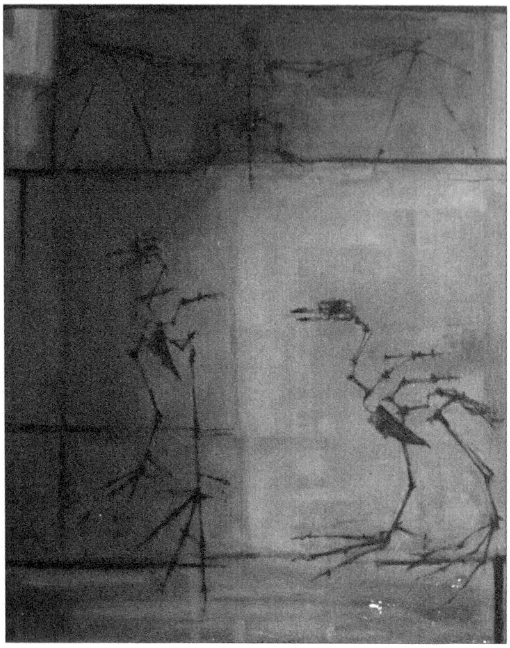

1.4 Isabel Rawsthorne, *Two Birds and a Bat*, 1947–1948, oil on canvas, 65 × 80 cm, private collection, by kind permission of the Estate of Isabel Rawsthorne.

gaunt, threadbare sculptures of that year, *La Chauve-souris* and *La Grande Mante* (The Bat and The Great Mantis) gained great success in London and Paris.[83] The British and Natural History Museums were favourite haunts of avant-garde artists, as was the Left Bank taxidermist shop Deyrolle.[84] Rawsthorne introduced it to Paolozzi and he bought her a bat skeleton. Together they devoured the 1942 reprint of D'Arcy Wentworth Thompson's venerable 1917 text on evolution, *On Growth and Form*, which he in turn showed to Turnbull and Henderson.[85] Rawsthorne's paintings between 1947 and 1948, 'bird, bat and fish skeletons walking, enclosed in rectangles of varying sizes, in colours of umber, ochres, siennas', as she described them to Giacometti, echoed his sculptures and the gridded skeleton diagrams in Thompson's book.[86] She exhibited twelve, including *Two Birds and a Bat* (Figure 1.4), at the new Hanover Gallery in February 1949.[87] The motifs set up resonances as restless as Paolozzi's collages. The grisaille grid evoked newsprint, or the scaffolding of the post-war city (it covered her hotel) and the look of a cage or glass vitrine. Like Giacometti and Richier's figures, the dead/alive birds resembled X-rays, popularly associated with atomic science and shorthand for the silhouetting of the skeleton when the body is penetrated by radiation.[88]

In 1947 Paolozzi's two gouaches named *Shooting Gallery* (Scottish National Gallery of Modern Art, private collection), inspired by the fair next to his Paris studio, added fairground targets to birds and grids to create a brighter, ironic post-military format. His main project was a set of skeletal shapes in the manner of Giacometti's table sculptures: *Paris Bird* (1948, Tate), for example, intended for an exhibition devoted to *On Growth and Form*.[89] Another of these, *Growth* (1947, Scottish National Gallery of Modern Art), based on the bone structure of a fish, was shown with two reliefs in *Les mains éblouies* with Dubuffet and other *brut* artists at the Galerie Maeght in October 1949, before being exhibited at the ICA in London in 1950.[90] Turnbull explored zoological imagery in a more abstract way, in monotypes such as *Sea Forms* (1949, Tate) and sculptures such as *Heavy Insect* (1949, Scottish National Gallery of Modern Art), *Mobile Stabile* (Tate) and *Playground* (1948, Leeds Museums and Art Gallery) shown in a joint show with Paolozzi at the Hanover Gallery in February 1950.[91] They can be seen in a photograph of the display taken by Nigel Henderson (Figure 1.5). In 1950, Slade student Richard Hamilton's *Chromatic Spiral* (Tate), also inspired by *On Growth and Form*, presented a more abstracted equivalent for Rawsthorne's monochromes, and in 1952

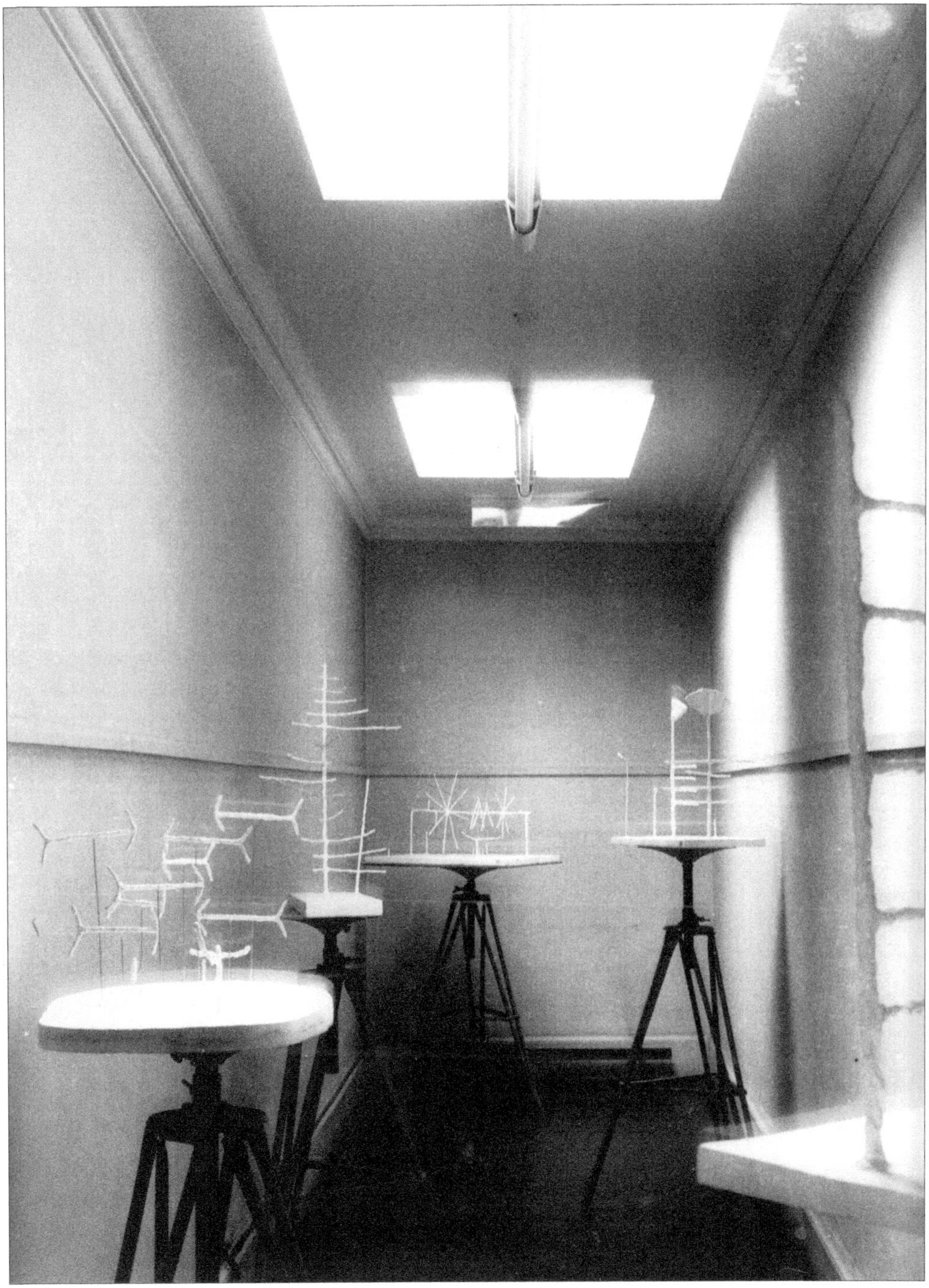

1.5 Nigel Henderson, *William Turnbull Exhibition, Hanover Gallery*, 1950, private collection © Estate of Nigel Henderson.

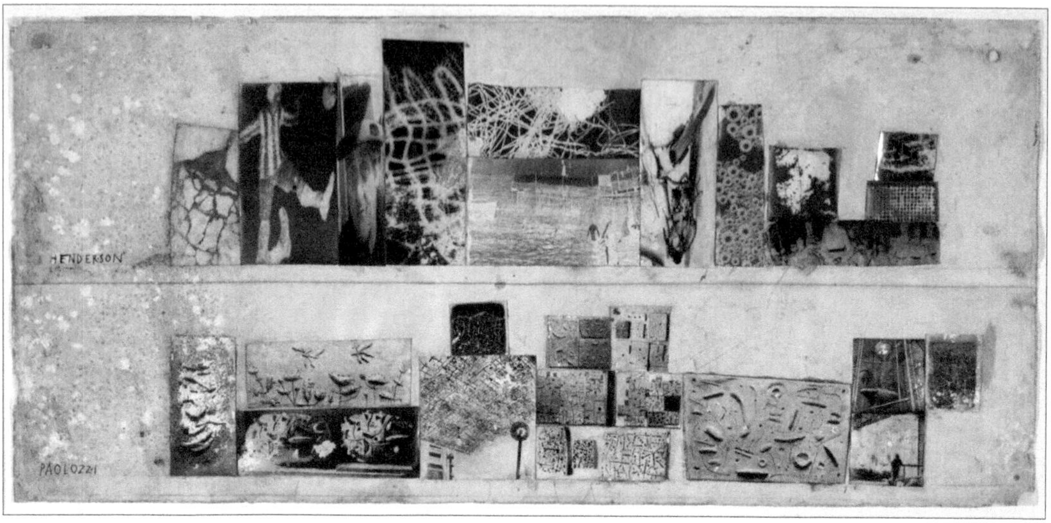

1.6 Nigel Henderson and Eduardo Paolozzi, *Untitled (Study for Parallel of Life and Art)*, 1952, photographs, black and white, on paper, pen and graphite on paper, 35.8 × 77.9 cm, Tate © Estate of Eduardo Paolozzi and Estate of Nigel Henderson.

X-rays and biological, zoological and geological motifs were featured in his exhibition *Growth and Form* at the ICA.[92]

The use of Thompson's *On Growth and Form* was part of a broader interest in the primordial present, manifested in the exhibition *40,000 Years of Modern Art: A Comparison of Primitive and Modern* at the ICA in 1948 (including Picasso, Giacometti, Rose Pulham, Bacon and Paolozzi; Ayer spoke at the discussion), and Siegfried Giedion's talk 'Palaeolithic Art and Modernism' given at the ICA in 1950.[93] The idea had a specifically nuclear dimension: in his seminal essay, 'Explosive Propositions: Artists React to the Atomic Age', Stephen Petersen observes that nuclear annihilation brought to mind primal origins; in other words ground zero 'paralleled or reflected in reverse the Stone Age'.[94] 'Just think of the prehistoric man who invented the bow and arrow to kill animals', Giacometti remarked. 'Now they've invented the atomic bomb, which can kill a hundred thousand people at once.'[95] Orwell went further: 'For forty or fifty years past', he wrote in 'You and the Atomic Bomb', 'Mr H.G. Wells and others have been warning us that man is in danger of destroying himself with his own weapons, leaving the ants or some other gregarious species to take over. Anyone who has seen the ruined cities of Germany will find this notion at least thinkable.' Pieces like Rawsthorne's *Two Birds and a Bat*, Turnbull's *Heavy Insect* and Paolozzi's terracotta reliefs, embedded with a montage of fossil and fish-like shapes, reverberated with these ideas (see Figure 1.6).

In the Cold War context, hybrid beings transformed from surreal dreams to real permutations of the evolutionary past and future, anticipating the trope of radioactive mutation. Sarah Wilson sees Richier's bat and insect hybrids, for instance, as: 'a reversal of evolution, the degenerating of the human through mammal and bat to bird and insect forms … regression to a more bestial universe in the aftermath of the atrocities of the war years.'[96] The skeleton was potent in this respect because it was understood to be the part of the body that retained the clearest link to earlier states. The bones of birds were thought to retain traces of dinosaurs. Rawsthorne's arrangement of bird skeletons in pairs referenced comparative drawings in Thompson's *On Growth and Form,* or Gerhard Heilmann's *The Origin of Birds* (1926) and sequential arrangements in natural history displays.[97] Their shadowy appearance prefigured Giacometti's apocalyptic description of the Galerie de Paléontologie et d'Anatomie Comparée: 'scored with their dark armature of bars, shining with the last light of day, still disoriented, lost in the desolation of the room of nameless snakes'.[98] Contemporary reviewers recognized this bleak natural history. Rawsthorne's 'breath on glass' effect, as she put it, was interpreted as the cases or cabinets of museum displays.[99]

The vitrine aesthetic also evoked the aquarium.[100] In 1948, Paolozzi and Rawsthorne discovered William Beebe's account of deep-sea exploration, *Half Mile Down* (1934), which represented a different kind of primal wasteland.[101] In 1948 she began to paint in inky blues inspired by Beebe's descriptions of his descent; as she wrote to Giacometti, 'the glare of an unimaginable blue which appeared to be brilliant but that was too dark to distinguish a blank page from a printed one' (author's translation).[102] She created *Night* (private collection, 1948–1949), and sooty *in camera* figures in gridded spaces. Bacon adopted a similar palette in 1951, developing what Harrison has described as the 'deep space of ... internal framings, like museum specimens displayed in vitrines' of his so-called 'blue' paintings, culminating in 1953 with *Untitled (Half-length Figure in the Sea)* (private collection).[103] Sylvester attributed equivalent effects to Paolozzi and Turnbull's 1950 exhibition: its 'principal source is the aquarium', he wrote. To enter it was 'to fly like a bird among branches or to swim under water among the inhabitants, mobile or stationary, of the sea'.[104] Henderson conveyed this effect by photographing the display with a blurred after image, a technique he used in his 'distressed' photographs such as *Stressed Photograph of a Bather* (c. 1950, Tate).[105]

In 1948 Rawsthorne applied the underwater appearance to a cityscape called *The Desert* (Figure 1.7), contemporary with Giacometti's *La Place.* It was a different kind of wet wasteland, the bombsite and scaffolded roofs outside her studio window. A walking woman in the form of a 'forked animal' echoes Paolozzi's contemporary impression of post-war London: 'moving through the fog all these lost souls somehow ... terribly austere ... that kind of atmosphere of damp astringency'.[106] *The Desert* was exhibited at the ICA in 1952. In the same exhibition Rose Pulham showed *Ville Entière* (Entire City) (Colour plate 1).[107] Afterwards they hung together in Rawsthorne's apartment, where Paolozzi admired them. Rose Pulham's far more distant cityscape reworked Ernst's Surrealist *La Ville Entière* (1935, exhibited at the Redfern Gallery in the 1930s and acquired by the Tate in 1941) on a post-nuclear scale, a Cold War Golgotha. Giacometti and Rawsthorne's scaffolding became black uprights equally distributed through a panoramic perspective, suggesting both blasted trees and the supports of disintegrated buildings, equating to the chaotic shadows of Satsuo Nakata's grainy photograph on the front page of the *Daily Express*.[108] The surface has the texture and colour of charcoal. Rose Pulham

1.7 Isabel Rawsthorne, *The Desert*, 1948–1951, oil on canvas, 75 × 98 cm, private collection, by kind permission of the Estate of Isabel Rawsthorne.

enhanced tension by dividing Ernst's single city in two, which, with the title, encourages thoughts of Hiroshima and Nagasaki and the unthinkable entirety of their destruction. The small pointed feature in the distance confuses these cities alarmingly with London and Paris, summoning up the silhouette of an English steeple or the Eiffel Tower, reminding us of the analogous blasted cities of *One World or None*.

The word 'Entire' in Rose Pulham's title referred to an explosive aesthetic, as well as subject, connected to the skeletal, calligraphic shape. In 1948 Giacometti's famous letter to Pierre Matisse described his 'skeleton in space' as a search for stylistic 'entirety'.[109] His abbreviated forms aspired to an all-at-once impact, 'sudden communication' as Leiris described it in 1929.[110] On the eve of the war in 1939, Walter Benjamin had defined human experience similarly, paraphrasing Paul Valery, as a series of 'shocks and collisions'.[111] For him the work of art aspired to the effect of 'catastrophe', the sudden 'technique of the putsch'.[112] In the same year Artaud asserted a 'convulsive' aesthetic.[113] Neither Benjamin nor Valery survived the war but Artaud went on to celebrate Van Gogh's convulsive '*peinture linear*'.[114] From 1945, French *brut* art was characterized by the spontaneity of 'hazard' and the scuffed succinctness of graffiti, an 'attempt to dissect contemporary pain down to its tragic ideograms', as André Malraux summarized it.[115] Artaud described his own linear technique: 'staggering blows given in all directions of chance, possibility, hazard or destiny'.[116] For Richier the 'perforations' in her sculptures 'lead like lightning bolts into the matter'.[117] British avant-garde painting and sculpture shared this convulsive aesthetic. Turnbull recalled being struck by Siegfried Giedion's elaborations of equivalent ideas on cave-drawing walls in his talk, 'Palaeolithic Art and Modernism'.[118] The graffiti shapes and battered surfaces of his and Paolozzi's sculpture aspired to this (see Figure 1.6). Both young men were also interested in the famous photograph which introduced Thompson's *On Growth and Form*, capturing the explosive shape of a splash of milk (see Figure 1.8). Its lobed projections appeared in many of their pieces, as well as Hamilton's *Chromatic Spiral*. Rawsthorne, perhaps most aware of French writings, called her calligraphic style '*peinture ecriture*' and wrote to Giacometti: 'I cannot ... paint little by little ... because I would like the object on the canvas to be sudden.'[119] In *The Desert*, the birds startled by the walking woman are rough smears of movement; the figure comprises little more; three strokes set her striding, the furthest leg a scratch in the grey pavement, and breasts two clots of impasto. The method was best represented by Bacon's famous bravura brushing, however, especially after he began working on raw canvas, which caught spontaneous marks, in 1947. He often described the effect he was after as 'violent', paraphrasing Valery.[120] His calligraphic paint stood in for the calligraphic figure. Its 'very substance', he wrote in 1953, 'can make such a direct assault on the nervous system.'[121]

This language of impromptu allegory and technical 'putsch' chimes with the atomic aesthetic emerging more broadly in the West. On 5 July 1946 the Parisian designer Louis Réard responded to a new two-piece swimsuit launched by his rival called the 'atom' with a skimpier garment, which he named the 'bikini' after the connotations of shock of the Bikini test.[122] Lyn Gamwell argues for an 'Atomic Sublime' based on sudden, disintegrative forms, and Petersen's 'Explosive Propositions: Artists React to the Atomic Age' finds these qualities linking Jackson Pollock in New York to Italian groups based in Milan, Arte Spaziale and Arte Nucleare (founded 1947) and Tachisme in Paris (announced in 1952).[123] The *brut* post-war sculptors and painters that preceded and informed Tachisme provide practical and theoretical precedents for these later trends.

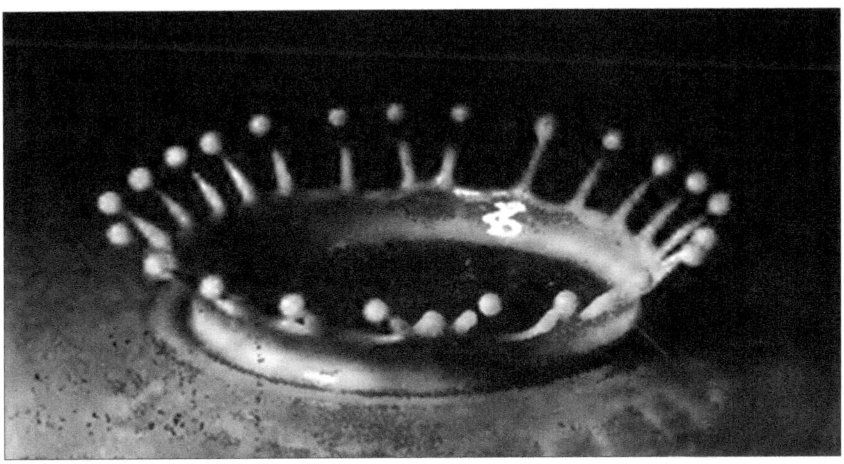

1.8 An instantaneous photograph of a 'splash' of milk from Harold E. Edgerton, Massachusetts Institute of Technology, Thomson, *On Growth and Form*, title page.

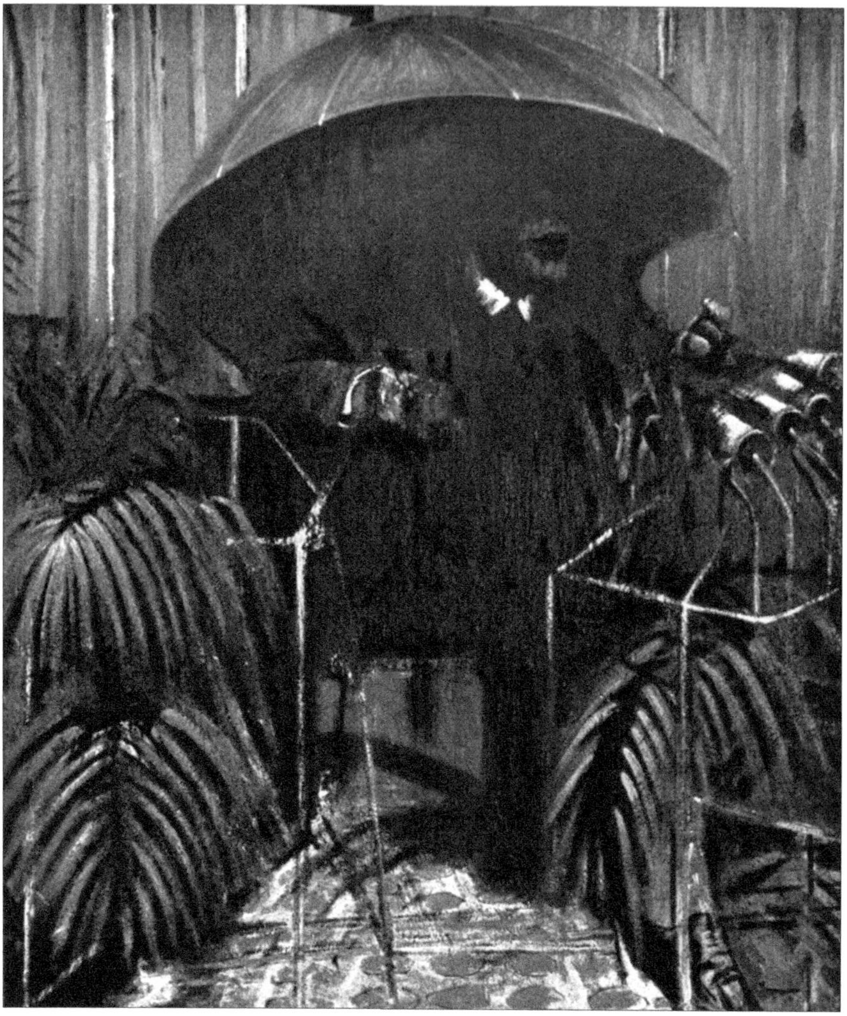

1.9 Francis Bacon, *Study for Man with Microphones*, 1946 (later *Gorilla with Microphones*, c. 1947–1948, partially destroyed), oil on canvas, 145 × 128 cm, by kind permission of the Estate of Francis Bacon.

To conclude with Bacon, his extraordinary technique was orientated towards portrait-like subject matter parallel to French artists such as Dubuffet and Artaud. During the war he had, like Rose Pulham, painted surreal *in camera* bodies owing a debt to Picasso. From 1945 they became more frankly figurative. Rose Pulham's *grisaille* nude in a studio, *Figure in a Room* (c. 1945, private collection) drew on Gruber and Giacometti's austere figures. Bacon developed more unsettling single bodies crying or shouting, such as *Painting 1946* (1946, MOMA) and *Study for Man with Microphones* (Figure 1.9) (later *Gorilla with Microphones*, 1946 onwards, Dublin City Art Gallery, The Hugh Lane) exhibited at the Lefevre Gallery and the Anglo-French Art Centre in 1946.[124] Both men experimented with Giacometti-esque space-frames. Bacon's often suggested broadcasting booths or rostrums.[125] Ambiguous, multivalent motifs drew on the strategy of allegory. The black umbrella in *Painting 1946* and *Study for Man with Microphones*, for example, synthesized many sources.[126] One of these may have been the umbrella that accompanied Wilfred Burchett to Hiroshima, a trope of British urbanity linking rainy London to the journalist's experience of radioactive rain in Japan. Bacon was, like Paolozzi, using montaged, printed images accumulated in his studio. Rather than collaging directly, however, he employed them in place of preliminary sketches.

In November 1949, he showed a new series of *grisaille, in camera* figures, *Heads*, to shock and acclaim at the Hanover Gallery.[127] Some explored the post-nuclear theme of the zoological hybrid.[128] *Head II* (1949, Ulster Museum) featured ape-like characteristics inspired by Robert Yerkes' 1920s research into humans and chimpanzees. *Fragment of a Crucifixion* (1950, Stedelijk Van Abbemuseum) combined an owl with a human mouth and genitals based on illustrations in ornithological and medical texts, in front of forked walking figures in a cityscape reminiscent of *The Desert* or *La Place*.[129] Bacon's purchase, sometime in the 1940s, of Kathleen Clark's book of X-rays, *Positioning in Radiography*, might also be placed in the context of the nuclear aesthetic.[130]

Bacon used analogy to interrogate power. Works such as *Study for Man with Microphones* have been linked to news pictures of Fascist leaders on podiums haranguing crowds, most extensively by Martin Hammer in his book *Francis Bacon and Nazi Propaganda*.[131] This war imagery was adapted to later Cold War conditions. In 1950 Sam Hunter photographed scraps in Bacon's studio that included photographs of a crowd, of marching troops and of Goebbels shouting from a rostrum.[132] Harrison has identified one of these images of Goebbels, found in Bacon's studio at his death, as a post-war rather than a wartime document, a page torn from a *Picture Post* dated 26 April 1947.[133] The article, headed 'What is De Gaulle Up To?', dealt with the French general's controversial attempt in 1947 to create a super-party, the Rassemblement du Peuple Français (Rally of the French People, or RPF). It placed yelling Nazi leaders and extracts from their speeches next to chillingly similar recent images of de Gaulle and his statements. The ambitions and style of 'The Sublime Figure Perched on the High Rostrum', as de Gaulle, once the hero of the Resistance, is described in the text, keyed into Cold War frictions. *Study for Man with Microphones* was Bacon's first use of the grey curtain.[134] The motif was informed by several fine art sources, but also coincides with the establishment of the term 'iron curtain' – popularized by Goebbels in 1943 and reiterated by Churchill in a speech in August 1945 – as the overarching image of Cold War stand-off between the Soviets and the West.[135]

In 1946, Bacon began to base some *Heads* on Popes, notably reproductions of Velasquez's *Pope Innocent X* (1650, Galleria Doria Pamphilj). The enthroned sitter was another variant of 'The Sublime Figure Perched on the High Rostrum' and his name, Innocent, introduced a provocative irony. The first of these to be exhibited, *Head IV* (Arts Council), appeared in the 1949 show. Like many other versions, it featured round-framed glasses.[136]

A famous source for these was a film still of the blood-spattered 'Woman with Pince-nez' in Eisenstein's *Battleship Potemkin*. The juxtaposition of her face with the Pope's brought a figure of catastrophe together with a figure of power. The glasses also set up associations with the modern Pope Pius XII. One of his stern official post-war portraits, enthroned like *Pope Innocent X*, was published in October 1945 as the full-page 'Picture of the Week' in *Life* magazine.[137] This lay-out was a ready-made montage of power and catastrophe as the photograph was positioned opposite a page headed in large letters 'The Bomb', one of the first editorials to appear after Hiroshima and Nagasaki; an arrow (a favourite Baconian motif) pointed from text to image. Even more significantly, on the reverse of the photograph was an array of pictures of Nazi commanders illustrating an article announcing their trial for war crimes. Bacon collected images of the trials, and his possession and retention of this particular page is supported by the appearance of Pius' features in *Pope 1 – Study after Pope Innocent X by Velasquez* (1951, Aberdeen Art Gallery & Museums) and his folded hands and skull cap in *Study (Imaginary Portrait of Pius XII)* (1955, Sainsbury Centre). Thus, a connection arises in Bacon's work, as it did with de Gaulle, between recent and current authority figures in the context of the Cold War that can be dated to ten days after Orwell coined the term.

The last months of 1949 and the early 1950s brought Britain directly into the Cold War and confirmed the divide with Russia. In 1950 Paolozzi adapted the format of the figure in a framework to a Russian leader. The collage *Mr Gromyko* (Figure 1.10) used the dark lines and cut divisions of the grid to section a face featuring Gromyko's chin and the heading 'Russia's Gromyko – What would be the judgement of history?' from a 14 August 1947 edition of *Time* magazine, obtained in Paris.[138] The round glasses and hairline add a resemblance to Bacon's Popes, or perhaps Pope Pius himself. Andrei Gromyko was Minister of Foreign Affairs for the Soviet Union and a well-known Cold War personality, instrumental in rejecting the Baruch Plan. Paolozzi's retrieval and use of the three-year-old image in 1950 may have been prompted by the Soviet Union's unexpected achievement of its nuclear test, RDS-1, in August 1949.

Britain conducted its first nuclear test in October 1952. This hardening of positions on the part of Britain and Russia was fraught for left-orientated British and French artists. The avant-garde circle had direct contact with the political debates behind the accelerating confrontation through their contacts with commentators, journalists and politicians. Between 1949 and 1952, Bacon and Rawsthorne often spent weekends at the country house of Tom Driberg, for example, now a Member of Parliament. Visitors included fellow Labour MPs Jenny Lee, who was violently opposed to the nuclear deterrent, and her husband Aneurin Bevan, Minister for Health, who supported it. Rawsthorne remembered 'fast, pointed, conversations'.[139] In 1952, Bacon adapted the format of *Study for Man with Microphones* to create *Untitled (Marching Figures)* (Colour plate 2), which replaced the figure on the rostrum-like framework

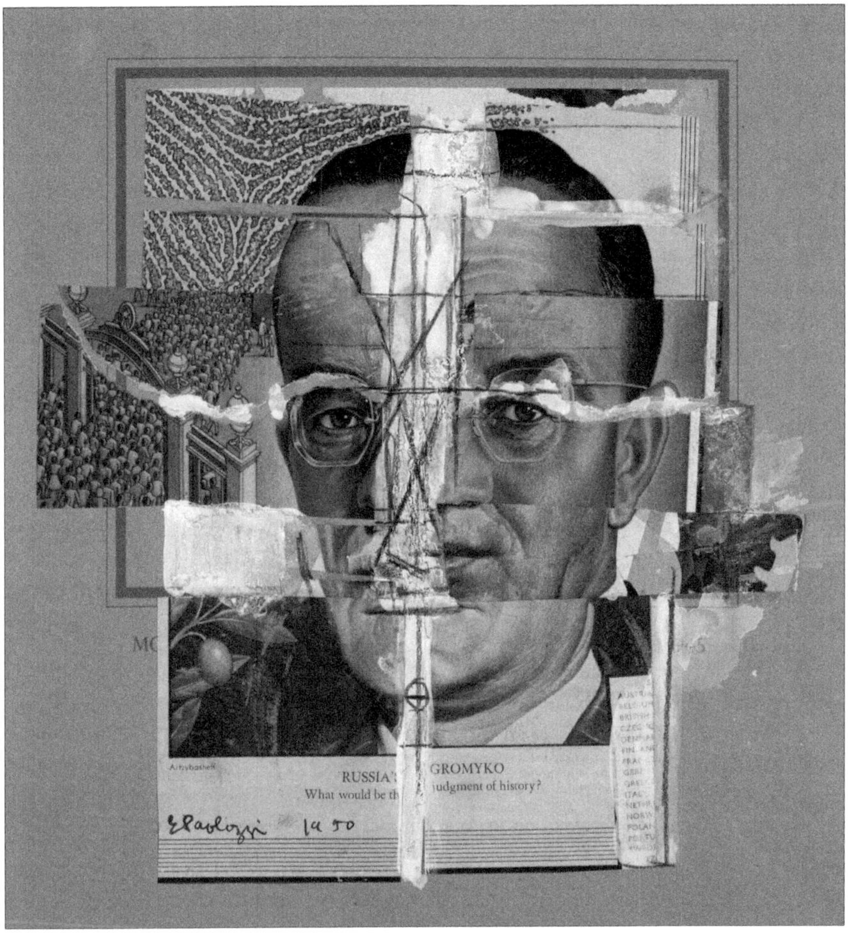

1.10 Eduardo Paolozzi, *Mr Gromyko*, 1950, collage, 33 × 25.5 cm, British Council © Estate of Eduardo Paolozzi 2005. All rights reserved, DACS.

with an imperious bear, inevitably evoking Russia for which it was the well-known symbol. Behind it, explosive blue brushstrokes radiate over a smutty, scumbled recess. The bear overlooks more forked, walking figures, this time aligned as though marching. The image was derived from Amédée Ozenfant's 1912 illustration in *Art* juxtaposing ancient Egyptian model troops with Egyptian and British armies in Egypt.[140] Marches and demonstrations against British occupation of Egypt had been widely covered in *Picture Post* and elsewhere from 1951, resulting in Mohamed Naguib's expulsion of Britain from Egypt in July 1952.[141]

Paolozzi's modification of *Time* magazine's 'Russia's Gromyko' of 1947 to the many-faced figure in 1950 evoked a new phase of the Cold War in which dissembling and personal paranoia joined global threat. The surprise success of the 1949 Russian test was partly due to espionage by agents, some of whom were part of the artists' circle. Rawsthorne knew the intelligence community well through her covert war work and friendships with others involved in espionage, such as Sefton Delmer, A.J. Ayer and Ian Fleming (at that time completing the first James Bond spy thriller, *Casino Royale*).[142] According to her memoir, their circle finally discovered the involvement of their friend Donald Maclean in May 1951 when he defected to Russia with Guy Burgess.[143] Bacon's *Untitled (Marching Figures)* coincided with Fleming's publication of Cyril Connolly's book, *The Missing Diplomats*, which described the events,[144] and the revelation that Maclean had been passing nuclear secrets.[145] The restless, multivalent motifs used throughout the group fitted an age in which multivalent identities and secret signs had become an everyday reality. Communists, spies and those bound by the Official Secrets Act had this in common with homosexuals (to remain illegal until 1967) like Bacon and Driberg. Refugees living strangers' lives and friends who had survived under Occupation, especially those working in the Resistance or with Jewish connections, had experienced this most powerfully. The art of this period was adept at layering covert significances.

To conclude, this group of British artists found an anti-ideal alternative to the specificities of history painting and the universals of Abstraction. Their variations on an austere, abbreviated style corresponded to the *brut*, calligraphic and 'convulsive' quality being simultaneously explored by French artists and writers, the 'sort of basic art' that Paolozzi described. They eschewed narrative and narrow historical reference, and instead advanced multiple, transhistorical motifs which did not so much blend as syncopate.[146] These montage techniques (visual and literal) moved on from Surrealism's concern with the mysteries of the mind to an existential engagement with the real. The art that resulted, though not illustrational in any conventional sense, nonetheless rose out of the materialist approaches of the circle, their military awareness, interest in zoological and anthropological science, and their Benjaminian treatment of found images, for instance. Furthermore, the transmedia ubiquity of Cold War thinking, a natural consequence of its real and inescapable relevance to every individual and branch of life, catalysed a cross-cultural dialogue and imagery intensified by the war and post-war interdependence of the London and Paris avant-gardes. Outsider painters and sculptors who were neither established nor official war artists had spent the conflict in other roles, often secret. This inflected and expanded their post-war vision and made them sensitive to a 'cold war feeling'. The work that resulted was a clearly conceived alternative to British Neo-Romanticism that did not idealize nuclear nature.

By the mid-1950s concepts of holocaust and apocalypse had been superseded by annihilation. The global inevitability of radiation spread was dramatized memorably in Nevil Shute's novel, *On the Beach*, prefaced by T.S. Eliot's words from *The Hollow Men*: 'This is the way the world ends.'[147] In 1956, the Hungarian and British artists Peter and Joan Foldes made a shocking six-minute animation of a nuclear end of the world which would receive international acclaim. It was called *A Short Vision* and can be viewed at the British Film Institute site.[148] Peter Foldes had been one of the first to exhibit at the Hanover Gallery, showing mobiles in skeletal and insect forms in 1948. His and Joan's film synthesized many of the features of the early Cold War aesthetic. Its starkly unsentimental image of nature, its depiction of a wide dark city and its dramatization of faces and bodies melting down to their skeletons were all executed in a 'sudden', 'basic' graphic style. Fascinatingly, its final image, before finishing in a dying flame and darkness, is entirely abstract, the colourful all-over dots and dashes that were used to represent the nuclear explosion in the 1946 film *One World or None*, by 1956 associated with Pop and Abstract Expressionist styles.

Acknowledgements

With gratitude to Tate, the Estates of Francis Bacon, Isabel Rawsthorne and Patricia Lambert, and Daily Express Newspapers for their generous support through the writing and publication of this essay, and to Biddy Noakes, Martin Harrison, Véronique Wiesinger and Martin Guy for their inspiration and patient advice.

Notes

1 Eduardo Paolozzi, interview with Frank Whitford, National Life Story Collection: Artists' Lives (10 June 1993; 4 May 1994; 30 December 1994; and 5 January 1995), British Library, C466/17: 249–50.

2 Selected to represent British art in *London, Paris, New Trends in Painting and Sculpture: An Exhibition of the Work of 16 Artists*, Institute of Contemporary Arts, New Burlington Galleries, Old Burlington Street, London W1, 7 March–4 April 1950; London was represented by Isabel Rawsthorne, Francis Bacon, Lucian Freud, John Craxton and Peter Lanyon, with the sculptors Robert Adams, Reg Butler and Fred McWilliam. Paris sent older artists, some working in an abstract style, and Germaine Richier.

3 Initially called the 'Young Group' meeting first 'early in 1952;, at the ICA arranged by the Deputy Director Dorothy Moreland: Anne Massey, *The Independent Group: Modernism and Mass Culture in Britain, 1945–1959* (Manchester: Manchester University Press, 1995), 46.

4 David Mellor, 'Existentialism and Post-War British Art', in *Paris Post War: Art and Existentialism 1945–55*, ed. Francis Morris (London: Tate, 1993), 53–54.

5 Martin Harrison, *Transition: The London Art Scene in the Fifties*, London: Barbican Art Gallery, 2002 – for the Cold War flavour, see page 25; Catherine Jolivette, 'Science, Art and Landscape in the Nuclear Age', *Art History*, vol. 35, no. 2 (April 2012), also published in *British Art in the Cultural Field, 1939–69*, ed. David Peters Corbett and Lisa Tickner (Oxford: Wiley-Blackwell, 2012).

6 Isabel Rawsthorne, autobiographical notes, Tate Archive, 9612.2.4.1.

7 Carol Jacobi, 'Cat's Cradle: Francis Bacon and the Art of Isabel Rawsthorne', *Visual Culture in Britain*, vol. 10, no. 3 (December 2009): 294–296.

8 Lee Richards, *The Black Art: British Clandestine Psychological Warfare against the Third Reich* (Peacehaven: Psywar, 2010), 132.

9 Paolozzi was in the Alien Pioneer Corps, working with the Ordinance Corps.

10 This idea is explored in detail by Gregor Dallas, *1945: The War that Never Ended* (London and New Haven: Yale University Press, 2006).

11 George Orwell, 'You and the Atomic Bomb', *Tribune*, 19 October 1945.

12 Isabel Rawsthorne, unpublished autobiography, private collection, 71; an earlier manuscript version is held in the Tate archive with slightly different page numbering.

13 Edward Ward on Buchenwald, BBC Home Service, 1 April 1945; Richard Dimbleby on Belsen, 19 April 1945; Alan Moorhead, an associate of Rawsthorne, published 'Glimpses of Germany II, Belsen', *Horizon*, July 1945, 26–34.

14 Rawsthorne, autobiography, 79; she and Giacometti lived at his studio off the Rue d'Alesia and Rose Pulham at Hotel Bisson on the Quai des Grand Augustins.

15 British Neo-Romanticism was formally identified and labelled by the artist and critic Robin Ironside, *Painting Since 1939* (London: British Council, 1946).

16 Lawrence Alloway, *Nine Abstract Artists: Their Work and Theory* (London: Tiranti, 1954), 2.

17 'Tragic Painters', *The Times*, 18 June 1938, 12.

18 Paolozzi, interview, 106–107.

19 Sarah Wilson, 'Paris Post War: In Search of the Absolute', in *Paris Post War*, 26–27.

20 Jean-Paul Sartre, *L'existentialisme est un humanisme*, Club Maintenant, 29 October 1945, published as *L'existentialisme est un humanisme* (Paris: Editions Nagel, 1946).

21 Simone de Beauvoir, *La force des choses 1944–51* (1963); *Force of Circumstance*, trans. Richard Howard (Harmondsworth: Penguin, 1968), 47.

22 Sartre, *L'existentialisme est un humanisme.*

23 Jean-Paul Sartre, 'La République du Silence', *Les Lettres Français*, 9 September 1944.

24 Albert Camus, *Le Mythe de Sisyphe: Essai sur l'absurde* (Paris: Gallimard, 1942); *The Myth of Sisyphus*, trans. Justin O'Brian (1955) (Harmondsworth: Penguin, 2000), 11.

25 Michel Leiris, Journal, 4 November 1945, in *Journal: 1922–1989*, ed. Jean Jamin (Paris: Gallimard, 1992), 422.

26 Rawsthorne, autobiography, 88; Patty Lambert, 'What I Remember about Isabel', manuscript, private collection, 6; Tabou opened and became a venue for American bebop on 17 April 1947, Antony Beevor and Artemis Cooper, *Paris After the Liberation, 1944–1949* (London: Doubleday, 1994), 246.

27 Andree Collie, 'Souvenirs Sur Soutine', *Le Spectateur des Arts*, vol. 1 (December 1944): 14–18; Raymond Cogniat, *Soutine* (Paris: Editions du Chene, 1945); see Martin Hammer, 'Found in Translation: Chaim Soutine and English Art', *Modernist Cultures*, vol. 5 (2010): 218–242; Antonin Artaud, *Van Gogh Le Suicidé de la Société* (February 1947) (Paris: K éditeur, 1947), in *Artaud Anthology*, ed. Jack Hirschman (San Francisco: City Lights Books, 1965), 140; trans. 'Van Gogh The Suicide Provoked by Society', *Horizon*, vol. 17, no. 97 (January 1948): 46–49.

28 Jean-Paul Sartre, 'La Recherche de l'Absolu', in *Alberto Giacometti, Exhibition of Sculptures, Paintings, Drawings*, catalogue (New York: Pierre Matisse Gallery, 1948), 4; also published in French: *Les Temps Modernes*, 28 January 1948, 1153–1163.

29 Theodora Fitzgibbon, *With Love: An Autobiography, 1938–46* (London: Century, 1982), 136; David Mellor and Val Williams, *Too Short a Summer: The Photographs of Peter Rose Pulham, 1910–1956* (York: Impressions Gallery of Photography, 1979), 10; George Melly remembers them as friends and fellow painters during and after the war, *Don't Tell Sybil: An Intimate Memoir of ELT Mesens* (London: Heinemann, 1997), 91; Patty Lambert records that Rawsthorne and Bacon came across each other during the war, notebook, private collection, 3v; and Rawsthorne recorded that she got to know Bacon 'in Paris with Peter', autobiographical notes, Tate Archive, MS 9612.2.4.3; he is a constant presence in Rawsthorne's correspondence after his return to London in November 1948.

30 Robert Melville, 'Eduardo Paolozzi', *Horizon* (September 1947): 212–213; Paolozzi, interview, 61.

31 Isabel Rawsthorne, postcard to A. Giacometti, 8 May 1947, FAAG, 20003-1890, *Correspondances*, 88–89; David Sylvester, oral testimony, 1988, cited in Laurence de Buzon-Vallet, *Alberto Giacometti: la collection du Centre Georges Pompidou, Musée national d'art moderne: Saint-Etienne, Musée d'art moderne*, catalogue (Paris: Centre Pompidou, 1999), 124.

32 James Hyman, *The Battle for Realism: Figurative Art in Britain During the Cold War 1945–1960* (New Haven and London: Yale University Press, 2001), 2.

33 *Samedi Soir*, 3 May 1947, 1, translated by the author.

34 Paolozzi, interview, 79.

35 Jean-Paul Sartre, 'Qu'est-ce que la litterature?', *Les Temps Modernes*, 17 February 1947–22 July 1947; a reprint (Paris: Gallimard, 1978) was in Bacon's studio at his death, Bacon's Books, Donation 2, Dublin City Art Gallery, The Hugh Lane.

36 Ministry of Information, *Make Do and Mend* (London: HMSO, 1943); the campaign to recycle was launched in 1941 and 'make do and mend' became a widespread slogan, Ina Zweiniger-Bargielowska, *Austerity in Britain: Rationing, Controls, and Consumption, 1939–1955* (Oxford: Oxford University Press, 2000), 120; Paolozzi linked his collages to a childhood milieu in which 'you didn't throw anything away', interview, 8.

37 Paolozzi, interview, 94.

38 Henry Ford, interview with Charles Wheeler, 'Fight to Disarm His Life's Work, Henry Ford Vows', *Chicago Tribune*, 25 May 1916, 10.

39 Walter Benjamin, 'On Some Motifs in Baudelaire', *Zeitschrift fur Sozialforschung* (1939), 8, 1–2, reproduced in *Illuminations Walter Benjamin*, ed. Hannah Arendt (London: Pimlico, 1999), 189.

40 The best known is T.S. Eliot's 'mythical method': T.S. Eliot, 'Ulysses, Order, and Myth', *The Dial* (November 1923), in *Selected Prose of T.S. Eliot* (London: Faber and Faber, 1975), 177–178, but see also Carola Giedion-Welcker, *Moderne Plastik: Elemente der Wirklichkeit, Masse und Auflockerung* (Zürich: H. Girsberger, 1937), trans. *Modern Plastic Art: Elements of Reality, Volume and Disintegration* (Zurich: H. Girsberger, 1937), 18.

41 Amédée Ozenfant, *Art* (Paris: Budry, 1928); Amédée Ozenfant, *The Foundations of Modern Art* (1928), trans. John Rodker (New York: Brewer, Warren and Putnam, 1931); Paolozzi, interview, 35, 73, 97.

42 'Sigmund Freud taking his first airplane flight', 1926, Robin Spencer, *Eduardo Paolozzi: Writings and Interviews* (Oxford and New York: Oxford University Press, 2000), 22; Freud was the cover story in *Time*, 26 June 1939. Einstein arriving at Newark Airport on 23 March 1939, 'Einstein Story', *Life*, 3 April 1939, 20; 'Adolph Hitler takes Prague and admires the castle view', *Life*, 3 April 1939, 18–19.

43 George Soule, 'The Lessons of Last Time', *The New Republic*, vol. 106 (2 February 1942): 163–184.

44 Paolozzi, interview, 18–23; Paolozzi remembered 10 June 1940, when Italy declared war on Britain: 'my father listening to this speech, and knowing what the consequences were, and we had framed in the back shop, even a picture of Mussolini, and we had a big map of the Abyssinia, which had been supplied by the Fascist club, and every day from the radio he would put the positions where the Italian army advanced from the north and up from the south ...'

45 Most notably, Brooke Rapaport, 'The Greater Mystery of Things: Aspects of Vital Forms in American Art', in *Vital Forms: American Art and Design in the Atomic Age 1940–1960*, ed. Brooke Kamin Rapaport and Kevin Stayton (New York: Brooklyn Museum of Art, 2001), 78–121, catalogue for exhibition 12 October 2001–6 January 2002; see also Denise M. Rompilla, *From Hiroshima to the Hydrogen Bomb: American Artists Witness the Birth of the Atomic Age* (published PhD thesis, University of New Jersey) (ProQuest, 2008).

46 Civilian, 'The New Weapon', *The Times*, 28 September 1939, 9.

47 BBC Home Service, 6 August 1945.

48 *The Evening Telegraph*, 8 August 1945, 1.

49 These are now collected in the Hiroshima Peace Memorial Museum, opened in 1955; Kyo MacLear, *Beclouded Visions: Hiroshima-Nagasaki and the Art of Witness* (Albany, NY: SUNY Press, 1999), 41.

50 *Mainichi Shimbun*, 21 August 1945, 1.

51 'War's Ending', *Life*, 20 August 1945, 25–27.

52 Accompanying the report, *The Atomic Bombings of Hiroshima and Nagasaki by the Manhattan Engineer District*, 29 June 1946, National Archive Washington DC, United States Air Force Strategy Bombing Survey, RG 243 Series H.

53 See rare colour image of 'First Atomic Bomb', *Life*, 19 November 1945, 107; Barbie Zelizer, *Remembering to Forget: Holocaust Memory Through the Camera's Eye* (Chicago: University of Chicago Press, 2000), 33. For another good example, see 'The drawing shows more graphically than aerial photographs (pp. 26, 27) effect of atomic bomb hit on Hiroshima, smoke billows 40,000 feet', *Life*, 20 August 1945, 25.

54 'Dresden Ablaze in First RAF Raid', *Daily Sketch*, 15 February 1945, 1; J.R. Eyerman, 'Atom Bomb Victims' in 'Tokyo Express Arrives at Hiroshima', *Life*, 8 October 1945, 34; a rare image of the destroyed city itself is Bernard Hofmann's 'The head of a statue of Christ severed by the bomb blast at Nagasaki, lies before the ruins of a Roman Catholic cathedral', but little is to be seen of the wider city, *Life*, 15 October 1945, 37.

55 John Hersey, *Hiroshima* (London: Penguin, 1946), v–vi; 'Editorial', *Life*, 20 August 1945, 30; David Pace, 'Old Wine – New Bottles: Atomic Energy and the Ideology of Science in Post-War France', *French Historical Studies*, vol. 17, no. 1 (Spring 1991): 49.

56 The Acheson-Lilienthal and Baruch Plans, presented to the United Nations Atomic Energy Commission (UNAEC), 14 June 1946. For a summary of these narratives in the USA, see Robert Jacobs, *The Dragon's Tail: Americans Face the Atomic Age* (Amherst: University of Massachusetts Press, 2010), 4–5.

57 Mary (Freda) Kirchwey, 'One World or None', *The Nation*, 18 August 1945; see also Norman Cousins, 'Modern Man Is Obsolete', *Saturday Review of Literature*, 14 August 1945, 5.

58 Dexter Masters and Katharine Way, eds, *One World or None: A Report to the Public on the Full Meaning of the Atomic Bomb* (New York: McGraw-Hill Book Company, 1946).

59 *One World or None*, 1947, Philip Ragan and the National Committee on Atomic Information; see Spencer Weart, *Nuclear Fear: A History of Images* (Cambridge, MA: Harvard University Press, 2009), 115–116; it can be viewed at <http://publicdomainreview.org/2012/11/23/one-world-or-none-1946/>.

60 John Hersey, 'Hiroshima', *The New Yorker*, 31 August 1946; broadcast and published shortly afterwards, John Hersey, *Hiroshima* (New York: Alfred A. Knopf, Inc., 1946).

61 'Comment', *Horizon* (December 1945): 365; see also Noel Busch, 'Occupation of Japan', *Horizon*, vol. 16, no. 32 (September 1947): 159–181; Joe Alsop, 'American Foreign Policy', *Horizon*, vol. 16, no. 93–94 (October 1947): 68–80.

62 Gavin Parkinson, *Surrealism, Art and Modern Science* (London and New Haven: Yale University Press, 2008), 201–202; Jean Duché, 'Andre Breton nous parle', *Le Figaro literaire*, 5 October 1946.

63 Rawsthorne, autobiographical notes, 19; Fitzgibbon, *With Love*, 69; Giacometti, letter to Rawsthorne, Thursday, January 1940, *Correspondances*, 56.

64 25 September 1941, Tube Alloys Committee, KGB Archives file number 13676, vol. 1, Jeffery Richelson, *A Century of Spies: Intelligence in the Twentieth Century* (Oxford: Oxford University Press, 1997), 137; Michael Goodman, *Spying on the Nuclear Bear: Anglo-American Intelligence and the Soviet Bomb* (Stanford: Stanford University Press, 2007), 69–70.

65 'La Premiere Bombe Atomique', *Combat*, 7 August 1945, 1; Albert Camus, 'Editorial', *Combat*, 8 August 1945.

66 Jean-Paul Sartre, 'La Libération de Paris: une semaine d'apocalypse', *Clarté*, vol. 9 (24 August 1945): 1; trans. Richard McClearly, *The Prose Writings of Jean Paul Sartre*, ed. Michel Contat and Michael Rybalka (Evanston: Northwestern University Press, 1974), 162–164.

67 A.J. Ayer, *Part of My Life* (London: Collins, 1977), 281–289; Rawsthorne, autobiography, 84–85; Bertrand Russell published his best-known work, *History of Western Philosophy* (London: George Allen & Unwin) in 1945; Rawsthorne's admiration was confirmed by Corinna MacNeice, in conversation with the author, 10 April 2011, to whom she gave a copy; a battered 1981 reprint was in Bacon's studio at his death, Bacon's Books, Donation 1, Dublin City Art Gallery, The Hugh Lane.

68 Bertrand Russell, 'The Bomb and Civilization', *Forward*, vol. 39, no. 43 (August 1945); 'The present atomic bomb', speech given in the House of Lords, *Hansard*, 28 November 1945, columns 87–92; 'What I believe', BBC Third Programme, 1947, published as *Atomic Challenge: A Symposium* (London: BBC, 1947).

69 Cover, *Picture Post*, 6 August 1945; on 29 November 1945, *Life* magazine discussed the possibility of a 36-hour war, 27–38; first atomic bomb second epoch in history, 107.

70 Wilfred Burchett, 'The Atomic Plague: "I Write This as a Warning to the World"', *Daily Express*, 5 September 1945, 1; a full account appeared in Wilfred Burchett, *Shadows of Hiroshima* (London: Verso Books, 1983).

71 Isabel Rawsthorne, autobiography, private collection, 63; For Rawsthorne's involvement in the Spanish Civil War, for example, see Sefton Delmer, *Trail Sinister: An Autobiography Volume 1* (London: Secker & Warburg, 1961), 263–268; a full account is forthcoming in Carol Jacobi, *Out of the Cage: The Art of Isabel Rawsthorne* (London and New Haven: Yale University Press, 2014), chapter 6.

72 *Effects of the Atomic Bomb*, Nichiei Kyo Productions, premiered Tokyo, July 1952; see MacLear, *Beclouded Visions*, 42–43; 'When Atom Bomb Struck – uncensored', *Life*, 29 September 1952, 19–25.

73 See, for example, the gaudy mushroom cloud and headline, 'Britain's 1st Atom Bomb Test', *Picture Post*, vol. 57, 11 October 1952, 2.

74 *The Family of Man*, Museum of Modern Art, 24 January–8 May 1955.

75 In Paris, Andre Fougeron's *Hommage à Andre Houiller* of 1949 heroized an anti-nuclear bill-sticker, while Bernard Lorjou's *L'Age Atomique* (1950, Centre Pompidou) heralded the nuclear future; see especially Italian Arte Spaziale and Arte Nucleare group, founded in 1948, Voltolino et al., *Manifesto dell'Eaismo* (Livorno: Società editrice italiana, 1948); Stephen Petersen, 'Explosive Propositions: Artists React to the Atomic Age', *Science in Context*, vol. 17, no. 4 (2004): 587–589.

76 Parkinson, *Surrealism, Art and Modern Science*, 206.

77 A. Giacometti, letter to Pierre Matisse (1947), *Alberto Giacometti, Exhibition of Sculptures, Paintings, Drawings*; M. Leiris, 'Alberto Giacometti', *Documents*, vol. 4 (1929); Patrick Elliott, 'Drawings by Alberto Giacometti. Paris', *The Burlington Magazine*, vol. 143, no. 1178 (May 2001): 315; Julia Kelly has shown that the pairing of Giacometti and Sartre for the artist's first two major solo post-war exhibitions had a striking impact on the way that his work and his life was perceived, 'Alberto Giacometti, Michel Leiris and the Myths of Existentialism', in *Giacometti: Critical Essays*, ed. Julia Kelly and Peter Read (Farnham: Ashgate, 2009), 151–169; Véronique Wiesinger, *The Studio of Alberto Giacometti* (Paris: Centre Pompidou, 2007); for example, *Art of the 20th Century* heads its section on this period: 'Giacometti and the Threat to Humanity', *Art of the 20th Century*, vol. 1 (London: Taschen, 2000), 287.

78 William Shakespeare, *King Lear*, Act 3, Scene 4, M. Leiris, 'Thoughts around Alberto Giacometti', trans. D. Cooper, *Horizon*, vol. 114 (June 1949): 411.

79 Andrew Richie, *Sculpture of the Twentieth Century* (New York: Museum of Modern Art, 1952), 33; see also Francis Ponge, 'Reflections sur les statuettes, figures et peintures d'Alberto Giacometti', *Cahiers d'Art* (1951).

80 Rawsthorne, diary, 1947, TGA 9612.2.7, written on the pages for March.

81 Jean Fautrier, *Otages*, Foutrier Galerie René Drouin, November 1945; *Mirobolus, Macadam & Cie, Hautes Pâtes de Jean Dubuffet*, Galerie René Drouin, June 1946; Antonin Artaud, *Portraits et Dessins par Antonin Artaud*, Galerie Pierre, June 1947.

82 Eduardo Paolozzi, *Study for Sculpture*, n.d, lot 167, 20th Century British Art, Bonhams, 29 November 2005; Robin Spencer, conversation with the author, 16 March 2009; other examples are *Fish*, collage, 1946, private collection, developed into *Fish*, plaster, 1948.

83 *Germaine Richier*, Anglo-French Art Centre, 1947. Germaine Richier was considered a major figure and crucial influence by David Sylvester, *Germaine Richier*, catalogue (London: Hanover Gallery, 1955); see Sarah Wilson, 'Germaine Richier, Disquieting Matriarch', *Sculpture Journal* (December 2005): 51–70; Darrell Davisson presents *La Grande Mante* (The Great Mantis, 1946) as a post-nuclear image, *Art After the Bomb: Iconographies of Trauma in Late Modern Art* (Bloomington, IN: Authorhouse, 2008), 51.

84 Diane Kirkpatrick, *Eduardo Paolozzi* (London: Studio Vista, 1970), 18.

85 Paolozzi, interview; D'Arcy Wentworth Thompson, *On Growth and Form* (1917), revised edition (Cambridge: Cambridge University Press, 1942); although it is an English text, it is known that Paolozzi discovered it in Paris, and Rawsthorne had a history of reading zoological texts inherited from her father, an amateur zoologist, unpublished autobiography, private collection, 1–2; Nigel Henderson, *Nigel Henderson: Paintings, Collages and Photographs*, catalogue (London: Anthony d'Offay Gallery, 1977).

86 Isabel Rawsthorne, letter to Peter Rose Pulham, late 1948, Tate Archive 9612.1.3.13.

87 *Isabel Lambert Recent Paintings*, February–March 1949; Isabel Rawsthorne, letter to Peter Rose Pulham, late 1948, Tate Archive 9612.1.3.13.

88 For example, 'Science: Atomic X Ray', *Time*, 24 September 1945.

89 Paolozzi would have known Giacometti's *Table* (1933, then in the collection of Don du Vicomte Charles de Noailles); Kirkpatrick, *Eduardo Paolozzi*, 18.

90 ICA, *Aspects of British Art, 1950–51*, 13 December 1950–12 January 1951, as 'Standing Sculpture'.

91 *Eduardo Paolozzi Drawings and Bas-Reliefs*, Mayor Gallery, 10–31 May 1949; *Kenneth King, Eduardo Paolozzi, William Turnbull*, 21 February–18 March 1950.

92 *Growth and Form*, ICA, 4 July–31 August 1951.

93 *40,000 Years of Modern Art: A Comparison of Primitive and Modern*, ICA, 20 December 1948–29 January 1949, *Discussion Forum: Is Art an Essential or an Accessory to Society?* December 1948; Siegfried Giedion, 'Palaeolithic Art and Modernism', ICA, 1950, later written up in Siegfried Giedion, 'Space Conception in Prehistoric Art, in *Explorations in Communication*, ed. Edward Carpenter and Marshall McLuhan (Boston: Beacon Press, 1960), and Siegfried Giedion, *The Eternal Present: The Beginnings of Art* (New York: Pantheon Books, 1962).

94 Petersen, 'Explosive Propositions', 591.

95 Alberto Giacometti, in conversation with Grazia Livi, 'Interrogghiamo gli artisti del nostro tempo: Che cosa ne pensano del mondo d'oggi? Giacometti', *Epoca*, no. 643 (Milan, 20 January 1963): 58–61, trans. Reinhold Hohl, *Alberto Giacometti: Sculpture, Painting, Drawing* (London: Thames and Hudson, 1972), 283.

96 Wilson, 'Germaine Richier: Disquieting Matriarch', 53.

97 For example, 'Comparative Illustration of Archaeopteryx and Pigeon', in Gerhard Heilmann, *The Origin of Birds* (London: H.F. & G. Witherby, 1926). The theory received impetus in 1942 from William Beebe's 'Evolution of a Lizard into a Bird', *Proceedings of the Eighth American Scientific Congress*, vol. 3 (1942): 296, but Rawsthorne does not seem to have discovered Beebe until 1948, after conceiving the skeleton paintings.

98 Alberto Giacometti, *Paris Sans Fin (1957–65)* (Paris: Editions Verve, 1969), plates 80–84, text reproduced in *Alberto Giacometti Ecrits: Articles, Notes et Entretiens* (Paris: FAAG/Hermann, 2007), 159, trans. Elizabeth Cowling, 'Elective Affinities: Giacometti in the Palace of Bones', *Apollo* (October 2003): 34–42.

99 *The Times*, 8 March 1949, 7.

100 Rawsthorne visited the Aquarium of the Stazione Zoologica in Naples in 1949, and Paolozzi in 1952, Rawsthorne, letter to Giacometti, October 1947, FAAG, inv. 2003 1891, *Correspondances*, 90; Robin Spencer, Eduardo Paolozzi, *Fish* (1949, private collection), Lot Notes, *20th Century British Art*, 8 June 2007, Christies, London, King Street.

101 William Beebe, *Half Mile Down* (New York: Harcourt, Brace and Company, 1934); Rawthorne asked Rose Pulham to tell Paolozzi that she had ordered Beebe's last book, *The Book of Bays* (New York: Harcourt, Brace, 1942), letter to P. Rose Pulham, early 1949, TGA 9612.1.3.65; 37.

102 I. Rawsthorne, letter to A. Giacometti, 1948, FAAG, inv. 2003-1897, 99; Beebe, *Half Mile Down*, 162–166.

103 N.G., *The Observer*, 22 February 1959, 18; Martin Harrison, *In Camera: Francis Bacon: Photography, Film and the Practice of Painting* (London: Thames and Hudson, 2005), 106; Martin Harrison, *Francis Bacon: A Terrible Beauty* (Dublin: Dublin City Art Gallery, The Hugh Lane, 2008), 75.

104 David Sylvester, 'Paolozzi and Turnbull's Sculptures', in *Kenneth King, Eduardo Paolozzi, William Turnbull*, catalogue (London: Hanover Gallery 1950); the exhibition ran from 21 February to 18 March 1950.

105 Wallis Neville, 'Nightmare', *The Observer*, 20 November 1949, 6; Percy Wyndham Lewis, 'Round the London Art Galleries', *The Listener*, vol. XLII, no. 1086 (17 November 1949): 860.

106 Paolozzi, interview, 49.

107 I. Rawsthorne, letter to P. Rose Pulham, late 1952, TGA 9612.1.3.74; *Recent Trends in Realist Painting*, ICA, 2 July–2 August 1952.

108 For example, *Looking east from inside the Hiroshima Military Area, near the junction of the street car line*, United States Air Force Strategy Bombing Survey, National Archive Washington DC, RG 243 Series H.

109 Giacometti, letter to Pierre Matisse.

110 Michel Leiris, 'Alberto Giacometti', *Documents*, vol. 4 (September 1929): 129.

111 Paul Valéry, *Analecta* (Paris: Gallimard, 1935), 264, quoted in Walter Benjamin, 'Uber einige Motive bei Baudelaire' (On Some Motifs in Baudelaire), *Zeitschrift fur Sozialforschung*, vol. 8 (1939): 1–2, reproduced in *Illuminations: Walter Benjamin*, ed. Arendt, 158.

112 Benjamin, 'Uber einige Motive bei Baudelaire', 171, 166.

113 Antonin Artaud, *Le Théâtre et son Double* (Paris: Gallimard, 1938), re-published in Antonin Artaud, *Le Théâtre de la Cruauté*, 5–6 spécial Artaud, *Revue 84*, Paris, 1948, translated in *The Theory of the Modern Stage*, ed. Eric Bentley (Harmondsworth: Penguin, 1968), 66.

114 Artaud, *Van Gogh Le Suicidé de la Société*, 140.

115 André Malraux, preface, *Jean Fautrier Otages*, catalogue (Paris: Galerie René Drouin, 1945).

116 Antonin Artaud, 'Le Visage Humain', *Portraits et Dessins par Antonin Artaud*, catalogue (Paris: Galerie Pierre, 1947).

117 Germaine Richier, quoted in Luca Barbero, *Germaine Richier*, catalogue (Venice: Peggy Guggenheim Collection, 2006), 24.

118 William Turnbull, interview with Martin Harrison, *Transition: The London Art Scene in the Fifties*, catalogue (London: Barbican Arts Gallery, 2002), 112.

119 Rawsthorne, diary, 1947; Isabel Rawsthorne, letter to Alberto Giacometti, 1948, Véronique Wiesinger, *Alberto Giacometti, Isabel Nicholas, Correspondances* (Paris: Fondation Alberto et Annette Giacometti, 2007), 97; also I. Rawsthorne, letter to P. Rose Pulham, 14 June 1949, TGA 9612.1.3.26.

120 Francis Bacon, quoted in David Sylvester, *The Brutality of Fact: Interviews with Francis Bacon* (London: Thames and Hudson, 2008), 65, 82.

121 Francis Bacon, 'Matthew Smith – A Painter's Tribute', in *Matthew Smith: Paintings from 1909 to 1952*, catalogue (London: Tate Gallery, 1953), 12.

122 Louis Réard, *Bikini*, unveiled at the Piscine Molitor, 5 July 1946.

123 Lyn Gamwell, *Exploring the Invisible: Art, Science, and the Spiritual* (Princeton: Princeton University Press, 2002), chapter 12; Petersen, 'Explosive Propositions', 571, 587–589.

124 *British Painters Past and Present*, Lefevre Gallery, July–August 1946; *Seventh Exhibition*, Anglo-French Art Centre, November–December 1946.

125 Giacometti, *Nude in the Studio* (1945, Centre Georges Pompidou); *Standing Nude* (1946, private collection), AGD 136.

126 As early as 1978, Hugh Davies connected the umbrella to various precedents, including that in Ford Madox Brown's *The Last of England* (1852–1855, Birmingham Museums and Art Gallery) or the umbrella that became the emblem for the pre-war Prime Minister Neville Chamberlain and his appeasement policies, *Francis Bacon: The Early and Middle Years, 1928–1958* (PhD thesis, Princeton University, 1975) (New York and London: Garland, 1978), 72.

127 Robert Melville, 'Francis Bacon', *Horizon* (December 1949): 419–422.

128 Biddy Noakes noted Isabel Rawsthorne's enthusiasm for Robert Yerkes and Konrad Lorenz, in conversation with the author, November 2007; M. Harrison, 'Francis Bacon Extreme Points of Realism', in *The Violence of the Real*, ed. Armin Sweite (London: Thames and Hudson, 2006), 54, n. 27; a paint-splattered copy of Robert Yerkes and Ada Yerkes, *The Great Apes: A Study Of Anthropoid Life* (1929) (London: Yale University Press, 1945) was in Bacon's studio at his death, Bacon's Books Donation 1, Dublin City Art Gallery, The Hugh Lane. Bacon was later a fan of J.Z. Young's *Introduction to the Study of Man* (Oxford: Oxford University

Press, 1971), shown to Hugh Davies, 'Interviewing Bacon, 1973', in *Francis Bacon: New Studies*, ed. Martin Harrison (Göttingen: Steidl, 2009), 111.

129 Eric Hoskings and Cyril Newberry, *Birds in Action* (London: Collins, 1949), plate 76; Ludwig Grunwald, *Atlas-Manuel des maladies de la bouche* (1894), translated from the German by Georges Laurens (Paris: Bailliere, 1903); Francis Bacon, interview, 17 March 1973, quoted in Davies, 'Interviewing Bacon, 1973', 97.

130 Kathleen Clark, *Positioning in Radiography* (London: Heinemann Medical Books, 1939); Francis Bacon, interview with David Sylvester (1962), *Interviews*, 32.

131 Martin Hammer, *Francis Bacon and Nazi Propaganda* (London: Tate, 2012), 52–53.

132 Sam Hunter, 'Anatomy of Horror', *Magazine of Art* (January 1952): 12; Bacon's Nazi interests are explored in detail by Chris Stephens and Martin Hammer in '"Seeing the Story of One's Time": Appropriations from Nazi Photography in the Work of Francis Bacon', *Visual Culture in Britain*, vol. 10, issue 3 (2009): 315–351, and in Hammer, *Francis Bacon and Nazi Propaganda*.

133 'What is De Gaulle Up To?', *Picture Post*, 26 April 1947, 26; Harrison, *In Camera*, 66.

134 Harrison, *In Camera*, 102–103.

135 Joseph Goebbels, 'Das Jahr 2000', *Das Reich*, 25 February 1945, 1–2; Winston Churchill, *Hansard*, House of Commons, c84, 16 August 1945, vol. 413.

136 David Seymour's 1949 photograph of Pius in his *sedia* was in Bacon's studio, Hunter, 'Anatomy of Horror', 12.

137 *Life*, vol. 19, no. 18, 29 October 1945, 37–38.

138 'Russia's Gromyko – What would be the judgment of history?', from 'United Nations: Negative Neanderthaler', *Time*, 18 August 1947, cover story.

139 Rawsthorne, letter to P. Rose Pulham, June 1949, TGA 9612.1.3.31.

140 Ozenfant, *Art*; Ozenfant, *The Foundations of Modern Art*; Harrison, *In Camera*, 12–13.

141 See, for example, *Picture Post*, vol. 53, 27 October 1951, 4; the *coup d'etat* was extensively covered in *Life*, 25 August 1952, 102–114.

142 Ian Fleming, *Casino Royale* (London: Jonathan Cape, 1952); Fleming associated with Delmer as a journalist in Russia and with Rawsthorne and Delmer socially and in intelligence during the war, Delmer, *Trail Sinister*, 387; Rawsthorne, unpublished autobiography, 53–54.

143 Rawsthorne, unpublished autobiography, 19; also, letter to Rose Pulham, 21 October 1953, in which she discusses the defection of Maclean's wife and speculates that he must therefore still be alive, TGA 9612.1.3.97.

144 Cyril Connolly, *The Missing Diplomats* (London: Queen Ann Press, 1952).

145 Goodman, *Spying on the Nuclear Bear*, 72–75.

146 Lee Fahnestock, 'Introduction', Francis Ponge, *The Nature of Things*, trans. Lee Fahnestock (New York: Red Dust, 1995), 8–9.

147 Nevil Shute, *On the Beach* (London: Heinemann, 1957).

148 *A Short Vision*, directed by Peter and Joan Foldes, British Film Institute, 1956: <http://www.screenonline.org.uk/film/id/442000/>, accessed 16 April 2013.

2

Geometries of Hope and Fear: The Iconography of Atomic Science and Nuclear Anxiety in the Modern Sculpture of World War and Cold War Britain

Robert Burstow

In the 1952 monograph of the sculptor Barbara Hepworth, her friend and apologist Herbert Read observed that her work since 1934/1935 had developed 'a dialectic ... between the antitheses of Geometry and Grace'. For Read, the geometrical aspect was especially evident in her drawings, above all those from 1936 to 1942 (Colour plate 3). He identified the origin of her formal geometries in the Constructive art of Naum Gabo and her former husband Ben Nicholson, which aimed to create a 'new reality' where art would 'enhance life, assert it and assist its further development'.[1] Like them, Hepworth signalled her desire for a new social and political order by alluding to the geometrical forms associated with modern science and mathematics. The reintegration of artistic and scientific endeavour that they imagined was exemplified by their friendship with the atomic physicist John Desmond Bernal. In common with many in the late 1930s and early 1940s they presumed that the advance of science, especially of the new field in which Bernal worked, would bring about a more enlightened, equitable and peaceful world. By the time Read's monograph was published, the geometric imagery of simplified atomic diagrams had become a ubiquitous symbol of cultural and social progress, conspicuously appropriated by advertising and interior decoration at the 1951 Festival of Britain.[2]

Just as the optimistic symbolism of atomic geometry acquired popular currency, Read provocatively redeployed the term to characterize a very different kind of modern sculpture. In the summer of 1952 – between the Hepworth monograph going to press and publication – he published an introduction to the eight 'young' British sculptors representing Britain at that year's Venice Biennale in which he famously described their sculpture as embodying 'the geometry of fear'.[3] His metaphorical phrase responded to their invention of new expressive forms that had 'a linear, cursive quality', exemplified by Reg Butler's *Woman* (1949; Figure 2.1). Coming at a moment

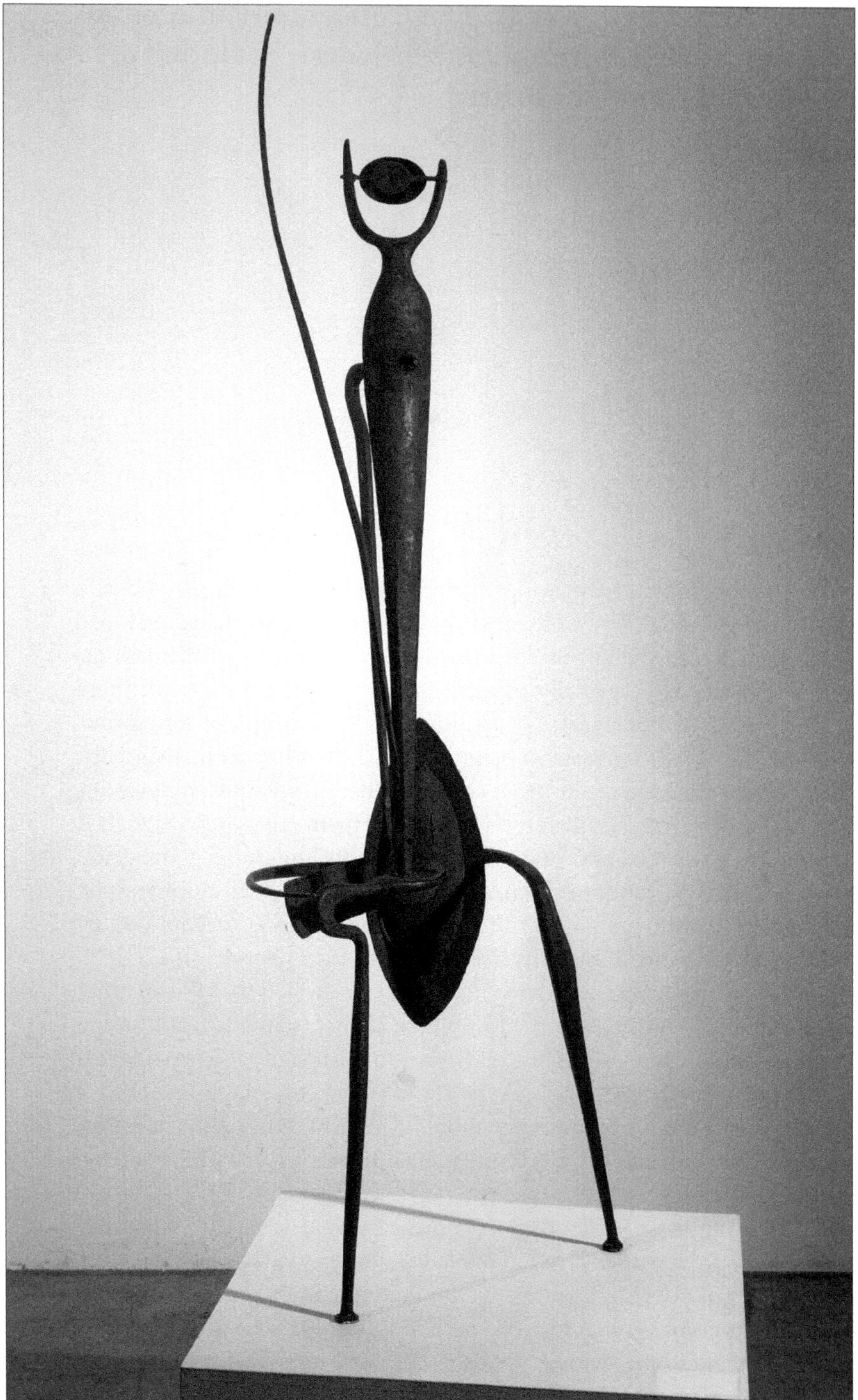

2.1 Reg Butler (1913–1981), *Woman*, 1949, forged steel, 221 × 77 × 75 cm, exhibited British Pavilion, Venice Biennale, 1952, coll. Tate Gallery, photo © Tate, London 2013.

when the ideological conflict of the Cold War had been transformed into military confrontation in Korea and increased risk of nuclear warfare, Read's memorable phrase slipped into the discourse of post-war British sculpture, shifting the semantic associations of 'geometry' away from the Constructive art of utopian Socialism (familiar to regular Biennale visitors from Hepworth's 1950 exhibit) toward an Expressionist art of dystopian anxiety. Given the former associations of geometry with proportion, perfection and purity, Read's application of the term to an art of deformation, despair and hybridity had the gloomy implication that formal, social and political unities were no longer imaginable, let alone attainable. Although Hepworth and Gabo – the most committed Constructive sculptors – continued to make geometrical works that represented their positive vision of the future, contemporary fears of Cold War enmity and global catastrophe ensured that 'the geometry of fear' overshadowed the geometry of hope.

Unlike the corresponding literature of American and continental European art,[4] comparatively little has been published on modern British sculpture's complex and changing relation to the 'atomic' or 'nuclear' age. In tracing their relation from the mid-1930s to the mid-1960s, this chapter takes Read's provocative allusion to 'geometry' in 1952 as a pivotal moment of recognition. The first half of the chapter draws on recent research into the influence of science and mathematics on Hepworth's and Gabo's optimistic Constructive art of the later 1930s and early 1940s, investigating the far-reaching but little recognized impact on them both of Bernal's atomic research. The second half of the chapter examines the more familiar association of British post-war 'humanist' sculpture with anxieties over nuclear warfare in the early years of the Cold War. It demonstrates that some sculptors, such as Henry Moore and the naturalized Hungarian émigré 'Peter' (Laszlo) Peri, made works that both overtly celebrated atomic science and opposed nuclear weaponry, while those now routinely identified as 'Geometry of Fear sculptors'[5] invented a more coded sculptural language to address the perils of the nuclear age. Throughout the chapter, a premium is put on Read's influential writings as they were critical to shaping the meaning of modern British sculpture, both during the period and after. The chapter concludes with an analysis of how Read himself and others have understood his seminal description of post-war sculpture.

Hope: Constructive Sculptors and Atomic Science (c. 1936–1945)

The friendship between Hepworth, Gabo and Bernal was characteristic of the London Constructivist circle, which brought artists, architects, designers and writers on art and design into contact with scientists and writers on science. Encouraged by Socialist convictions, they devised a Modernist programme for art and design that sought to reintegrate art and science, and to challenge the intellectual compartmentalizations of bourgeois culture. As the architect and co-editor of their 'manifesto' publication *Circle: International Survey of Constructive Art*, Leslie Martin observed: 'In science as in art, "appearance" has been jettisoned in favour of a world discovered only through the penetration of appearances.'[6] Among their scientific friends were zoologist Julian Huxley, anatomist Solly Zuckerman, biologist C.H. Waddington, geneticist C.D. Darlington and science journalist Jim Crowther. But as other art historians have demonstrated, it was the brilliant, charismatic polymath Bernal who most encouraged Hepworth's and Gabo's dialogue with science and mathematics.[7] In 1937 Bernal became Professor of Physics at Birkbeck College, University of London, having established his reputation as a leading crystallographer at the Royal Institution and Cavendish Laboratory. Known to many friends and colleagues as 'Sage', Bernal refined the practice and theory of X-ray crystallography pioneered by William Henry Bragg and his son William Lawrence Bragg, inventing cameras and photographic methods to analyse the sub-microscopic arrangement of atoms in complex molecules. As a Marxist, a pacifist and an admirer of modern art and architecture, he was especially interested in the social function of art and science.[8] In 1937, he lectured on 'Architecture and the Scientist' at the Royal Institute of British Architects and, at Hepworth's invitation, contributed an essay on 'Art and the Scientist' to the sculpture section of *Circle* and the foreword to an exhibition catalogue of her own work.[9] His enthusiasm for modern art deviated from the official Communist Party doctrine of 'socialist realism', and he argued for the social utility of Hepworth's sculptures by likening them, paradoxically and prophetically, to Neolithic menhirs.[10] By 1952, he owned two works by Hepworth and at least one by Gabo.[11]

Hepworth was introduced to Bernal in the early 1930s by their mutual friend, the writer and collector Margaret Gardiner (then Bernal's lover), who witnessed their fascination with each other's professional work.[12] Bernal's biographer, Maurice Goldsmith, records that Hepworth 'loved the visits of Bernal, who would examine her works and explain their mathematical and geometrical forms'.[13] According to their mutual friend Marcus Brumwell and Hepworth herself, they exchanged ideas excitedly and drew things for one other on the studio floor, while Bernal occasionally wrote mathematical equations in delible pencil on her sculpture (rather as mathematical models were sometimes inscribed).[14] Sculptor and crystallographer shared a common interest in the nature of matter and three-dimensional form. A few years after meeting Bernal, Hepworth's interest in crystal structures led her to accompany a mineralogist to the National Geology Museum.[15] Meanwhile, Bernal became sensitive to parallels between abstract geometric sculpture and crystallography, noting in his writing on Hepworth that: 'By reducing the traditional forms of sculpture it is possible to see the geometry which underlies it.'[16] His understanding of formal structure, including characteristics such as reflection and symmetry, is evident in his essay in *Circle*, where he likens a sculpture's internal formal relationships to the forces of attraction and repulsion between atoms, comparing Hepworth's *Two Forms* (1935) to a scientific illustration of the 'equi-potential surface of two like charges'.[17] In their fruitful correspondence, Hepworth addresses him respectfully, even deferentially, accepting his 'scientific' criticisms of her sculpture. While she acknowledges important differences between art and science, she assumes they share a rational, Constructive search for aesthetic and political truth:

> ... science, surely, does not work as emotionally as ptg or sculpture. You always seem to me to be searching for, discovering + applying basic laws + principles ... I know no one more fitted to make these laws clear to other people. Your criticism is most stimulating because you know exactly where a law has been broken + can apply the principle so that you can make the solution clear. It seems to me that we know those laws emotionally + intuitively, learning by our experience + trying to live the sort of life that will most easily enable us to express the laws – that is construction.[18]

Hepworth was deeply attracted to Bernal's political ideas: she joined the Hampstead-based anti-Fascist group, For Intellectual Liberty, in which Bernal and Gardiner played leading roles, and considered joining the Communist Party.[19] As Hepworth confided to Gardiner, his intellectual and humanitarian ideals seemed to offer a solution to the growing threat of Fascism:

> I can't see how science and the human spirit can be reconciled unless there is a clear majority of men of the calibre of Desmond ... I think there will be a new form of ethics – social and political – very *much* to the good and what we had hoped for ... I cannot see any hope of stopping this suicidal impulse unless Art and Science stand firm together.[20]

Hepworth's admiration for and contact with Bernal remained strong until the end of the Second World War, during which time the influence of crystallography grew from a means of understanding her work into an initiating subject.

Gabo's relationship with Bernal was never as close as Hepworth's, tempered no doubt by his own unfavourable first-hand experience of post-Revolutionary Russia.[21] He and Bernal became acquainted after Gabo's move from Paris to London in 1936, meeting in London and later St Ives.[22] Yet, as Martin Hammer and Christina Lodder have shown, Gabo's knowledge of science and mathematics was considerable, far greater than Hepworth's, following his early study of physics and chemistry at Munich University and his continuing appetite for scientific literature.[23] In his 1949 essay 'On Constructive Realism', Gabo used his knowledge of atomic science and relativity to defend the right of the artist to produce innovative responses to 'a world which is pictured to us as a conglomeration of oscillating electrons, protons, [and] neutrons', reasoning:

> ... If the scientist is permitted to picture to us an image of an electron which under certain conditions has less than zero energy ... and if he is permitted to see behind this simple common table, an image of a curvature of space – why, I ask, is not the contemporary artist to be permitted to search for and bring forward an image of the world more in accordance with the achievements of our developed mind ...[24]

His awareness that atomic physics undermined simplistic distinctions between solids and space has been compared to his method of constructing sculpture in sheets of clear plastic to render its form open and transparent.[25] Of Gabo's *Construction in Space: Crystal* (1937–1939), for example, Jacky Klein observes that its 'sense of dynamic forces and centrifugal energies relates to his interest in modern physics, in particular to discoveries which pictured the physical universe as a continuous field of forces'.[26] It was only after his arrival in London and meeting with Bernal that his aim of expressing 'the dynamic interior of objects' was fully realized by fabricating sculpture entirely from transparent planes.[27]

As sculptors in diverse materials, Hepworth and Gabo were understandably interested in the atomic structure of matter, which determines many of its most important characteristics, such as hardness, texture, strength and colour. But their interest in Bernal's research went beyond a mutual fascination with the parallels between cutting-edge science and art, and the shaping of their philosophical views of 'reality'. Their frequent use of crystalline forms and of the term 'crystal' in the titles of their works, together with a recognition of this in contemporary criticism, confirms that the science of crystallography had a more specific influence on their work in this period.

Gabo's awareness of crystallography is widely acknowledged in the literature of the artist.[28] Despite his disavowals of its influence, it manifested itself in the forms and titles of several works produced in the later 1930s and early 1940s, most overtly in the multiple versions of *Construction in Space: Crystal* and *Construction in Space with Crystalline Centre* (c. 1938–40; Figure 2.2). Of the latter, Hammer and Lodder declare that it 'evokes a scientific imagery

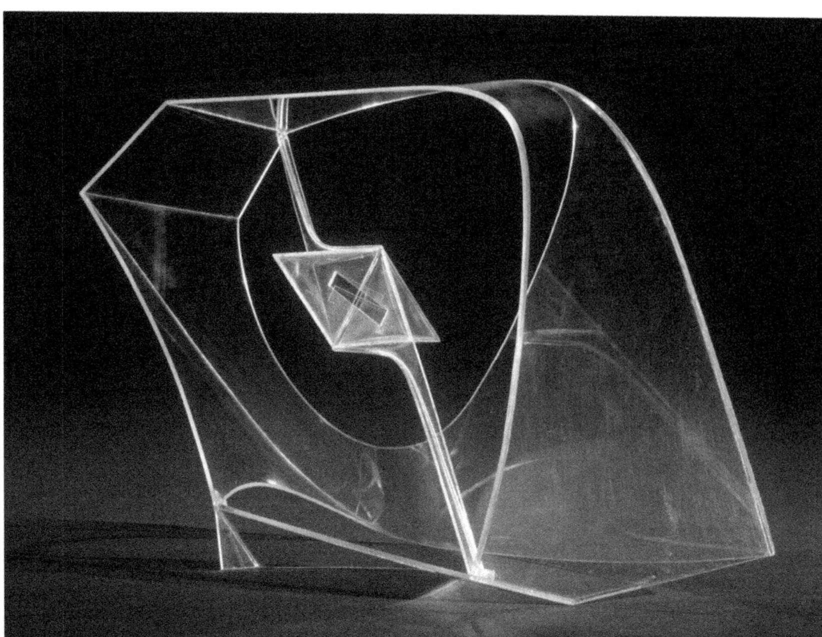

2.2 Naum Gabo (1890–1977), *Construction in Space with Crystalline Centre*, 1938–1940, perspex and celluloid, 32.4 × 47 × 22 cm; coll. Tate Gallery, photo © Tate, London 2013.

of crystalline or atomic structure'.[29] But its connection to crystallography may go deeper: the basic formal structure – a curvaceous shallow 'tunnel' enclosing a central octahedral crystal, which several commentators note appears to have the potential to rotate[30] – bears a conceptual affinity with the apparatus Bernal invented and used in his analytical technique of 'single crystal rotation'. The X-ray goniometer – a staple of laboratory practice from the 1930s to the 1950s – directs a beam of rays at a pinhead-sized crystal mounted on a rotating spindle inside a cylindrical camera, enabling its atomic structure to be calculated from the photographic diffraction pattern obtained.[31] Small wonder that Hepworth, who particularly admired Gabo's construction, specifically recommended Bernal to view it, and that Gardiner subsequently purchased it.[32] Another work from this period which may have some relation to Bernal's crystallographic methods is the entirely transparent plastic construction *Spiral Theme* (1941), commissioned by Gardiner as a birthday present for Bernal.[33] Hammer and Lodder suggest it was partly inspired by the artist's own photographs of reflected light patterns,[34] connecting its genesis to a source similar to the goniometer's photographic diffraction patterns.

The influence of crystallography on Hepworth's art has been shown by Anne J. Barlow to be widely acknowledged, though its exact nature is disputed.[35] Each of the 20 or so pencil and gouache drawings of 1936–1942 that Read identified as exemplary of the geometrical tendency in her work depicts one or more irregular-shaped, polygonal figures on an orthogonally divided ground. The majority are traversed by converging and crossing straight lines, some of which intersect to create parabolic curves, giving an overall impression of spatially ambiguous figures that resemble multi-faceted, crystalline forms. Hepworth's biographer A.H. Hammacher observed that the crystal is their 'leading motif' and that 'the poetry of the crystal' marks '[t]his episode of her career',[36] having previously noted her understanding of 'mathematics and geometry'.[37] The poet Kathleen Raine, who commissioned four of the drawings for her poetry collection *Stone and Flower* (1943)[38] – the frontispiece, *Crystal*, and three others entitled with variations of the book's title – believed 'Barbara's hard crystals and solid geometry' expressed their shared understanding of the natural world and 'sense of the interior landscape of contemporary science'.[39] Although curator and critic Alan Bowness (Hepworth's son-in-law) has asserted that the drawings are 'an examination of crystal structure',[40] Barlow is sceptical of 'superficial comparisons between "scientific" form and Hepworth's work', reiterating Hammacher's view that the drawings emerged 'emotionally, not mathematically'.[41] Even if not 'scientific', the visual and circumstantial evidence confirms that Hepworth's drawings from this period – and, as we shall see, her sculpture – were informed, like Gabo's constructions, by crystallographic science. That Bernal owned one of Hepworth's crystal drawings confirms their interest to him.[42]

While we should not expect to find exact sources for Hepworth's drawings in crystallographic imagery, her contact with Bernal makes it likely that she

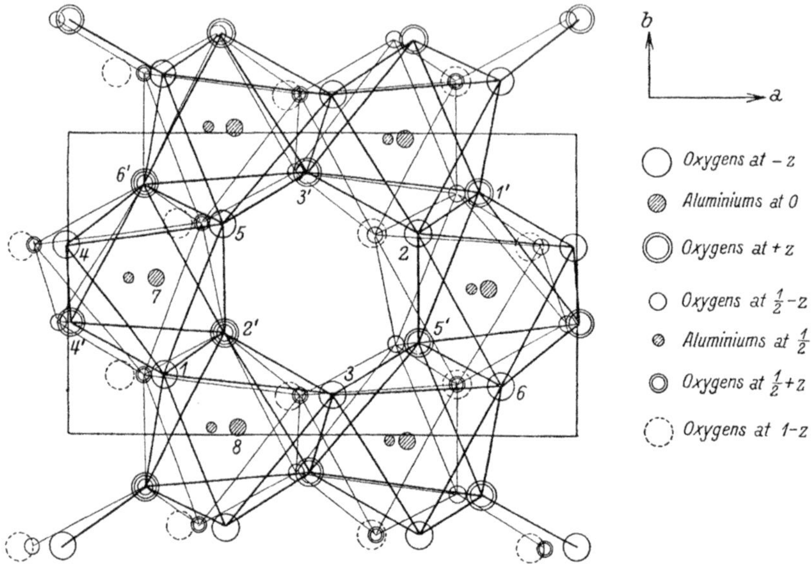

FIG. 6—Structure of Al(OH)$_3$ projected on (001), showing two layers of octahedra

2.3 Crystal structure diagram of alumium hydroxide (hydrargillite), from J.D. Bernal and Helen Megaw, 'The Function of Hydrogen in Intermolecular Forces', *Proceedings of the Royal Society*, Series A, 151: 873 (2 September 1935): 400, fig. 6, photo © Proc.R.Soc A.

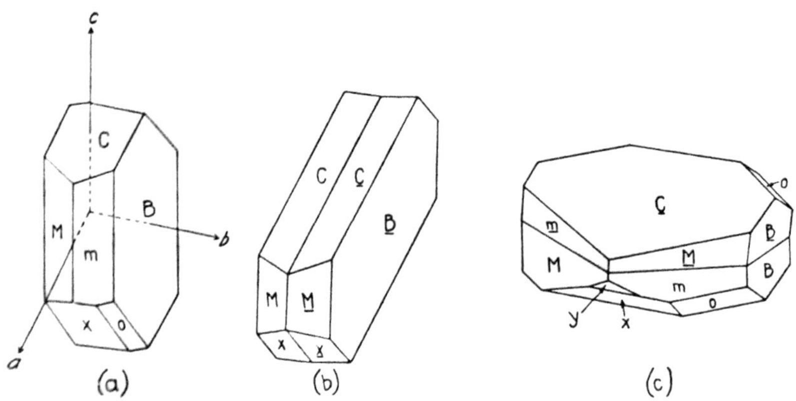

FIG. 136.—Twinning of albite (after Miers). (a) A single crystal of albite. B(010), C(001), m(110), M(1$\bar{1}$0), x(10$\bar{1}$), O(11$\bar{1}$). (b) Albite twinned on (010) (Albite law). (010) is a reflection plane of orthoclase. (c) Albite twinned on axis (010). (010) is a rotation axis of orthoclase.

2.4 'Twinning of albite', from William L. Bragg, *Atomic Structure of Minerals* (Ithaca, NY: Cornell University Press; London: Humphrey Milford; Oxford: Oxford University Press, 1937), 249, fig. 136.

was familiar with his method of drawing crystal structure diagrams and may have seen published examples. Indeed, her drawings bear some resemblance to diagrams of hydrogen-related molecules that Bernal co-published with his doctoral research student Helen Megaw in the *Proceedings of the Royal Society* in 1935, notably to the individual unit cells within the atomic lattices of naturally occurring minerals such as aluminium hydroxide (Figure 2.3).[43] Hepworth's fascination with Bernal's crystallographic ideas raises the possibility that she may have seen other published sources or attended public lectures on the subject by other crystallographers, such as the British pioneer of X-ray crystallography (and Bernal's former research director at the Royal Institution), William L. Bragg. Unlike Bernal's and Megaw's lattice diagrams, Bragg's *Atomic Structure of Minerals* (1937) contains illustrations of single, multi-faceted crystals drawn in three dimensions, which especially resemble some of Hepworth's drawings (Figure 2.4).[44] While neither Bernal's nor Bragg's publications were necessarily known to the sculptor, they contain the types of crystalline diagram she is likely to have known. And just as the parabolic curves seem to derive from, but not copy, the mathematical models she (and Gabo) had studied in science museums, so the geometric networks of straight lines seem to allude to crystal diagrams, without exactly imitating them.[45] This fusion imbued Hepworth's work with a scientific resonance, evoking the complex modern experience of the natural world.

In Hepworth's 1952 monograph, Read acknowledged the connection between these drawings and her contemporary sculpture, and Hepworth herself described them as 'my sculptures born in the guise of two dimensions'.[46] Both this monograph and William Gibson's earlier one juxtapose reproductions of one of the crystal drawings with a version of *Sculpture with Colour* (Colour plate 4),[47] suggesting that this work exemplifies Hepworth's increasingly self-conscious engagement with crystallography in the early 1940s. Forced by wartime restrictions to work on a smaller scale than usual, what more fitting subject than the crystal? The first and smallest version of the sculpture was the only sculpture Hepworth took with her when she moved from London to St Ives in 1939, and during the following three years she made five larger versions. Collectively, they transform the converging lines, parabolic curves and bright colours of the crystal drawings into plaster or wood, introducing two significant innovations in Hepworth's *oeuvre*, painted colour and string. The external form of the five later versions, with their more sharply abutting planes, recall the simple octahedral crystals pictured in some of Hepworth's drawings. If read as a hollowed-out crystal, the stringing may be taken to signify the internal atomic lattice. Given that one version was mounted on a turntable, the sculpture invites comparison with Gabo's almost concurrent *Construction in Space with Crystalline Centre*, and in turn with Bernal's technique of rotating a crystal to expose its atomic 'interior'.[48]

Hepworth's and Gabo's enthusiasm for X-ray crystallography extended their circle's longstanding interest in the structure of natural form, which had previously attracted it to a genre of micro-photography found in books such as D'Arcy Wentworth Thompson's *On Growth and Form* (1916 and 1941) and Karl Blossfeldt's *Art Forms in Nature* (English edition, 1929).[49] Yet however fascinated they were by atomic science, their purpose and understanding were aesthetic and imaginative rather than analytical and 'scientific'. Above all, as Hammacher noted, it was the symbolism of the crystal that mattered most to them: 'The geometry of the crystal, the laws it obeys, its clarity – all satisfy the requirements as to form which the anti-surrealists demanded during these years. Crystals were almost symbols of the ideal they pursued.'[50] As Hammacher hints, the formal perfection of the crystal – its order, precision, predictability, universality, and so on – could be taken to symbolize their utopian social ideals. The molecular organization of atoms invited comparison with the strength and coherence of a unified society composed of a mass of equal and interconnected members. Likewise, crystallography and atomic physics were associated with the 'red science' movement of the 1930s,[51] through figures such as Bernal and his predecessor as Birkbeck Professor of Physics, Patrick Blackett.

The symbolic value attached to the crystal by British Constructivists is made clear in the final chapter of Herbert Read's only novel, *The Green Child* (1935), written in the summer of 1934 when Read is almost certain to have been aware of Bernal's crystallographic work. His fantasy novel traces the journey of a former political dictator to an apolitical, peace-loving society which lives in an underground network of crystalline grottoes. Read's narrator comments that: 'The science which we call crystallography ... was the most esteemed of all sciences in this subterrestrial country; indeed, it might be regarded as science itself, for on it were based, not only all notions of the structure of the universe, but equally all notions of beauty, truth and destiny.'[52] For the 'workmen' in this harmonious community, the highest 'stages of existence' are to be first a maker of crystals – 'suggested' by natural forms but 'not exact imitations' of them – and then, when judged to have 'constructed' five perfectly formed crystals, a 'crystal-polisher' or 'sage' (Bernal's nickname, as it happens). The elite group of sages spend their final days engaged in what is regarded as 'the most difficult problem in their philosophy', contemplating the precision, perfection and order of artificial crystals. With flattering implications for the practitioners and theorists of Constructive art, Read's utopian allegory identifies the artistic construction and philosophical contemplation of crystalline sculptural forms as the path to absolute sensual and intellectual enlightenment.[53]

When Constructive art enjoyed a critical resurgence in the 1940s, its crystalline forms came to embody the hope of post-war social transformation. Hepworth's *Sculpture with Colour*, Gabo's *Spiral Theme* and several of their related drawings were selected by curators Margot Eates and Hartley Ramsden for inclusion in the London Museum's wartime exhibition *New Movements in Art: Contemporary Work in England*.[54] According to Gabo, the exhibition received about 10,000 visitors and his own work a very positive critical response.[55] As Chris Stephens has noted, *Spiral Theme* acquired a particularly high status among Constructive artists and their supporters: Read, who had not previously reviewed Gabo's work, hailed it as 'the highest point ever reached by the aesthetic intuition of man'. Two years later it was reproduced alongside a Constructivist work by Hepworth in *This Changing World* (1944), an influential collection of essays edited by Marcus Brumwell, who had become a friend of Gabo, Hepworth and Read through their work for the Design Research Unit (DRU), which he directed, or for his advertising agency. Many of the essays were written by supporters of Constructive art (two of them former contributors to *Circle*), including Bernal, Crowther, Darlington, Raine, Ramsden, Read and Waddington.[56] Drawn from articles published in *World Review* – a journal edited by Brumwell and owned by the liberal-leftist publisher Edward Hulton – the anthology laid out a progressive post-war programme for art, architecture, science, education, politics, and so forth, applying the utopian spirit of Constructivism to plans for the physical and social reconstruction of Britain.[57]

That crystal structure patterns featured extensively in post-war 'Festival-style' designs for fabrics, wallpaper, products and interiors is a direct consequence of the influence of Bernal and Hepworth. Lesley Jackson has shown that the Festival Pattern Group's promotion of 'crystal design' had its origins in a proposal made by Brumwell in 1946 for the DRU to publish a book on the subject by Bernal's former doctoral student Helen Megaw. Brumwell had met Bernal through Read, and witnessed Bernal's enthusiasm for Hepworth's sculpture at first hand.[58] Revealingly, he sought an opinion on the proposed book from Hepworth, who had responded enthusiastically:

> I think it's a marvellous idea to use these 'designs' for fabrics ... The main point seems to me to produce them as suggested in series – with their proper names – exactly as they really are. To me they are more beautiful than any man-made pattern [emphasis in original].[59]

When the idea was taken forward in 1949 by the recently founded Council of Industrial Design, it was because its Chief Industrial Officer, Mark Hartland Thomas, attended a lecture on crystallography given to the Society of Industrial Artists by Bernal's former colleague Kathleen Lonsdale (using slides prepared by Megaw).[60] With Megaw acting as scientific advisor to the Festival Pattern Group, the pre-war utopian symbolism of Constructivism's crystalline geometries was pressed into post-war service to signify the coming of a progressive, cooperative and equitable, social-democratic new age.

Fear: Humanist Sculpture and Nuclear Anxiety (c. 1945–1970)

In the year that the imagery of atomic science achieved ubiquity at the Festival of Britain, the Nobel Prize in Physics was belatedly awarded to J.D. Cockcroft and Ernest Walton, who had famously 'split' the atom in 1932 while working alongside Bernal at the Cavendish Laboratory under Ernest

Rutherford.[61] Following their momentous breakthrough, the technology of the atom bomb was developed during the Second World War by the joint Anglo-American-Canadian Manhattan Project and deployed to devastating effect at Hiroshima and Nagasaki. The Soviet Union achieved atomic capability in 1949 and Britain followed in 1952. As the world fractured into ideologically antagonistic spheres of American and Soviet influence, the likelihood of Cold War escalating into nuclear warfare was increased by a series of international political crises and military interventions, most notably the Korean War of 1950–1953, when Chinese and Soviet-backed forces supported the Communist North against the UN-backed South. From 1953 to 1958, Britain joined the nuclear arms race by conducting H-bomb tests in the Pacific region.[62]

The proliferation of nuclear weapons and growing risk of their use meant that the positive and optimistic attitudes toward atomic science of the pre-war era were increasingly eclipsed by anxiety and pessimism. Distinguished scientists in the United States – most famously Albert Einstein, Linus Pauling and veterans of the Manhattan Project, Robert Oppenheimer, Hans Bethe and Joseph Rotblat – opposed the development of nuclear arms by founding the Emergency Committee of Atomic Scientists and initiating the international Pugwash conferences. Eminent scientists in Britain, such as Huxley, Waddington and Zuckerman, expressed public concern and joined campaign groups for peace and disarmament. Among them were atomic physicists and crystallographers, including Bernal, Blackett and another of Bernal's former research students, Dorothy Crowfoot Hodgkin. They supported British protest organizations that emerged in the 1950s, including the Christian, pacifist Toldas group and the larger, non-aligned National Campaign for the Abolition of Nuclear Weapons Tests (NCANWT) and its successor, the Campaign for Nuclear Disarmament (CND). Bernal and Zuckerman were especially vocal, driven by longstanding concern for the social responsibility of science and perhaps by the findings of their wartime military research into the impact of explosive blasts on people and buildings. Bernal brought Communist and non-Communist campaigners together in Scientists for Peace and the Cambridge Scientists Anti-War Group, and joined the CND's Easter marches to London from the government's Atomic Weapons Research Establishment at Aldermaston.[63]

Just as many of the scientists associated with British Constructive art became opponents of nuclear weapons, so did many of the artists. As early as 1950, Henry Moore signed a letter to *The Times* opposing the use of atomic weapons in Korea.[64] By 1958, he and Hepworth had become sponsors of the CND,[65] and its forerunner the NCANWT.[66] They were among the 300 artists and writers who publicly supported the Marxist critic John Berger's call for an immediate end to the 'testing, storing and manufacture of nuclear weapons'.[67] Like many on the Left, Hepworth had felt compelled by the events of 1956 – the Soviet invasion of Hungary and the Suez Crisis – to re-examine her political conscience. Having distanced herself from Bernal and his politics,[68] her political sympathies now focused on pacifism, disarmament and internationalism. As she told Read:

> I became a pacifist two years ago but all this has pushed me into trying to do more by joining the Labour Party, the Toldas Group & United Nations Ass[ociation of Great Britain and Northern Ireland], etc. One can scarcely look 'earlier sculptures' in the face if one remains politically and socially inactive now.[69]

Her commitments are likely to have been galvanized by the loss of her first son, an RAF pilot, in the Korean War, while her troubled sense of personal responsibility may have been heightened by her own pre-war fascination with atomic science. That her optimism endured, however, is evident from her continued commitment to a broadly Constructivist aesthetic: her former secretary David Lewis recalls that 'in the post-war years of nuclear arms and the Cold War, she once again sought to make a counter-statement of serenity'.[70]

Unlike Hepworth, Henry Moore produced sculpture in the 1950s and 1960s that was widely seen as a direct response to the prevailing climate of nuclear anxiety. Even Berger, an unlikely champion of Moore, proposed that his *Falling Warrior* (1956–1957) become an emblem for the CND and the preliminary studies the subject of a CND-sponsored exhibition.[71] However, it was Moore's ambiguously entitled *Atom Piece* (1964–1967) that was to become the best-known British sculpture to confront the twin subjects of atomic science and nuclear warfare (see Figure 2.5). Commissioned by the University of Chicago, it was intended to commemorate Enrico Fermi's contribution to the Manhattan Project by marking the site of the nuclear pile where, in 1942, the Italian physicist initiated the first self-sustaining nuclear chain reaction. The commissioners hoped it would be 'a monument to man's triumphs, charged with high hope and profound fear'.[72] On Moore's own account, his proposal provocatively conflated the forbidding forms of 'a mushroom cloud' and 'a skull – a death's head'.[73] Given its resemblance to his helmet-head sculptures (conceived in the Second World War and developed during the Korean War) and to a contemporaneous CND poster, Iain A. Boal concludes that Moore's monument deliberately undermined his patrons' nuclear triumphalism, leading them to rename it *Nuclear Energy* so as to distance it from the military purpose of Fermi's experiment.[74] But Moore's intentions remain unclear and his later efforts to associate the work with organic forms and cathedral architecture have been cited to suggest that he had developed a more ambivalent stance to nuclear arms by the later 1960s.[75] Such equivocation was certainly characteristic of the time, as nuclear power became commonplace, the Windscale accident faded from memory and other radical causes increasingly preoccupied the Left. It may also have been a response to the commissioners' ambivalent desire for a monument that would instil both hope and fear.

The same dichotomy between the benefits and dangers of atomic science was clearly manifest in two works by the sculptor 'Peter' Peri. In 1960,

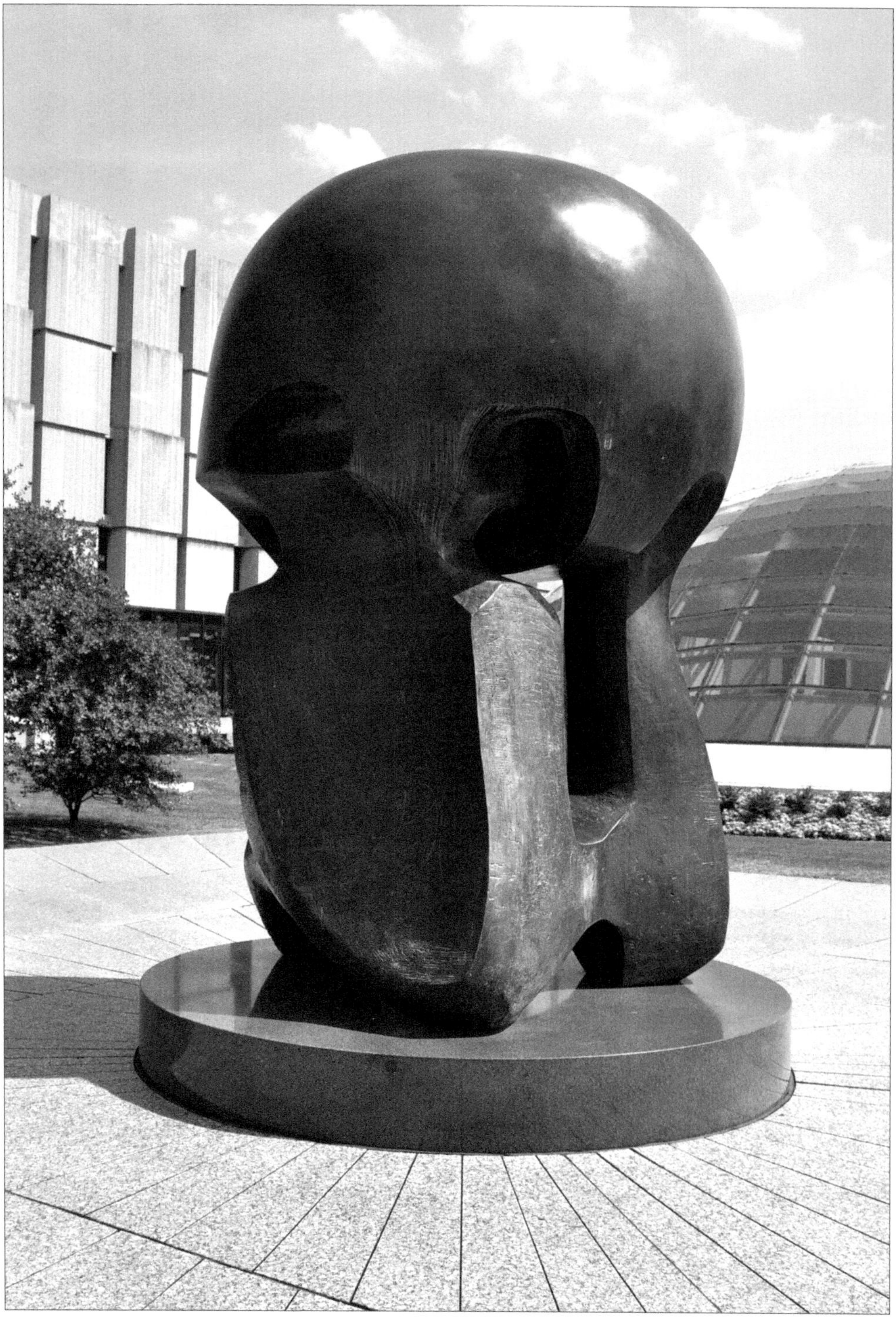

2.5 Henry Moore (1898–1986), *Nuclear Energy*, 1964–1966, bronze, h. 366 cm, University of Chicago, photo reproduced by permission of The Henry Moore Foundation.

he completed a wall sculpture for a Leicestershire school of a celebratory symbolic figure holding an atomic ball-and-spoke structure, entitled *Man's Mastery of the Atom = Self Mastery* (1959–1960).[76] In the same year, he made a small sculptural group in his characteristic social-realist style, entitled *Aldermaston Marchers* (1960; Figure 2.6), which, according to Helga László, was 'inspired by an actual peace march'.[77] As a Quaker (and Communist), Peri was committed to pacifism and disarmament.[78] The work constituted a rare and belated response to the call from his friend and supporter John Berger in *Marxist Quarterly* for an art explicitly 'protesting against the H-bomb'.[79] A preliminary drawing for the sculpture depicts a line of seven or eight anonymous marchers, with one carrying a CND pennant and another pushing a pram.[80] The final sculpture simplified and intensified the original conception and reflected the iconography of the burgeoning disarmament campaign by reducing the composition to a single family group: the man now holds the CND symbol more purposefully aloft like a military flag-bearer, while the woman is poignantly held back by a dawdling and uncomprehending child.[81] Peri's and Moore's direct engagement with themes of atomic science and nuclear warfare was, however, extremely rare in the 1960s and, paradoxically, even rarer in the preceding decade when Britain had witnessed a more intense period of nuclear testing and disarmament campaigning. In the 1950s, as the possibility of nuclear apocalypse increasingly haunted the popular imagination, British sculptors developed a subtler, metaphorical repertoire of forms, materials and subjects to express the collective unease.

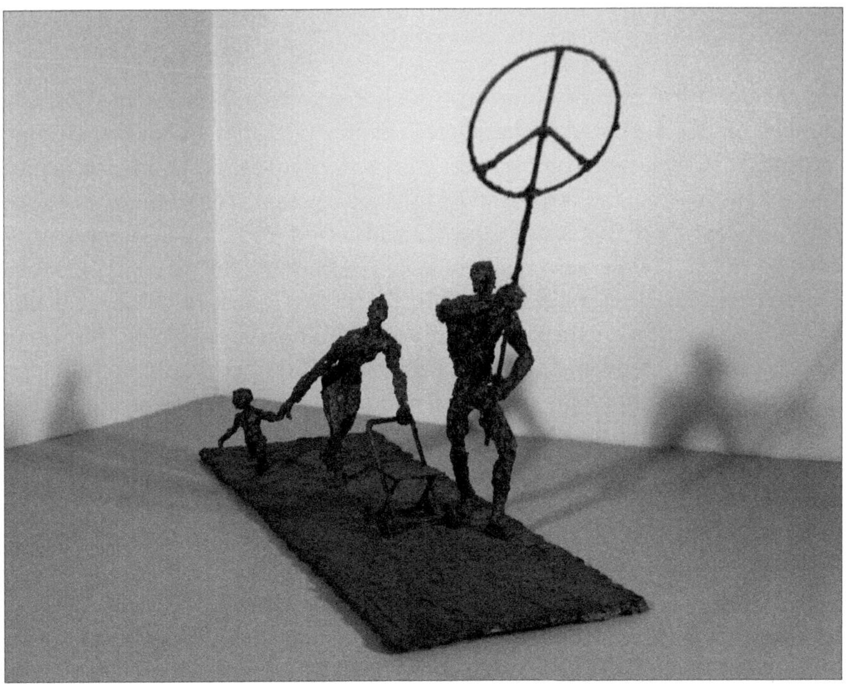

2.6 'Peter' (Laszlo) Peri (1899–1967), *Aldermaston Marchers*, 1960, gilt wire and polyester resin, 36 × 53.5 × 19.5 cm, photo © Religious Society of Friends in Britain.

Read's identification of the work of a younger generation of sculptors with 'the geometry of fear' signalled his recognition of the transformation that British sculpture was undergoing. He aligned the work of the eight represented at Venice – Robert Adams, Kenneth Armitage, Reg Butler, Lynn Chadwick, Geoffrey Clarke, Bernard Meadows, Eduardo Paolozzi and William Turnbull – with the Surrealism of Moore, Picasso, Calder and Giacometti (and, in Butler's case, Lam and Masson), claiming it was 'close to the nerves, nervous, wiry'. Contrasting their purposes with the 'transcendental values' of Hepworth and the Constructivists, he revealed a new scepticism toward the earlier movement's attempt 'to create the images of a civilisation not yet born, perhaps never to be born'. Employing ideas derived from nineteenth-century German aesthetics, he insisted that: 'They are all involved in some wider manifestation of the creative will, some general extension of consciousness.' He used the timing of the sculptors' birth – 'for the most part ... during or immediately after the First World War' – to explain their lack of classical calm and serenity, and to associate them with the bleak and disturbing imagery of T.S. Eliot's Modernist poetry of those years: 'They have seized Eliot's image of the Hollow Men and given it an isomorphic materiality. They have peopled the Waste Land with their iron waifs.' In the most striking and most frequently cited passage, Read described their work in terms of psychological repression and Surrealist fantasy, still drawing his foreboding imagery from Eliot's poetry:

> These new images belong to the iconography of despair, or of defiance; and the more innocent the artist, the more effectively he transmits the collective guilt. Here are images of flight, of ragged claws 'scuttling across the floors of silent seas', of excoriated flesh, frustrated sex, the geometry of fear.[82]

His metaphorical imagery hinted at the horrors of the Bomb – the ultimate symbol of the alienated technological civilization that Read increasingly lamented. It conjured a vision of post-war Man as desperate, culpable, vulnerable, injured and impotent. Read wrote these words in March or April of 1952,[83] less than a year after the US had tested a hydrogen bomb and as a tense military stalemate held in Korea. Although the hawkish General MacArthur had been relieved as UN Supreme Commander, the US had made preparations for the possible use of atomic weapons. With the rise of the disarmament movement, Read's pacifism and abhorrence of nuclear arms would soon lead him, like Hepworth and Moore, to support the NCANWT and Berger's protest, and later, as an anarchist, to sponsor the radical Direct Action Committee Against Nuclear War and its successor, the Committee of 100, which advocated non-violent, civil disobedience.[84]

Read's view of the new sculpture derived from an eclectic mix of cultural, psychological and political theory. As I have argued elsewhere, his conception of 'the geometry of fear' owes much to his longstanding attachment to the ideas of Wilhelm Worringer, whose 'psychology of style' associated geometric abstraction with the art produced in hostile environments where Man is alienated from nature.[85] (The crystallographers' discoveries of the geometry underlying natural materials did not deter Read's commitment to the paradigm.[86]) While the new sculpture was neither fully abstract nor literally 'geometric', Read's defining phrase effectively aligned it with the art of alienation. Although he ascribed a unity to it that was clearly an over-simplification and less obviously suited to the work of Adams, Paolozzi and Turnbull, its longstanding identification with his soubriquet reveals how well the language used in his introduction – which he described himself as 'a bit poetic'[87] – matched the metaphorical character of their sculpture. In 1963, he acknowledged that his earlier account of these sculptors had been over-generalized but once again asserted their unity, writing that 'with one or two exceptions, such as Robert Adams, they are all expressionists'.[88] His new characterization demonstrates the stylistic eclecticism of this sculpture which combines features of Expressionist art with those more typical of Surrealism (as he had emphasized in 1952) and of contemporaneous 'modernist realism'.[89] The amalgamation of emotionally charged content and figurative imagery has led many commentators to describe it as 'humanistic'.[90]

Among the formal tendencies that distinguished the new sculpture from the Modernist sculpture that preceded it were its open assembled forms and metamorphic imagery, which abandoned both Constructive geometry and Surrealist biomorphism. Sculptors favoured forged and welded metal over carved stone and wood, and casting from directly carved plaster rather than from the malleable forms of clay or wax. Instead of massiveness and volume, its forms are often skeletal, planar and spiky. An effect of weightlessness and suspension is produced by slender legs and open-form bases, conveying a sense of frailty and immateriality. In place of smoothly polished forms, surfaces tend to be pitted, scarred or scorched. Their battered and ruined 'skin' has been associated with post-war perceptions of the political or existential human predicament,[91] and in retrospect may remind us of the charred bodies of the victims of atomic and napalm bombing in Japan and Korea. As sculptors used these innovative and evocative materials and forms to create an imagery of attenuated and deformed bodies, insubstantial limbs and diminutive heads, it implied the fragility and insignificance of the individual, and a permanent loss of the social cohesion and natural harmony that underpinned idealist and organicist abstraction.

The subject matter of this sculpture is dominated by birds, animals, insects, crustaceans and humans. There are animals that have been startled, injured and killed, like Meadows' cockerels, birds and crabs (see Figure 2.7), and humans who have been blasted by strong winds, felled, impaled, mutilated and martyred. Of the many winged creatures – some drawn from myth, others from pre-history – most have been deprived of the power of flight. Numerous 'hybrids', as Read described them in 1952,[92] surrealistically metamorphose animal, human and sometimes vegetal forms. Margaret Garlake has compared them to the genetically mutated victims of

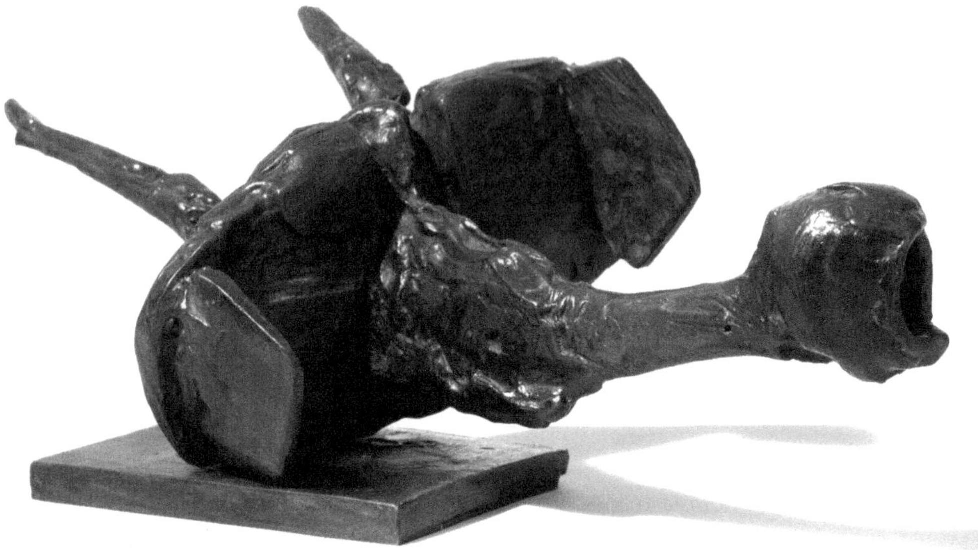

2.7 Bernard Meadows (1915–2005), *Fallen Bird*, 1958, bronze, 17 × 46 × 23 cm, coll. Tate Gallery, photo © Tate, London 2013.

irradiation featured in contemporaneous science fiction literature and film, and plausibly suggested that they represent 'the negation of identity, of selfhood'.[93] By contrast, there are symbols and personifications of cruel and domineering methods of persuasion and oppression. Among the animals are predatory beasts, insects and crustaceans armed with menacing beaks, teeth, claws or pincers, and protective shells or armour (Colour plate 5). Among the human figures, invariably male, are works entitled Politician, Orator, Director, Inquisitor, Dictator, Tyrant and Big Brother, sometimes equipped with armour and threatening weapons or 'machines'. Butler and Meadows, who were responsible for many such figures, frequently drew their subjects from history and science fiction, but their relevance to the adversarial politics and propaganda of the Cold War is self-evident.[94] Butler, a pacifist and wartime conscientious objector who was active in the disarmament campaign as an Aldermaston marcher and member of the Committee of 100, created the menacing *Manipulator* (1954), who controls the skies with a sinister apparatus.[95] Whether animal, human or hybrid, these creatures are often animated by prominent and vigilant eyes (or by a single cyclopean eye) that suggest, in Garlake's words, a 'control mechanism',[96] enforcing their role as a 'watcher' or 'sentinel'.

Much of the allegorical iconography deployed in this sculpture, from archaic helmet heads and fallen warriors to dominant and distressed animals possessed respectively of unfeeling or fearful eyes, recalls the prototypical Modernist image of aerial bombardment and civilian massacre, Picasso's *Guernica* (1937). This huge and renowned mural painting had toured London and several British cities before the war to enthusiastic acclaim from Modernist artists and critics.[97] Using the language of the bullfight, its juxtaposition of the overbearing bull and injured horse had long been interpreted as representing the brutality of the Spanish dictator and the suffering of his people.[98] Attuned by *Guernica*, contemporary critics recognized the unhappy creatures which populated British sculpture in the Cold War as 'visual metaphors':[99] albeit aggressor or victim, hunter or hunted, oppressor or oppressed, here were competitors in a vicious struggle for power.

As I have observed before,[100] many of these creatures embody within themselves a strange ambiguity, being predatory, threatening and aggressive but also vulnerable, anguished and frightened. This internalized dialectic denies them any simplistic identification with either good or evil, virtue or vice, just as Moore's doubly titled *Atom Piece/Nuclear Energy* monument possesses a 'duality' which enables it to evoke both the benefits and dangers of atomic science.[101] The origin of this 'ambiguity, or duality of meaning' has been plausibly located by Chris Stephens in Moore's helmet-head sculptures.[102] It was certainly consistent with Read's belief in the dialectical nature of great art, which he reasserted in an essay of 1950 by quoting Schelling's observation that: 'To be drunk and sober not in different moments but at one and the same moment – this is the secret of true poetry.'[103] The dialectic may even be found in those sculptures representing figures of authority, which also invariably possess an implicit weakness. Meadows' armed figures may have a commanding presence, with helmets, shields, armour and extended blades, but their top-heavy bodies, spindly legs and exposed flesh suggest their possible downfall through the vanities of high office – pomposity, hubris, narcissism, and so forth (see Figure 2.8). This duality may also be applied to the fixed and penetrating stares of these creatures, which are equally the hallmark of predator or prey; they may gaze imperiously or peek timidly, just as the piercing eyes of Moore's helmet heads may be those of binoculared observer or frightened casualty. Similarly, the many sculptures of Armitage, Butler and Chadwick that allude to surveillance technology – from primitive observation towers to modern radio and radar masts, detectors, listeners, monitors, and so on – are neither simply defensive nor offensive. Butler combined the watchfulness of the oppressed with the shadowing methods of the Cold War oppressor in his design for a Monument to the Unknown Political Prisoner (1952–1953) by placing anxious female 'Watchers' at the foot of a menacing communications tower.[104] In short, the

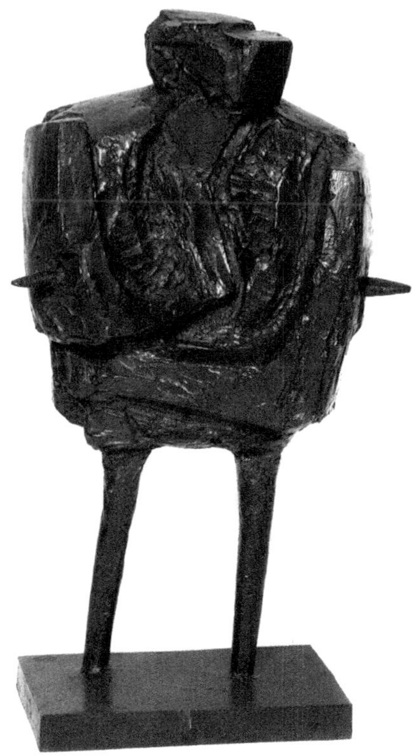

2.8 Bernard Meadows (1915–2005), *Maquette for Large Standing Armed Figure*, 1962, bronze, h. 57.2 cm, coll. Oldham Art Galleries & Museum, photo © Oldham Art Galleries & Museum.

new sculpture transfigured the imperatives of Cold War enmity – strength, alertness, deterrence, defence, surveillance, dissemblance, and so on – in a metaphorical and de-historicized language of anthropomorphic and zoomorphic forms. As with *Guernica*, Surrealist symbolism transformed specific political fears into broader psychological anguish. The duality and ambivalence of the imagery confounded a simple apportioning of rights and wrongs, reflecting the difficult moral choices facing Western liberals and leftists in the ideological battle between capitalism and Communism.

In the decade after Read coined the 'geometry of fear' phrase it established itself in the critical lexicon of modern sculpture, repeated many times by Read himself and by those sympathetic to his views.[105] In reviewing Chadwick's solo exhibition at the Venice Biennale in 1956, Bernard Denvir proclaimed it 'one of the happiest phrases coined by an English critic during the past twenty-five years'; however, ignoring Read's emphasis on the historical conditions that had shaped the new sculpture, Denvir argued that: 'Fear is no new thing in art, and it would be tedious to try to relate Chadwick's sculpture to any anxiety about the atom bomb.'[106] When Read himself later reiterated the phrase, he also attempted to reinterpret its meaning. In 1958, in a survey of post-war British art, he aligned the sculpture and phrase more explicitly with existentialist philosophy.[107] Elsewhere, in the same year, he reworked it in light of his growing allegiance to Jungian ideas, explaining that he had not intended to refer to fear of 'physical peril (in a battle or an air-raid, for example)' but to 'metaphysical' or 'unconscious fear', the 'demonic force that is pent in the unconscious' as the 'primary source of creative activity', 'released' when it enters an 'archetypal form'.[108] Six years later, writing in *A Concise History of Modern Sculpture*, he expanded further:

> It is temptingly easy to interpret the diverse phenomena of modern art as expressions of an 'Angst' or despair induced by the alienation prevailing in our technological civilisation – a 'geometry of fear', as I once expressed it. ... Anxiety [however] ... is seldom unmixed with hope ... The artist unconsciously projects the anxieties of his age, but he would have no creative energy if he were completely filled with despair.[109]

His backtracking was prompted by an exhibition in Darmstadt entitled *Evidences of Anxiety in Modern Art*, which included Armitage, Butler, Chadwick and Moore, and elaborated the meaning of their work in relation to his 1952 catalogue introduction.[110] Read's retrospective efforts to re-shape his interpretation of this sculpture was accepted by at least one commentator, his friend, the Arts Council's Director of Art, Philip James, who advised visitors to an exhibition in 1960 that these sculptors 'express the *ecstatic* [my emphasis] "*Angst*" of our age in what has been aptly called a "geometry of fear"'.[111]

More recently, David Hulks has coaxed more positive meanings from this sculpture and Read's evocative phrase by subjecting them to psychoanalytical examination. He argues that the sculpture '*acknowledged* psychic fragmentation under Cold War conditions, but ... showed a way of turning despair into exaltation'.[112] For Hulks, images of affirmation were allowed to prevail over those of despondency by achieving a 'double inflection', exemplified in the work of Armitage and Meadows. Although his argument rightly acknowledges the duality of this sculpture and shows how Read's later reinterpretation of the phrase allowed him to invest it with optimistic associations similar to those surrounding Constructive art, it accords less well with Read's original text, which attributes emphatically negative meanings to the sculpture – Armitage's figures are described as 'a sardonic commentary on the stretched agony of human relationships', while the 'virtue' of Meadows' cockerels and crabs is 'caught as in a snare'.[113] Notwithstanding Hulks' and Read's own attempts to bring a more nuanced psychological dimension to this sculpture and Read's account of it, most commentators who have connected them over the years have done so to assert the sculpture's troubled mood.

Almost from the moment that Read ascribed a unity to this sculpture focused on its 'iconography of despair', more formalist-minded critics voiced their criticisms of it. When in January 1953 the painter and critic Patrick Heron reviewed an exhibition of maquettes by British entrants to the Unknown Political Prisoner competition, including five of the sculptors shown in Venice and others working in a similar style, he dismissed them (with a few exceptions) as 'emptily expressionist' and devoid of 'any genuine, compelling *Angst*', concluding that 'the repetitive spikiness of all these iron thorns and cacti … are a thin cliché by now'.[114] By the mid-1960s, when Read's critical authority had waned, formalist and phenomenological methods of criticism were ascendant, and new forms of colourful abstract sculpture were emergent, a younger generation of sculptors and critics – many of them former students of Anthony Caro at St Martin's School of Art – increasingly denigrated the sculpture of the 1950s.[115] Reflecting on 'A Decade of British Sculpture' in 1964, David Thompson reiterated Heron's accusations that the 'Angst-ridden' new sculpture 'that Read in a much repeated phrase called "the geometry of fear" [had] degenerated almost flagrantly into cliché'.[116] In 1971, the critic Charles Spencer wrote still more scathingly:

> These sculptors were part of what might be called the 'mutilated' or 'tortured' school. They were humanists, concerned both with the human body and the human situation and expressed their tragic understanding of life in emotional clichés which amounted to a kind of blackmail – headless torsos, torn limbs, scabrous, pitted skin surfaces …[117]

In the changing aesthetic and political landscape of the 1960s, which saw the peaceful outcome of the Cuban missile crisis, the implementation of the first Nuclear Test Ban Treaty and US involvement in Vietnam, younger artists expressed their opposition to nuclear weapons more often through a satirical Pop aesthetic, resembling the black humour of Stanley Kubrick's 1963 film, *Dr Strangelove or How I Learned to Stop Worrying and Love the Bomb*. That year Colin Self created his *Leopardskin Nuclear Bomber No. 2* (1963), which conflated military and sexual bravado by 'tarting-up' the phallic fuselage of a model US fighter-plane with tight-fitting leopard-skin fabric.[118] At the turn of the next decade, Eduardo Paolozzi wryly associated napalm and nuclear warfare with the excesses of capitalist production and consumption by exhibiting stock-piled *Bombs* (1971) – which he described as 'my answer to [Warhol's] Brillo Boxes'[119] – and a replica skip entitled *Waste (Cloud Atomic Laboratory)* (1971). Unlike the sculpture of the 1950s, such works were marked by an overt anti-Americanism. With the emergence of this new aesthetic, 'geometry of fear' sculpture and Read's commentary on it received scant historical or critical attention for nearly two decades, until interest was reawakened in 1981 by the Whitechapel Art Gallery's landmark exhibition, *British Sculpture in the Twentieth Century*.[120]

For more than two decades in the middle of the century, some of the most critically acclaimed modern sculpture produced in Britain had expressed social ideas related to developments in atomic physics. Before 1945, the mathematical geometry of crystal structures supplied Constructive sculptors with a recognizable signifier of the utopian hopes accompanying scientific advance. Gabo and Hepworth introduced formal and technical innovations to their work, inspired by the pure research of the crystallographers. After 1945, new forms of humanist sculpture expressed the anxieties created by the deadly inventions of the nuclear technologists through a metamorphic, metaphorical and dialectical language pioneered by Moore and given meaning by Read. The 'geometry of fear' sculpture adopted new images, forms and materials to confront themes of oppressive power, intense surveillance and identity erasure. Despite Read's retrospective efforts to reconcile the fear that he identified in this sculpture with timeless and universal aspects of the human condition and collective unconscious, the poetic resonance of his original insightful description ensured its indissoluble tie to the historical realities of the Cold War and the Bomb.

Acknowledgements

I would like to thank the Digital and Material Arts Research Centre at the University of Derby for funding my research, and Helen Burstow for her encouragement and support.

Notes

1 Herbert Read, 'Introduction', in *Barbara Hepworth: Carvings and Drawings* (London: Lund Humphries, 1952), ix–x. Hepworth's drawings are reproduced in large numbers in the monograph.

2 See Lesley Jackson, *From Atoms to Patterns: Crystal Structure Designs from the 1951 Festival of Britain* (Shepton Beauchamp: Richard Dennis in association with the Welcome Collection, 2008).

3 Herbert Read, 'New Aspects of British Sculpture', in *The XXVIth Venice Biennale: The British Pavilion*, exhibition catalogue (London: British Council, 1952), np.

4 See, for example, Mona Hadler, 'The Bomb in the Postwar Era: From the Sublime to Red Hot Candy', *Source: Notes in the History of Art*, vol. 21, no. 1 (Fall 2001): 38–43; Brooke Kamin Rapaport and Kevin Stayton, eds, *Vital Forms: American Art and Design in the Atomic Age, 1940–1960* (New York: Brooklyn Museum of Art in association with H.N. Abrams, 2001); Stephen Petersen, 'Explosive Propositions: Artists React to the Atomic Age', *Science in Context*, vol. 17, no. 4 (2004): 579–609; *Be-Bomb: The Transatlantic War of Images and all that Jazz, 1946–1956*, exhibition catalogue (Barcelona: Museu d'Art Contemporani de Barcelona; Madrid: Centro de Arte Reina Sofia, 2007). Some rare but over-generalized comments on British sculpture appear in Darrell D. Davisson, *Art After the Bomb: Iconographies of Trauma in Late Modern Art* (Bloomington, IN: Authorhouse, 2008).

5 See, for example, Ann Jones, in *Geometry of Fear: Works from the Arts Council Collection*, exhibition guide (London: South Bank Centre, 2007), np. This exhibition demonstrates how the group has been enlarged since 1952 to include others who worked in a similar manner and who had often been taught by one of the original eight, such as Robert Clatworthy, Elisabeth Frink, John Hoskin and Leslie Thornton.

6 J. Leslie Martin, 'The State of Transition', in *Circle: International Survey of Constructive Art*, ed. J.L. Martin, Ben Nicholson and Naum Gabo (London: Faber and Faber, 1971 [1937]), 217.

7 See Anne J. Barlow, 'Barbara Hepworth and Science', in *Barbara Hepworth Reconsidered*, ed. David Thistlewood (Liverpool: Tate Gallery, 1996), 95–107; Martin Hammer and Christina Lodder, 'Hepworth and Gabo: A Constructive Dialogue', in *Barbara Hepworth Reconsidered*, 109–133; Martin Hammer and Christina Lodder, *Constructing Modernity: The Art and Career of Naum Gabo* (New Haven and London: Yale University Press, 2000), 379–381.

8 On Bernal, see Maurice Goldsmith, *Sage: A Life of J.D. Bernal* (London: Hutchinson, 1980); Brenda Swann and Francis Aprahamian, eds, *J.D. Bernal: A Life in Science and Politics* (London: Verso, 1999); Andrew Brown, *J.D. Bernal: The Sage of Science* (Oxford: Oxford University Press, 2005).

9 J.D. Bernal, 'Architecture and the Scientist', *RIBA Journal*, 3rd series, vol. 44, no. 16 (June 1937): 805–812; J.D. Bernal, 'Art and the Scientist,' in *Circle*, 119–123; J.D. Bernal, Foreword to *Barbara Hepworth*, exhibition catalogue, Alex Reid and Lefevre Gallery, London, 1937, np. According to Hepworth, Bernal planned to write a book on art and science, Margaret Gardiner, *Barbara Hepworth: A Memoir* (London: Lund Humphries, 1994; Edinburgh: Salamander, 1982), 28.

10 Bernal, Foreword to *Barbara Hepworth*, np.

11 See *Barbara Hepworth: Carvings and Drawings*, xvi; and Hammer and Lodder, *Constructing Modernity*, 280 and 401.

12 Gardiner, *Barbara Hepworth: A Memoir*, 27.

13 Goldsmith, *Sage*, 87.

14 See Goldsmith, *Sage*, 225; and Brown, *J.D. Bernal*, 154. On the inscription of mathematical models, see Michael Compton, 'Gabo: Mathematics and Science', in *Naum Gabo: Sixty Years of Constructivism*, exhibition catalogue (London: Tate Gallery, 1987), 16.

15 Sally Festing, *Barbara Hepworth: A Life of Forms* (London: Viking, 1995), 148.

16 Bernal, Foreword to *Barbara Hepworth*, np.

17 Bernal, 'Art and the Scientist', 121, and Figures 5a and 5b.

18 Hepworth, letter to Bernal, no date [c. 1936/1937], University Library (Add. 8287/J84 [Box 83]), University of Cambridge.

19 Gardiner, *Barbara Hepworth: A Memoir*, 48–51.

20 Hepworth, letter to Gardiner, nd. [late 1930s], quoted in Gardiner, *Barbara Hepworth: A Memoir*, 28.

21 Gardiner, *Barbara Hepworth: A Memoir*, 44.

22 Hammer and Lodder, *Constructing Modernity*, 233 and 270.

23 For their fullest account, see Hammer and Lodder, *Constructing Modernity*, 379–401.

24 Naum Gabo, 'On Constructive Realism', in Katherine S. Dreier, James Johnson Sweeney and N. Gabo, *Three Lectures on Modern Art* (New York and London: Kennikat, 1971 [1949]), 76–77.

25 Martin Hammer and Christina Lodder, 'Naum Gabo and the Constructive Idea of Sculpture', in *Naum Gabo: The Constructive Idea*, exhibition catalogue (London: South Bank Centre, 1987), 42–45; Compton, 'Gabo: Mathematics and Science', 26.

26 Jacky Klein, 'Naum Gabo: Construction in Space (Crystal)', Tate Gallery, September 2002 <http://www.tate.org/art/artworks/gabo-construction-in-space-crystal-t06978>, Tate website, accessed 19 December 2012.

27 Klein, 'Naum Gabo: Construction in Space (Crystal)', accessed 19 December 2012.

28 For example, Hammer and Lodder, *Constructing Modernity*, 385.

29 Hammer and Lodder, 'Hepworth and Gabo', 128.

30 For example, Hammer and Lodder, *Constructing Modernity*, 262; Jacky Klein, 'Naum Gabo: Construction in Space with Crystalline Centre', Tate Gallery, September 2002 <http://www.tate.org/art/artworks/gabo-construction-in-space-with-crystalline-centre-t06977>, accessed 19 December 2012.

31 See Brown, *J.D. Bernal*, 61; and J.D. Bernal, 'My Time at the Royal Institution, 1923–27', International Union of Crystallography <http://www.iucr.org/people/crystallaphers/bernal2>, accessed 12 December 2012.

32 For Hepworth's recommendation, see Hepworth, letter to Bernal, 2 May 1939; cited in Hammer and Lodder, *Constructing Modernity*, 261, n. 251. Gardiner recalled purchasing the first completed version before September 1939, Hammer and Lodder, *Constructing Modernity*, 261, n. 253. Gardiner still owned it in 1978, see *Tate Gallery 1976–8: Illustrated Catalogue of Acquisitions* (London: Tate, 1979), 64, although it was not among the works gifted to the Pier Arts Centre, Stromness, Orkney, in 1979.

33 See Hammer and Lodder, *Constructing Modernity*, 280, nn. 79 and 80.

34 Hammer and Lodder, *Constructing Modernity*, 281 and plates 188–90.

35 Barlow, 'Barbara Hepworth and Science', 95–107.

36 A.H. Hammacher, *Barbara Hepworth* (London: Thames and Hudson, 1968), 89 and 86.

37 Quoted in Barlow, 'Barbara Hepworth and Science', 103.

38 Kathleen Raine, *Stone and Flower: Poems, 1935–43* (London: Nicholson & Watson, 1943).

39 Kathleen Raine, *Autobiographies* (London: Skoob, 1991), 266 and 224.

40 Alan Bowness, 'The Drawings of Barbara Hepworth', in *Barbara Hepworth: Drawings from a Sculptor's Landscape* (London: Cory Adams & Mackay, 1966), 17.

41 Barlow, 'Barbara Hepworth and Science', 106 and 103.

42 Goldsmith, *Sage*, 87.

43 J.D. Bernal and Helen Megaw, 'The Function of Hydrogen in Intermolecular Forces', *Proceedings of the Royal Society*, Series A, 151: 873 (2 September 1935): 384–420. See also boric acid, brucite, lepidocrocite and zinc hydroxide.

44 William L. Bragg, *Atomic Structure of Minerals* (Ithaca, NY: Cornell University Press; London: Humphrey Milford; Oxford: Oxford University Press, 1937), 249, fig. 136. See also blende (64, fig. 30), pyrites (72, fig. 39), 'octahedral' variants of spinel and magnetite (99, fig. 58), felspar (244, fig. 132), mica (207, figs 116 and 117) and orthoclase (236, fig. 131).

45 Hammer and Lodder, 'Hepworth and Gabo', 115–116.

46 Read, 'Introduction', in *Barbara Hepworth: Carvings and Drawings*, x; Hepworth, quoted in Chris Stephens, 'Drawings for Sculpture with Colour', in *Barbara Hepworth Centenary*, ed. Chris Stephens (London: Tate, 2003), 71.

47 For reproductions, see *Barbara Hepworth: Sculptress*, introduction by William Gibson (London: Faber & Faber for Shenval, 1946), 42–43; *Barbara Hepworth: Carvings and Drawings*, figs 61a and 61b.

48 Information on this sculpture is drawn from Matthew Gale and Chris Stephens, *Barbara Hepworth: Works in the Tate Gallery Collection and the Barbara Hepworth Museum St Ives* (London: Tate Gallery, 1999), 74–78; and Stephens, 'Drawings for Sculpture with Colour', 71.

49 See Barlow, 'Barbara Hepworth and Science', 102; Hammer and Lodder, 'Hepworth and Gabo', 128.

50 Hammacher, *Barbara Hepworth*, 85.

51 See Eric Hobsbawn, 'Preface' to Swann and Aprahamian, *J.D. Bernal: A Life in Science and Politics*, ix–xx.

52 Herbert Read, *The Green Child* (London: Robin Clark, 1989 [Heinemann, 1935]), 174. I am grateful to David Hulks for drawing my attention to this chapter of Read's novel.

53 See Read, *The Green Child*, 175–195.

54 Lancaster House, 18 March–9 May 1942. A version of the exhibition with a different catalogue toured seven British cities.

55 Hammer and Lodder, *Constructing Modernity*, 279–280. They note that Gabo's *Construction in Space with Crystalline Centre* may also have been exhibited.

56 Chris Stephens, '"The morrow we left behind": Landscape and the Rethinking of Modernism, 1939–53', *Art History*, vol. 35, no. 2 (April 2012): 219, 221, 224 and 232.

57 J.R.M. Brumwell, ed., *This Changing World: A Series of Contributions by Some of Our Leading Thinkers, to Cast Light upon the Pattern of the Modern World* (London: Routledge, 1944).

58 Goldsmith, *Sage*, 87.

59 Jackson, *From Atoms to Patterns*, 5.

60 See Jackson, *From Atoms to Patterns*, 8–10.

61 Brown, *J.D. Bernal*, 111.

62 See Lorna Arnold and Katherine Pyne, *Britain and the H-Bomb* (Basingstoke: Palgrave Macmillan, 2001).

63 Brown, *J.D. Bernal*, 414 and 454. The first march, in 1958, reversed the route.

64 B. Britten, E.M. Forster, H. Moore et al., 'Use of Atomic Weapons', *The Times*, 14 December 1950; cited in Chris Stephens, 'Henry Moore's *Atom Piece*: The 1930s Generation Comes of Age', in *Henry Moore: Critical Essays*, ed. Jane Beckett and Fiona Russell (Aldershot: Ashgate, 2003), 247, n. 9.

65 James Hyman, *The Battle for Realism: Figurative Art in Britain During the Cold War, 1945–1960* (New Haven and London: Yale University Press, 2001), 244, n. 76.

66 Richard Taylor, *Against the Bomb: The British Peace Movement, 1958–65* (Oxford: Clarendon, 1988), 9, n. 15.

67 John Berger, 'Artists and the H-Bomb', *Art News and Review*, vol. 10, no. 9 (24 May 1958): 2; for letters of response, see *Art News and Review*, vol. 10, no. 15 (16 August 1958): 2; cited in Hyman, *The Battle for Realism*, 244, nn. 76 and 79.

68 See Gardiner, *Barbara Hepworth: A Memoir*, 52–56.

69 Hepworth, letter to Read, 4 December [1956]; quoted in Gale and Stephens, *Barbara Hepworth*, 20; and Stephens, *Barbara Hepworth Centenary*, 246.

70 'Barbara Hepworth: A Memoir', in Stephens, *Barbara Hepworth Centenary*, 12.

71 John Berger, 'Round London', *New Statesman*, vol. 56, no. 1245 (July 1958): 15; cited in Hyman, *The Battle for Realism*, 171 and fig. 178.

72 William H. McNeill (University of Chicago), letter to Moore, nd [November 1963]; reprinted in David H. Katzive, 'Henry Moore's *Nuclear Energy*: The Genesis of a Monument', *Art Journal*, vol. 32, no. 3 (Spring 1973): 285.

73 Interview with Franz Schulze, *Chicago Daily News* (2 December 1967): 10; quoted in Iain A. Boal, 'Ground Zero: Henry Moore's *Atom Piece* at the University of Chicago', in Beckett and Russell, *Henry Moore*, 228.

74 Boal, 'Ground Zero', 228 and 235, and figs 10.4 and 10.5.

75 Stephens, 'Henry Moore's *Atom Piece*', 250–251.

76 See Terry Cavanagh and Alison Yarrington, *Public Sculpture of Leicestershire and Rutland* (Liverpool: Liverpool University Press, 2000), 11–12.

77 Helga László, 'The Realistic Sculpture of Laszlo Peter Peri' (MA dissertation, trans. John Lloyd, University of Budapest, 1990), vol. 1 (text), 83, and vol. 2 (plates), figs 66 and 99.

78 John Lloyd, 'Correspondence: Peter Peri's Politics', *Art Monthly*, no. 104 (March 1987): 25.

79 John Berger, 'The Necessity for Uncertainty', *Marxist Quarterly*, vol. 3, no. 3 (July 1956): 168–177; cited in Hyman, *The Battle for Realism*, 171.

80 László, 'The Realistic Sculpture of Laszlo Peter Peri', vol. 1, 'Supplement 6: sketches for the sculptures'.

81 On Moore's family groups and CND imagery, see Stephens, 'Henry Moore's *Atom Piece*', 248.

82 Read, 'New Aspects of British Sculpture', np. Read's introduction refers to three of Eliot's poems: 'The Lovesong of Alfred J. Prufrock' (1915), 'The Wasteland' (1922) and 'The Hollow Men' (1925).

83 Read sent the text to Lilian Somerville (British Council) on 2 April 1952, Margaret Garlake, 'Revisiting the Geometry of Fear', in *Rereading Read: New Views on Herbert Read*, ed. Michael Paraskos (London: Freedom Press, 2007), 135.

84 See Taylor, *Against the Bomb*, 9, n. 15, and 196, n. 20. On Read's support for Berger, see Hyman, *The Battle for Realism*, 244, n. 76.

85 Robert Burstow, '*The Geometry of Fear*: Herbert Read and British Modern Sculpture', in *Herbert Read: A British Vision of World Art*, ed. Ben Read and David Thistlewood (Leeds: Leeds City Art Galleries in association with The Henry Moore Foundation and Lund Humphries, 1993), 123–125.

86 See, for example, Herbert Read, 'Great Britain', in Marcel Brion et al., *Art Since 1945* (London: Thames and Hudson, 1958), 243.

87 Herbert Read, letter to Somerville, 2 April 1952; quoted in Garlake, 'Revisiting the Geometry of Fear', 135.

88 Herbert Read, 'Introduction' to *Sculpture in the Open Air*, exhibition catalogue (London: London County Council, 1963), np. The exhibition included all of those presented at Venice in 1952 except Paolozzi.

89 See James Hyman, 'Henry Moore and the Geometry of Fear', in *Henry Moore and the Geometry of Fear*, exhibition catalogue, James Hyman Fine Art, London, 2003, 6–11.

90 For example, Philip Hendy on Butler's work of 1949, cited in Hyman, 'Henry Moore and the Geometry of Fear', 8.

91 Hyman, 'Henry Moore and the Geometry of Fear', 10.

92 Read, 'New Aspects of British Sculpture', np.

93 Margaret Garlake, *New Art, New World: British Art in Postwar Society* (New Haven and London: Yale University Press, 1998), 194–200; Margaret Garlake, 'Identifying the Geometry of Fear', in *Henry Moore and the Geometry of Fear*, exhibition catalogue, 5; and Garlake, 'Revisiting the Geometry of Fear', 143.

94 See Margaret Garlake, *The Sculpture of Reg Butler* (Much Hadham and Aldershot: Henry Moore Foundation in association with Lund Humphries, 2006); Robert Burstow, 'Bernard Meadows', in *Sculpture in 20th Century Britain: Vol. 2: A Guide to Sculptors in the Leeds Collections*, ed. Penelope Curtis (Leeds: Henry Moore Institute, 2003), 232–235.

95 See Garlake, *The Sculpture of Reg Butler*, 16 and 25; Taylor, *Against the Bomb*, 197, n. 20.

96 Garlake, *New Art, New World*, 195.

97 See *Picasso and Modern British Art*, exhibition catalogue (London Tate, 2012).

98 See Carla Gottlieb, 'The Meaning of the Bull and Horse in *Guernica*', *Art Journal*, vol. XXIV, no. 2 (Winter 1964–1965): 106. (This interpretation of the bull was available in English by 1956.)

99 For example, David Thompson, 'A Decade of British Sculpture', *Cambridge Opinion*, no. 37 (January 1964): 24–32.

100 Robert Burstow, 'Aesthetics, 1950–75', in *Sculpture in 20th Century Britain: Vol. 1: Identity, Infrastructure, Aesthetics, Display, Reception*, ed. Penelope Curtis (Leeds: Henry Moore Institute, 2003), 170.

101 Katzive, 'Henry Moore's *Nuclear Energy*', 288.

102 Stephens, 'Henry Moore's *Atom Piece*', 251.

103 Herbert Read, 'Realism and Abstraction: A Footnote to the Preceding Essay', in Herbert Read, *The Philosophy of Modern Art* (London: Faber and Faber, 1982 [1951]), 104.

104 See Robert Burstow, 'Butler's Competition Project for a Monument to "The Unknown Political Prisoner": Abstraction and Cold War Politics', *Art History*, vol. 12, no. 4 (December 1989): 472–496; Robert Burstow, 'The Limits of Modernist Art as a "Weapon of the Cold War": Reassessing the Unknown Patron of the Monument to the Unknown Political Prisoner', *Oxford Art Journal*, vol. 20, no. 1 (Spring 1997), 68–80.

105 Read's introduction was given a new lease of life in 1956 when the British Council circulated the 1952 sculpture exhibition as an international touring exhibition. Among others who cited his phrase was Robert Melville in his introduction to the Lynn Chadwick exhibition at the British Pavilion of the XXVIIIth Venice Biennale in 1956.

106 Bernard Denvir, 'The Geometry of Fear', *Art*, no. 29 (June 1956): 3.

107 Read, 'Great Britain', 232–235.

108 Herbert Read, *Lynn Chadwick* (Amriswil: Bodensee-Verlag, 1958), 9–10.

109 Herbert Read, *A Concise History of Modern Sculpture* (London: Thames and Hudson, 1964), 218–225.

110 *Zeugnisse der Angst in der modernen Kunst*, exhibition catalogue, Mathildenhöhe Darmstadt, 1963.

111 Philip James, 'Sculpture Today', in *Sculpture in the Open Air*, exhibition catalogue (London: London County Council, 1960), np.

112 David Hulks, 'Despair, or Defiance: The Double Inflection in Herbert Read's Geometry of Fear', in Paraskos, *Rereading Read*, 144 and 151.

113 Read, 'New Aspects of British Sculpture', np.

114 Patrick Heron, 'Fruit or Thorn?', *New Statesman and Nation* (24 January 1953): 92.

115 See Burstow, 'Aesthetics, 1950–75', 171–175.

116 Thompson, 'A Decade of British Sculpture', 25.

117 Charles Spencer, 'Robert Adams', in *Robert Adams: Retrospective Exhibition*, exhibition catalogue (London: Arts Council, 1971), np.

118 See the artist's account of the work cited in Simon Martin, 'Art on the Eve of Destruction: Colin Self and the Nuclear Age', in *Colin Self: Art in the Nuclear Age*, exhibition catalogue (Chichester: Pallant House Gallery, 2008), 22.

119 'Speculative Illustrations: Eduardo Paolozzi in Conversation with J.G. Ballard and Frank Whitford', *Studio International*, 193 (October 1971), reprinted in *Eduardo Paolozzi: Writings and Interviews*, ed. Robin Spencer (Oxford and New York: Oxford University Press, 2000), 202.

120 See especially, Richard Calvocoressi, 'Public Sculpture in the 1950s', in *British Sculpture in the Twentieth Century*, ed. Sandy Nairne and Nicholas Serota, exhibition catalogue (London: Whitechapel Art Gallery, 1981), 134–153. The author quotes Read's catalogue introduction (143, n. 35).

3

'An Imagined Cataclysm Becomes Fact': British Photojournalism and Real and Imagined Nuclear War in *Picture Post*

Christoph Laucht

Britain's leading illustrated magazine, *Picture Post*, first covered real and imagined nuclear war in its 'Man Enters the Atom Age' number on 25 August 1945. In light of the recent bombings of Hiroshima and Nagasaki, the issue was dedicated entirely to atomic energy. Sir Edward Hulton, one of Britain's leading publishing tycoons, launched *Picture Post* in October 1938.[1] Aesthetically and stylistically 1920s German illustrated papers exerted a strong influence on *Picture Post*, and German-speaking émigrés played a crucial role in establishing its format.[2] Perhaps most notably, Stefan Lorant, *Münchner Illustrierte Presse*'s former editor-in-chief, became *Picture Post*'s first editor.[3] With the aid of his assistant Tom Hopkinson, Lorant ensured that *Picture Post* relied on the efficient use of typography – especially an appropriate balance between texts and images – to get the message across to readers.[4]

After its launch in 1938, *Picture Post* soon emerged as Britain's leading illustrated weekly magazine, holding the pole position until its cancellation in 1957.[5] The readership of Hulton's weekly magazine was particularly drawn from middle-class backgrounds.[6] Reports on the lives of 'ordinary' Britons, cutting-edge photography and the readiness to take political standpoints, in particular on social injustice, without advocating a particular party line were some of *Picture Post*'s chief characteristics.[7] In general, Hulton's weekly magazine advocated a social agenda situated left of the centre.[8] With its combination of state-of-the-art photography and texts, *Picture Post* represented not only a crucial medium through which many Britons formed their initial views of nuclear energy but an important vehicle of contemporary British art in the first decade or so of the atomic age.

While Mary Kaldor has broadly conceptualized the Cold War, which constituted the wider historical framework of most of the magazine's coverage of real and imagined nuclear war, as an 'imaginary war',[9] this chapter offers a more detailed analysis of the various layers of *Picture Post*'s coverage of the subject and their relationship and intersection of imaginary nuclear war with real nuclear war and its effects, especially Hiroshima, Nagasaki and atomic testing. These layers of representation include the conceptualization of real and imaginary nuclear war in the seminal 'Man Enters the Atom Age' issue, representations of the first post-war atomic tests at Bikini Atoll and the nuclear mushroom cloud image, as well as the impact of the hydrogen bomb on *Picture Post*'s coverage of imagined nuclear war. This chapter also looks at military strategy and scenarios of nuclear war, and at *Picture Post*'s reportage of the short- and mid-term effects of the atomic bombings of Hiroshima and Nagasaki.

Blurring the Boundaries between Imagined and Real Nuclear War: Three Levels of Imaginary War

Leafing through the pages of the 'Man Enters the Atom Age' issue, readers came across an article that was aptly titled 'An Imagined Cataclysm Becomes Fact' and addressed the first and most fundamental layer of imaginary nuclear war in *Picture Post*. Tellingly, it opened with the assertion that: 'The ruin produced by the atom bomb belongs to the world of the imagination. It is far more like the cataclysm foreseen by poets, painters and visionaries for generations than any piece of man-made devastation.' Spreading over an entire page, the article featured three apocalyptic drawings by Leonardo da Vinci. On the opposite page, under the heading 'Nature's Ruin – And Man's', it contained a blurry image of the mushroom cloud rising over Hiroshima on 6 August 1945 paired with an aerial photograph of columns of smoke billowing over Tokyo after the 1923 earthquake.[10]

These comparisons and apocalyptic references to nuclear weapons and their effects tapped into a longstanding tradition of prophecy belief. Prior to the arrival of the atom bomb, human beings often imagined Armageddon by referring to seemingly unprecedented natural events such as comets, volcanic eruptions, earthquakes, tsunamis, and so on. But the advent of nuclear arms fundamentally changed these assumptions because, as Paul Boyer notes, 'it seemed that man himself had, in the throes of war, stumbled on the means of his own prophesied doom'.[11] This first level of imagined cataclysm remained a recurring and underlying theme in *Picture Post*.

Another article in the same issue addressed a second layer of imagined nuclear war. As early as 1932, British politician, diplomat and writer Harold Nicolson had imagined the impact of atomic arms in his novel *Public Faces*. The *Picture Post* commentator placed Nicolson in line with science-fiction authors Jules Verne and H.G. Wells, who had conceived the idea of 'atomic bombs' being used by the 1950s in the novel *The World Set Free* (1914). 'But,' the article acknowledged, 'it is doubtful if anybody has ever hit off such an uncannily accurate advance description of any revolutionary

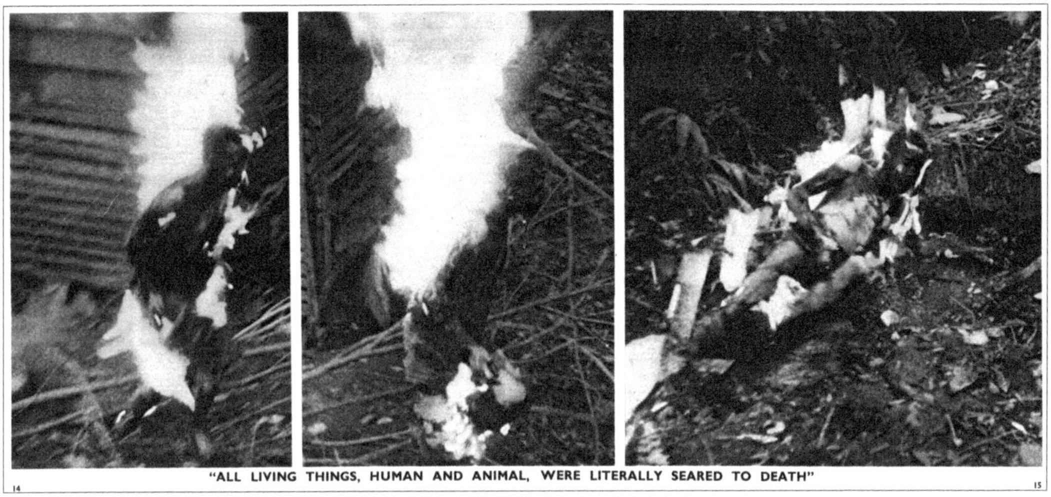

3.1 'All living things, human and animal, were literally seared to death', *Picture Post* (London, England), Saturday 25 August 1945, 14–15, Issue 8 © Mirror Group Newspapers and Getty Images. All rights reserved. Document Number PPO: 179560796.

discovery as Harold Nicolson' who 'even used the name "atomic bomb" for the weapon conceived by his imagination'.[12] This statement reveals the close connection between imaginary and real nuclear war because, as Bruce Franklin argues, 'our actual superweapons originated in their imagined history'.[13] In the pages of *Picture Post*, this applied to both atomic and thermonuclear weapons.

The *Picture Post* article on Nicolson then blurred the boundary between imagined and real nuclear war by linking the atomic explosion described in *Public Faces* with the recent nuclear attack on Hiroshima. It juxtaposed a report by one of the crewmembers of the *Enola Gay* – the American B-29 bomber that delivered the Hiroshima bomb – with Nicolson's description of the sight of an imagined atomic explosion by the fictitious crew of an American aircraft in the novel. In *Public Faces*, the British government acts responsibly and defuses the international conflict by using its nuclear deterrent to impose global disarmament. 'And so the atomic bomb of fiction became an instrument of peace,' *Picture Post* concluded, 'just as millions of men and women all over the world to-day are praying that it shall do in reality.'[14]

In the same 'Man Enters the Atom Age' number, a photo essay addressed a third layer of imagined nuclear war. It showed large pictures of a burning Japanese soldier who had been incinerated by a flame-thrower, 'a clumsy weapon, now utterly out of date' (see Figure 3.1). The article relied on conventional weaponry to explain the horrors of nuclear warfare to readers: 'The atom bomb which fell on Hiroshima had a very much more speedy, as well as very much more widespread, effect … that … was not much different from the one shown in the pictures.' To further 'authenticate' the impact of the attack, the essay featured several translated lines from a report by a local Japanese radio station. 'The atom bomb literally seared to death all living things – human and animal – in Hiroshima,' it opened, emphasizing: 'The dead are simply uncountable. It is not possible to distinguish the men from the women among the killed. The power of destruction of the atom bomb is beyond words.'[15] This interplay between images of victims' bodies and graphic language 'establishes a sense of relationship and compassion between reader and victim', as anthropologist Hugh Gusterson has generally demonstrated for the visual and textual representation of Hiroshima and Nagasaki survivors.[16]

Using current technology and recent experiences as a reference point, *Picture Post* strove to make the effects and the nature of real and imagined nuclear war comprehensible to readers. With its mapping of past (especially Second World War) experiences onto anticipated future scenarios of nuclear war, *Picture Post* resembled official British government propaganda on key policies and programmes, especially civil defence.[17] At the same time, references to the Second World War should also be seen against the background of Britain's direct exposure to bombing during the war, particularly during the Blitz or the notorious V-1 and V-2 attacks. To varying degrees, all three levels of imaginary nuclear war (nuclear cataclysm; the blurring of boundaries between imagined and real nuclear weapons; and the visualization of the effects of these weapons, particularly by referring to Second World War experiences) reappeared in later articles on nuclear war in *Picture Post* until its cancellation in 1957.

Symbolizing Nuclear War: Operation Crossroads and the Mushroom Cloud

After the atomic bombings of 6 and 9 August 1945 had introduced the British public to the new climactic weapon, British public opinion on nuclear arms turned more pessimistic by 1946.[18] Operation Crossroads, the United States' first post-war atomic tests in the Marshall Islands in the South Pacific in the summer of 1946, was crucial in this context. During the operation, two nuclear tests were fired, codenamed Able and Baker. 'I am convinced, however, that the witnessing of this impressive experiment will yield ample recompense for any inconvenience that may be experienced,' stressed Vice Admiral William Blandy, the military commander who oversaw the tests, regarding Operation Crossroads' significance in the official booklet issued to observers.[19] In the build-up to the tests, the *New York Times* published a graphic novel-style guide to the upcoming event in February 1946. Sidney Shalett's texts and Bertrand Zadie's illustrations explained the tests within their wider context, from Hiroshima to the (anticipated) discussion of the test results by the United States Joint Chiefs of Staff.[20] In October 1947, *Picture Post* informed readers in detail about Operation Crossroads. Martin Chisholm investigated the atom bomb's potential impact on Britain's naval planning and strategy.[21] That Chisholm's article relied heavily on photographs of the mushroom clouds of the two Bikini tests demonstrates that the mushroom cloud had arguably become both 'the most impressive of all nuclear symbols' and a household name after Operation Crossroads.[22] Gerhard Paul also terms it a major 'twentieth-century media icon' in light of its seeming omnipresence.[23]

Chief amongst the images featured in the *Picture Post* article was an enlarged photograph of the Baker test, subtitled 'The Climax of "Operation Crossroads": Ten Million Tons of Water Become Airborne'. Accompanying captions such as '"The Cauliflower" Blooms over a Doomed Fleet' and 'It Breaks Downwards and Spreads Destruction' added to the article's pessimistic message about the nuclear future.[24] This combination of mushroom cloud imagery and pessimistic caption texts was reminiscent of the more subtle meanings that observers often inscribed into this nuclear symbol by placing it, as Peter Bacon Hales has argued in the American context, 'in line with important modernist visions of science and warfare, and especially of their intersection'.[25] However, unlike many commentators in the United States, who viewed Operation Crossroads metaphorically as a decisive intersection for mankind in the atomic age, Chisholm's piece placed emphasis on doom and destruction.[26]

This pessimism reveals some fundamental differences in meaning that have been written into this symbol. Commentators such as Jane Caputi have argued that 'any attempt to assign definite symbolic significance to the mushroom cloud is haunted by ambiguity, paradox, and contradiction',[27] and Spencer Weart has noted that mushrooms bore ambivalent meanings and could be associated with either death and decay or magic and life.[28] In a global context, the mushroom cloud therefore represents a prime example of Victor Turner's concept of 'dominant symbols' which are classified according to three criteria: '(1) condensation; (2) unification of disparate meanings in a single symbolic formation; (3) polarization of meaning.'[29] While the image of the nuclear mushroom cloud arguably assumed more varied and diverse layers of meaning in the American context, oscillating between nuclear utopia and dystopia and even kitsch, its meanings were generally more negative and associated with destruction in the British context, as illustrated by Chisholm's article for *Picture Post*.[30]

The evolution of the term 'mushroom cloud' dates back to the world's first nuclear explosion – the Trinity test that took place in the New Mexican desert on 16 July 1945. Several British scientists, who were part of the British mission to the Manhattan Project and witnessed the test, played a part in its creation. Their first-hand descriptions are particularly evocative. Sir James Chadwick, head of the British mission, noted that: 'The ball of fire was intensely bright, equivalent to several suns by general estimate.'[31] The explosion initially looked 'somewhat like a raspberry,' recorded experimental physicist Otto Frisch, before it took on the shape of 'the trunk of an elephant'.[32] Another British scientist, James Tuck, reported that he saw 'a beautiful poplar tree shaped cloud, in turbulent motion' and offered what can perhaps be seen as the first artistic treatment of the mushroom cloud – three sketches of the explosion, including the mushroom-shaped cloud.[33] The *New York Times* journalist William Laurence, who was also the official chronicler of the Manhattan Project, later wrote in his best-selling book *Dawn over Zero* (1946) that: 'A great cloud rose from the ground and followed the trail of the great sun. At first it was a giant column, which soon took the shape of a supramundane mushroom.'[34] The same year that Laurence's book was published also saw the mushroom cloud take on more abstract shape and meaning in American artist Philip Evergood's painting *Renunciation* (1946) as a cynical commentary on Operation Crossroads.[35]

Alongside its semantic importance, Operation Crossroads marked the temporary climax of nuclear photography up to that point. Although – as film scholar Anthony Michaelis wrote retrospectively about Operation Crossroads in 1952 – 'undoubtedly the greatest achievement of cinematography in the field of scientific war research was the filming of the atomic bomb explosions at Bikini Island',[36] the relatively new field of atomic photography had gained momentum during the time of the Manhattan Project where special cameras had to be developed to document and assess experiments conducted as part of the development of the first atomic bombs. This pioneering work in atomic photography culminated in recording the Trinity test.[37]

Besides mushroom cloud images, Chisholm's *Picture Post* article included photographs of the ships that were destroyed in the tests and a map that visualized the effects of the two tests on navy vessels. It also featured a photograph of Hiroshima survivor Kiyoshi Kikawa, showing some of his burns. This inclusion can be seen as both a warning about the human cost of atomic warfare and a pessimistic outlook on the future.[38] But Operation Crossroads did not remain the only instance where *Picture Post* featured mushroom cloud images. Perhaps most explicitly, *Picture Post* bore a 'Nevada Mushroom' from an above-ground test at the Nevada Test Site in the United States on the cover of its 11 October 1952 issue to accompany an article that informed readers about Britain's first atomic bomb test in Australia.[39]

Changing the Game: The H-Bomb and Imagined Thermonuclear War

If *Picture Post*'s coverage of Operation Crossroads, with its extensive reliance on mushroom cloud imagery and dire commentary, had indicated an increasing pessimism with regards to the effects of real and imagined atomic war on the part of the magazine's editorial staff, the coming of the hydrogen bomb marked the real game-changer. Shortly after United States President Harry Truman had announced his administration's firm determination to develop thermonuclear arms in January 1950, *Picture Post* addressed this new weapon and its anticipated effects in an article under the graphic title 'Can Man Survive the Hydrogen Bomb?'[40]

Its author, Derek Wragge Morley, used two books by Austrian physicist Hans Thirring as his point of departure to discuss the H-bomb's design, yield and anticipated effects, including its use as a 'dust bomb.'[41] Since 'there is virtually … *no theoretical limit to the size of the bomb* [original emphasis]',[42] Morley argued that its yield 'would be limited only by the difficulties of

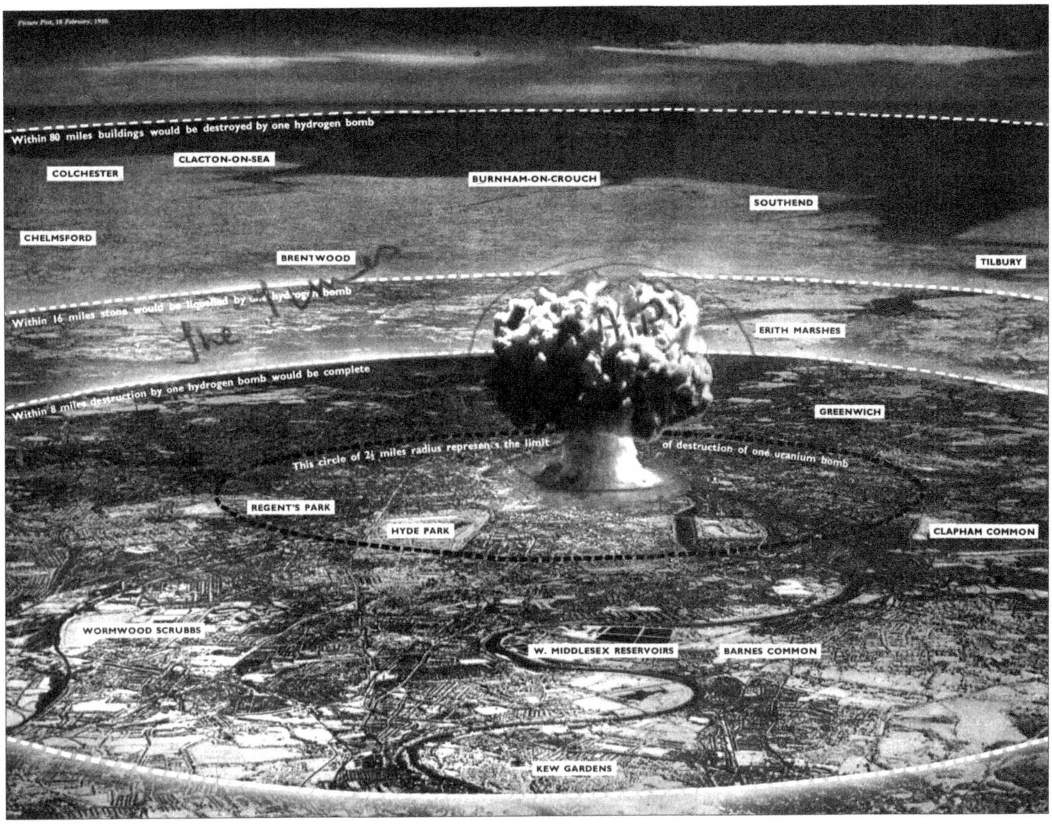

3.2 'This is what would happen if a hydrogen bomb were dropped on the British Houses of Parliament', *Picture Post* (London, England), Saturday 18 February 1950, 32–33, Issue 7 © Mirror Group Newspapers and Getty Images. All rights reserved. Document Number PPO: 179560770.

transport', pointing to 'the possibilities of water transport' and the beginning of nuclear war 'by means of a Pearl Harbour [*sic*] type of attack'.[43] This scenario was based on actual concerns by the United States authorities that Soviet agents might attempt to smuggle atomic bombs into Britain and other signatory nations of the newly founded North Atlantic Treaty Organization (NATO) to force their withdrawal from the military alliance.[44]

To illustrate the anticipated effects of thermonuclear warfare, the article featured two diagrams that detailed the impact of a hypothetical thermonuclear attack on London. With its focus on Britain's largest urban area, *Picture Post* adopted a technique from American popular media where, after Hiroshima, New York City had emerged as a chief target of imaginary nuclear warfare.[45] Similarly, Morley's article illustrates well Paul Saint-Amour's observation that 'the inhabitants of Cold War cities … became accustomed to a more overt and permanent variant of the uncanny frission felt in Hiroshima before the bombing, as a structuring condition of every-day life'.[46]

The first illustration in Morley's piece was an aerial photograph of the London area that stretched over one and a half pages with a superimposed mushroom cloud over Westminster as the hypothetical ground zero (Figure 3.2). Alongside the names of major London boroughs, landmarks and towns in the city's immediate environs, the picture featured superimposed concentric

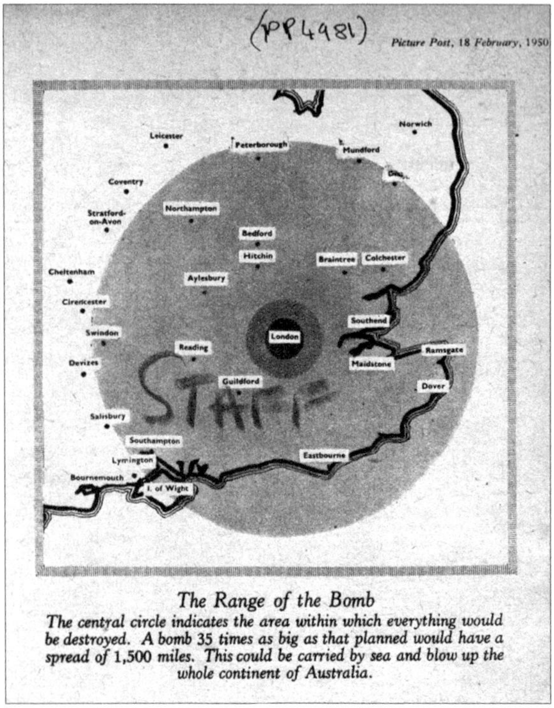

3.3 'The Range of the Bomb', *Picture Post* (London, England), Saturday 18 February 1950, 33, Issue 7 © Mirror Group Newspapers and Getty Images. All rights reserved. Document Number PPO: 179560770.

circles to visualize the imagined effects the attack would have on the entire area.[47]

Concentric circles played a major role in visualizing the impact of real and imaginary nuclear war. The report by a British team that had visited Hiroshima and Nagasaki in November 1945, entitled *The Effects of the Atomic Bombs at Hiroshima and Nagasaki*, is a key text in this context. Published on the eve of the Bikini Atoll tests in 1946, it featured, apart from a photo section that specified the damage caused by the bombs, highly abstract concentric circles that detailed the effects of the bombs in fairly generic terms.[48] *Illustrated London News* reprinted the diagram from the government report,[49] and it was later also included in the seminal Atom Train exhibition that the Atomic Scientists' Association (ASA), the chief organization of British nuclear scientists, organized and that toured the country in 1947–1948.[50]

Apart from using concentric circles, Morley's article imagined the effects of such an attack in graphic language: 'If a hydrogen bomb were dropped at Westminster, everything within 8 miles would be completely destroyed. No person or building would survive.' About 16 miles from ground zero 'temperatures would still be great enough to melt some stone' and 'Tidal waves would batter Calais and the French channel ports'. Using real nuclear warfare from the Second World War era as a reference point, he offered a comparison: 'Yet if the Nagasaki bomb had been dropped at the same point, little damage would have been done beyond Regent's Park, Hyde Park and Battersea Park'.[51]

A map of southeastern England placed the explosion's anticipated effects within the national context (Figure 3.3). 'The central circle indicates the area within which everything would be destroyed,' read the caption. 'A bomb 35 times as big as that planned would have a spread of 1,500 miles', cautioning in an alarmist manner: 'This could be carried by sea and blow up the whole continent of Australia.'[52] That *Illustrated London News* featured a similar diagram of an imagined H-bomb attack on London demonstrates the prominence that this scenario enjoyed in the popular media.[53]

Moreover, Morley looked into Hans Thirring's idea 'of war by radioactive dust'. Morley envisaged that an aircraft could deliver such a 'dirty bomb' that used multiple sets of explosives to disperse its radioactive particles. Two photographs relating to the experience of Nagasaki survivor and physician Paul Nagai, who had become an influential writer against atomic weaponry, humanized the imagined impact of this fairly abstract weapon.[54] And *Picture Post* once again mapped the experience of real nuclear war on to the effects of imagined thermonuclear warfare. The fact that scientists in the United States publicly discussed the possibility of using the H-bomb to disperse radioactive fallout into the atmosphere around the same time demonstrates the high importance that experts attached to this scenario.[55] Such a 'doomsday device' later gained particular fame through its inclusion in the plot of Stanley Kubrick's satire *Dr. Strangelove, Or: How I Learned to Stop Worrying and Love the Bomb* (1964).[56]

If earlier articles in *Picture Post* had imaged the H-bomb as a cataclysmic weapon of unprecedented force, the United States' first test of a thermonuclear device in 1952 eventually made it real. But it was the United States' Castle Bravo test in the South Pacific in March 1954 that sensitized people all around the world to the dangers emanating from the H-bomb. Particularly poignant were pictures of crew members of a Japanese fish trawler – the *Lucky Dragon* – who had fallen ill with radiation sickness after their ship had been hit by radioactive fallout from the test while fishing outside the designated danger zone. These caught much public attention, as they revealed the human cost of thermonuclear warfare.[57]

In *Picture Post* in October 1954 Tim Raison addressed the debate over civil defence that had erupted in Britain in the wake of the *Lucky Dragon* incident and had led the Coventry City Council to abandon its civil defence operations altogether.[58] In light of the H-bomb, Raison discussed the raison d'être of Britain's civil defence programme by imagining the impact of the Castle Bravo H-bomb on London, where 'it might have caused complete destruction over a four and a half mile radius, heavy to moderate damage over ten miles, and light damage for over twenty'. He emphasized 'three effects – blast, heat and radio-activity ... which people fear most'.[59]

These three chief effects of (thermo)nuclear arms figured high in official British government propaganda, such as the Home Office's first civil defence training manual that was exclusively dedicated to nuclear warfare. It contained familiar concentric circle diagrams to visualize the impact of the heat flash and the blast wave. In addition, it featured images of dosimeters, Geiger counters and a section with photographs of destroyed Hiroshima and Nagasaki that were reprinted from the British government report, *The Effects of the Atomic Bombs at Hiroshima and Nagasaki*.[60] Yet, the government continued to play down the effects of the H-bomb, even after the secret Strath Report (1955) had concluded that there was no effective defence against thermonuclear attack,[61] in order not to jeopardize public support for the country's aspiring H-bomb project.[62] Given the likelihood of a thermonuclear attack on Britain, Raison criticized the Home Office for planning to scrap the Civil Defence Corps' Mobile Column, the most effective asset in the event of an H-bomb attack. To get the point across and to suggest its readiness, his article included several photographs by Slim Hewitt of an exercise carried out by the Mobile Column.[63]

Civil defence formed an important part of imagined nuclear war.[64] To anticipate the effects of a nuclear attack and to optimize their operative capabilities, civil defence officials in the United States and elsewhere took to 'imagineering', as David Monteyne has demonstrated, and constructed post-nuclear-strike urban environments to train their personnel.[65] In October 1952, *Picture Post* featured a large photograph of two senior British civil defence officials studying an 'imagineered' model of a British urban centre post nuclear strike.[66] At the same time, civil defence programmes in Britain and elsewhere publicly relied on elements of performance and rehearsal to suggest an air of preparedness for and survivability of nuclear war.[67]

In July 1955, performance and rehearsal figured in a *Picture Post* article on the largest civil defence drill in United States history to date. Jack Winocour introduced the scenario in which 'nearly fifteen million Americans were killed and injured by Hydrogen bombs dropped on sixty of the biggest centres of population'.[68] The exercise formed part of a series of Operation Alerts in the United States between 1954 and 1960.[69] Winocour's piece marked the most sarcastic article on imaginary atomic war in *Picture Post*. 'Is this a dress rehearsal for doomsday?' he posed the (almost) rhetorical question. In the imaginary strike, an unnamed enemy nation deployed a five megaton H-bomb against New York City which 'dies in a flash, and a roar, at 2.5[0] this afternoon'. Winocour detailed the strike further: 'Ten minutes later I may be one of New York's 2,991,285 dead', quipping:

> The statisticians have calculated the figure to the nearest human digit with actuarial exactitude. Or luckier – or unluckier, depending on what you think about radiation sickness – I may be one of the city's 1,776,899 injured. If I escape I shall do so with 3,282,345 other New Yorkers, to live amid the ruin and rubble, or to disperse to the interior.[70]

Not only do casualty estimates form an integral part of civil defence planning (and imaginary nuclear war more widely) but projected casualty figures played a key part in nuclear decision making from the start, in particular US President Harry Truman's decision to authorize the use of the atomic bombs against Japan. Purely hypothetical discussions by US Secretary of War Henry Stimson and the United States Strategic Bombing Survey of alternatives to the atom bomb, especially estimated casualty figures of a planned invasion of mainland Japan, underline the highly abstract nature of this debate well.[71] The point raised by Barton Bernstein about this particular counterfactual argument, stating that 'operating in the penumbra of a hypothesized-but-never-lived future adds to the complexity of analysis', also applies to the study of imaginary nuclear war more broadly.[72]

Winocour's article was lavishly illustrated with photographs from the exercise, ranging from US President Eisenhower in his command bunker to images of evacuees and civil defence workers, and given evocative titles such as 'members of a New York decontamination squad appear on Broadway' and 'The Mass Retreat From the Pentagon'.[73] These photographs, along with Winocour's text, illustrate the theatrical nature of civil defence exercises.[74]

Winocour's account blended the hypothetical thermonuclear attack on New York City with his own personal British Second World War experience as 'the sirens wail, bringing back for me the unquieted memory of the war years', especially the 'never forgotten summer of 1944, when flying bombs first disclosed to all of us the shape of things to come'.[75] While Winocour's report reveals the degree to which civil defence operations generally led to a militarization of civilian life,[76] his remarks stress a fundamental difference in British and American experiences of imaginary nuclear war. Apart from the effects of the Hiroshima and Nagasaki bombings that figured high in British and US civil defence planning, British scenarios by and large also included experiences of the Blitz and mapped them onto imagined nuclear war.[77]

Alongside civil defence, fallout from nuclear tests was one of the key issues of the day and figured high in scenarios of imagined nuclear war. As the 1954 Castle Bravo test in the South Pacific had so fatally demonstrated in the *Lucky Dragon* incident, extensive US and Soviet thermonuclear tests drastically revealed that, by the mid-1950s, fallout was an international problem that could not be limited to a country or test site.[78] Shortly after the British government had announced that it would go ahead and develop its own thermonuclear weapons, *Picture Post* addressed the fallout issue in two articles in its 23 April 1955 issue. In the first article, the geneticist and evolutionary biologist John Burdon Sanderson Haldane focused on possible genetic effects of radioactive fallout on current and future generations. If calculations were accurate, 'the prospect is pretty serious,' he concluded, with 'literally hundreds of millions of deaths in future generations'. On its first page, the article featured a photograph of a worker from the Atomic Energy Research Establishment Harwell in radiation protection gear and a picture of the gigantic mushroom cloud from the first US thermonuclear test in 1952. 'Shall We, One Day, Have to Live Like This?' asked the accompanying caption and concluded that: 'If nuclear bomb tests … were to continue, on an unlimited scale, future generations might have to be clothed in similar garments.'[79]

A diagram of a stylized female body that showed 'How Excessive Radiation Might Attack the Human Body' followed Haldane's essay. Additionally, a close-up of one healthy and one irradiated germ cell completed this visualization of the key health effects of radiation. Below, under the cautionary title 'Science Must Warn Mankind', scientific experts including the president of the British Association for the Advancement of Science, Lord Adrian, and Joseph Rotblat, a senior ASA member, warned of the impact that nuclear testing and its radioactive fallout could have on human health.[80]

British popular culture and art, too, responded to the threat posed by radiation and mutation. Hammer Films, Britain's leading production company of horror films, addressed nuclear fears and offered a very critical commentary on the risks involved in unrestricted scientific and technological advance, exploiting the mutation theme in the TV serials *The Quatermass Experiment* (1953) and *Quatermass II* (1955), as well as in the feature film *X the Unknown* (1956). Like the (usually) invisible fallout from nuclear testing, these Hammer productions remained fairly vague about the origins of the featured monsters, even questioning – especially in the case of *X the Unknown* – the feasibility of locating and annihilating a threat whose origin and nature are not fully known.[81] In a similar fashion, British artist Francis Newton Souza later used distorted shapes and figures in his paintings as a form of anti-nuclear protest after 1960.[82]

Imagining the Unthinkable: Military Strategy and Scenarios of Nuclear War

While the development of the H-bomb severely impacted the ways in which *Picture Post* viewed the prospects of imaginary nuclear war in increasingly dystopian terms, the first Soviet nuclear test in 1949 and the start of the Korean War in June 1950 led the illustrated magazine to report on military strategy. Military expert Lieutenant-General Sir Brian Horrocks imagined the scenario of a Soviet invasion of the Federal Republic of Germany, including the use of atomic bombs. While he assessed NATO's strategic position as disadvantageous by comparison with the Soviet Union and underlined Britain's high vulnerability to attack by Soviet strategic nuclear weapons, Horrocks envisaged that tactical nuclear weapons might give NATO forces an advantage over numerically superior Red Army troops. A map representing 'A Diagrammatic View of How the West Must Organise Defence in Depth in Europe' accompanied the piece, to illuminate Horrock's key points.[83]

Not all experts on military strategy shared Horrock's optimistic assessment, particularly with regard to the use of tactical nuclear weapons. As early as July 1948, Liddell Hart had argued in *Picture Post* that: 'The atom-bomb is *not* [original emphasis] a tactical weapon designed to repel the penetrating advance of armies.'[84] Hart reiterated his opposition to tactical nuclear weapons immediately after the British government had publicly announced its decision to pursue its own thermonuclear weapons development programme in February 1955. He particularly feared an escalation of limited nuclear into all-out atomic war. Thermonuclear weapons served as a catalyst to amplify the effects of such an escalation dramatically. 'In the case of Britain, where the vital targets are closely grouped, five to ten hydrogen bombs would almost certainly suffice to annihilate all its main centres of industry, comprising half the population of the country,' Hart imagined the impact of a thermonuclear attack on Britain. Photographs of American delivery systems for nuclear warheads accompanied Hart's article, showing a large gun used to fire tactical nuclear weapons, an Honest John mobile missile launcher and an intermediate-range Matador rocket. In addition, the piece featured a map of Europe, with superimposed arrows indicating the distance between Moscow,

Rome and Western Germany, as well as Eastern Germany, London and Paris, which served to illustrate the distances that rockets as a delivery system for atomic warheads could cover.[85]

Missiles also played an important role in *Picture Post*'s scenarios of imaginary nuclear war in two articles in 1952 and 1956 respectively. As 'Rocket-fever has taken hold' of Britain and other nations, Trevor Philpott provided readers with a review of recent developments in rocketry in Britain, West Germany and the United States in 1952. The article was lavishly illustrated with photographs of the secret missile research facility on the Woomera Range in Australia, where British plans to develop and test its own nuclear-capable missiles were to be put into practice, and pictures of current US missiles: Tarzan, JB2, Sparrow and Matador. In addition, it contained graphics that explained the modus operandi of air-to-ground, surface-to-surface and interceptor missiles.[86]

This interest in rocketry corresponded with a wider cultural fascination with space exploration. Starting in April 1950, *Picture Post*'s publisher, Hulton Press, launched the highly successful *Eagle* comic books that featured a weekly strip about the adventures of 'Dan Dare: Pilot of the Future'.[87] The Dan Dare story 'The Red Moon Mystery' (1952) addressed both atomic and thermonuclear arms. When an asteroid is on collision course with the Earth, the Intraplanet Space Fleet deploys the last remaining twelve atom bombs of the 'U.N. Police strategic reserve' to divert the object – in vain. Since 'All Super Explosives were completely dismantled after the final peace congress in '65', including the production facilities, there are no H-bombs left to do the job.[88] This indicates how deeply thermonuclear arms pervaded popular culture.

Picture Post kept readers up-to-date about the latest developments in missile technology with an article in the 10 March 1956 issue. Woodrow Wyatt offered a survey of key advances in rocketry in Britain and especially the United States. 'Already in this century we have had the tank, the aeroplane, radar, and hydrogen and atom bombs,' he emphasized. 'Now, as though all that were not enough, comes the age of the guided missile.' A particular threat was the American Nike missile that 'can carry an atomic warhead which will blow up whole formations of enemy bombers at a time'. But in Wyatt's eyes it was the nuclear-tipped intercontinental ballistic missile (ICBM) that was about to revolutionize military strategy because of its capability of enabling the two superpowers to attack each other directly. Photographs of American Terrier and Nike missiles taking off, as well as a Nike missile intercepting a B-17 bomber, accompanied the piece.[89]

Apart from rocketry, *Picture Post* dedicated an article in its 26 November 1955 issue to 'The Shape of Wars to Come'. It consisted of two pictures. The first one showed an image of a US Army atomic cannon in the foreground and a mushroom cloud looming in the background. On the opposite side, a photograph depicted the test of a so-called washdown system on a US Navy aircraft carrier that could be used as a 'counter-measure' to decontaminate military hardware from radioactive fallout.[90]

Reversing the Effects of Real Nuclear War: Hiroshima, Nagasaki and Urban Renewal

While the tone of *Picture Post*'s coverage of nuclear war and atomic testing generally became more pessimistic over time, conversely, the magazine reported in an increasingly positive tone about the short- and mid-term consequences of the atomic bombings of Hiroshima and Nagasaki. In its issue of 24 August 1946, *Picture Post* ran a short article on the reconstruction of the two Japanese cities. Its title, 'After the Atom Bomb: An Astonishing Rebirth', used a procreational motif that was already indicative of its positive tenor. The piece featured a giant aerial photograph of newly built houses in Hiroshima bearing the declamatory caption 'How Man Comes Back to Hiroshima'. It rejected gloomy predictions about Hiroshima and Nagasaki's inhabitability that experts had put forth immediately after the atomic bombings and instead stressed the resilience of both humans and nature. A photograph showing a garden full of crops amongst the ruins of Nagasaki supported this claim. 'The atom bomb is not the Last Weapon after all', the piece concluded in an ambivalent tone: 'That may or may not be a source of consolation.'[91]

Although photographs of victims of the atomic bombings of Hiroshima and Nagasaki featured in several articles as a warning against the human cost of atomic war, *Picture Post* dedicated only very few articles exclusively to the effects of the nuclear attacks.[92] Indeed, it took about a decade until Hiroshima featured again in two full articles. In March 1955, *Picture Post* reviewed Kaneto Shindô's film *Children of Hiroshima* (*Genbaku no Ko*, 1952) which dealt with the impact of the bombing through Japanese eyes. Given Prime Minister Winston Churchill's recent H-bomb announcement, Sylvain Mangeot's essay marked a temporary return to a more pensive coverage of the bombings of the two Japanese cities, featuring stills from the film.[93]

On the occasion of the tenth anniversary of Hiroshima, *Picture Post* published a photo report by journalist Trevor Philpott and photographer John Chillingworth about life in the Japanese city ten years on. Their essay reached a new (and unrivalled) level of optimism about the effects of real nuclear war in *Picture Post*.[94] So optimistic was their portrayal that it appeared almost as a British version of 'the bright atomic future narrative' that Scott Zeman has identified in coverage of the atomic bomb in American popular science magazines in the period after Hiroshima until the first post-war nuclear tests at Bikini Atoll in July 1946.[95]

Philpott and Chillingworth's piece opened with two panoramic views of Hiroshima, one taken shortly after the atom bomb had devastated the city and the other one showing the same area in 1955. 'They are calling it the Atomic Desert, as they wander among the embers looking for relatives,' the caption for the former picture read. The second photograph, by contrast, depicted a modern, rebuilt city centre, with the subtitle 'They are calling it the Peace Commemoration City, and declaring that soon it will be the most systematic and progressive city in all Japan', a reminder to readers of the tremendous reconstruction effort that had taken place in the former nuclear wasteland.[96]

'In Hiroshima now,' the article read, 'there is hardly a sign that even a little bomb ever dropped there.' Philpott praised the city's modern buildings and infrastructure. Several photographs of scenes of everyday life visually supported the optimistic message. These included images of men playing on pinball machines (Pachinko), a traditional music parade, supposedly 'healthy' students at a Hiroshima school engaging in physical exercise, the Peace Memorial and the headquarters of the Atomic Bomb Casualty Commission.[97]

With its strong focus on urban renewal through atomic devastation, Philpott and Chillingworth's report followed in the vein of science-fiction literature that imagined 'nuclear Holocaust as urban renewal' as early as the 1930s.[98] This is especially evident where the city's Deputy Mayor, who was also one of the chief engineers in charge of planning Hiroshima's post-war reconstruction, is recorded as saying: 'We didn't rebuild the old town, we built a new city.' Drawing his first highly positive conclusion, Philpott asserted: 'a city can be an atomic desert in 1945, its debris littered with one hundred thousand peeled corpses; and ten years later it can be bigger and brighter and better than it ever was.'[99]

Concerning the medical and environmental effects of the atomic bombings, however, Philpott's text was more cautious than the 1946 article that had simply condemned experts for jumping to fast conclusions. Philpott cited Japanese and American medical experts who had not ruled out the possibility of genetic defects in future generations of Hiroshimans, before he reached his second conclusion 'that humanity isn't going to be so easily blotted out, even in the Atomic Age'. War itself should be banned, Philpott argued, 'not any particular one of its weapons'.[100] *Picture Post*'s coverage of the effects of the atomic bombings of Hiroshima and Nagasaki represented a rare case in which Hulton's illustrated magazine featured 'the Utilitarian view' of nuclear war (more common in American publications like *Life* magazine) which acknowledged the seriousness of any such conflict but questioned its capability of annihilating entire nations, let alone humankind.[101]

Conclusion

While *Picture Post* had an ambivalent take on the short- and mid-term effects of the atomic bombings of Hiroshima and Nagasaki which it viewed increasingly optimistically, its coverage of imagined scenarios of atomic war was in general pessimistic. Between the 'Man Enters the Atom Age' issue of 25 August 1945 and its final issue on 1 June 1957, *Picture Post* had addressed the issues of real and imaginary nuclear war with varying intensity and in ambiguous terms. Often the effects of conventional warfare during the Second World War and the atomic bombings of Hiroshima and Nagasaki were blended with scenarios of imagined nuclear war. While the Bikini Atoll tests of 1946 marked a turning point towards an increasingly dystopian tone in the magazine's reportage on real and imaginary nuclear war, it was the coming of the H-bomb that was the real game-changer.

In June 1957, *Picture Post* folded, partly as a consequence of the rise of television as a mass medium and an increase in the cost of paper.[102] Nevertheless, Hulton's illustrated magazine had served as an important forum through which many Britons shaped their initial knowledge and views on nuclear energy and real and imaginary nuclear war. And it represented an important manifestation of British art in the early nuclear age.

Notes

1 Tom Hopkinson, 'Introduction', in *Picture Post 1938–50*, ed. Tom Hopkinson (London: Chatto & Windus: Hogarth Press, 1984), 9.

2 Tom Hopkinson, 'Foreword', in *Picture Post 1938–50*, unpaginated; David Reed, *The Popular Magazine in Britain and the United States 1880–1960* (London: British Library, 1997), 185.

3 Irme Schaber, 'Pioniere mit Langzeitwirkung: Der Einfluss der fotografischen Emigration der NS-Zeit auf die englische Fotolandschaft und Bildpresse am Beispiel von Kurt Hutton, Felix H. Man, Wolf Suschitzky und weiteren Fotoschaffenden', *Yearbook of the Research Centre for German and Austrian Exile Studies*, 3 (2001): 78.

4 Reed, *Popular Magazine*, 185.

5 *A National Survey of Readership of Newspapers and Magazines on a Continuing Basis*, ed. Institute of Practitioners in Advertising (London: Institute of Practitioners in Advertising, 1956), 38; Siân Nicholas, 'Being British: Creeds and Cultures', in *The British Isles: 1901–1951*, ed. Keith Robbins (Oxford: Oxford University Press, 2002), 111.

6 *The Hulton Readership Survey 1948*, ed. J.W. Hobson and H. Henry (London: Hulton Press, 1948), 8, 11; *The Hulton Readership Survey 1953*, ed. J.W. Hobson and H. Henry (London: Hulton Press, 1954), 5, 11, 21; *A National Survey of Readership of Newspapers and Magazines on a Continuing Basis*, 38, 40, 42, 46, 48.

7 Tom Hopkinson, 'How Picture Post Began', *Picture Post*, 2 October 1948, 11–13, 47; Edward Hulton, 'The Last Picture Post', *Picture Post*, 1 June 1957, 3.

8 Cf. Sean Kenneth Smith, '*Picture Post*, 1938–1945: Social Reform and Images of Britain at War' (PhD dissertation, Stanford University, 1992).

9 Mary Kaldor, *The Imaginary War: Understanding the East–West Conflict* (Oxford: Basil Blackwell, 1990), 4.

10 'An Imagined Cataclysm Becomes Fact', *Picture Post*, 25 August 1945, 10–11.

11 Paul Boyer, *When Time Shall Be No More: Prophecy Belief in Modern American Culture* (Cambridge, MA: Belknap Press of Harvard University Press, 1992), 115.

12 'The Man Who Saw It Coming', *Picture Post*, 25 August 1945, 25; Ferenc Szasz, *Atomic Comics: Cartoonists Confront the Nuclear World* (Reno, NV: University of Nevada Press, 2012), 20. See also Paul Brians, *Nuclear Holocausts: Atomic War in Fiction, 1895–1984* (Kent, OH: Kent State University Press, 1987).

13 Bruce Franklin, *War Stars: The Superweapon and the American Imagination* (New York: Oxford University Press, 1988), 4.

14 'The Man Who Saw It Coming', 25. For the original passage from Harold Nicolson's novel, see *Public Faces* (London: Constable, 1932), 302–303.

15 'What the Atom Bomb Means If Mankind Should Now Fail to Control Its Use', *Picture Post*, 25 August 1945, 14–15.

16 Hugh Gusterson, *Nuclear Rites: A Weapons Laboratory at the End of the Cold War* (Berkeley, CA: University of California Press, 1996), 112.

17 Matthew Grant, *After the Bomb: Civil Defence and Nuclear War in Britain, 1945–68* (Basingstoke: Palgrave Macmillan, 2010), 5–6.

18 *The Gallup International Public Opinion Poll: Great Britain 1937–1975*, ed. George Gallup, vol. 1 (New York: Random House, 1976), 132, 183–184.

19 William Blandy, 'Introduction', in 'Operation Crossroads', n.d., the Papers of Sir James Chadwick, 1914–1974, Churchill Archives Centre, Churchill College, University of Cambridge, Cambridge, United Kingdom, CHAD IV/3/18, unpaginated.

20 Sidney Shalett and Bertrand Zadie, 'Operation Crossroads', *New York Times*, 17 February 1946, 87.

21 Martin Chisholm, 'The Navy in the Atom Age', *Picture Post*, 25 October 1947, 7–13, 32.

22 Spencer Weart, *Nuclear Fear: A History of Images* (Cambridge, MA: Harvard University Press, 1988), 401–402. See also Vincent Leo, 'The Mushroom Cloud Photograph: From Fact to Symbol', *Afterimage*, vol. 13, nos 1–2 (1985): 9.

23 Gerhard Paul, '"Mushroom Clouds": Entstehung, Struktur und Funktion einer Medienikone des 20. Jahrhunderts im interkulturellen Vergleich', in *Visual History: Ein Studienbuch*, ed. Gerhard Paul (Göttingen: Vandenhoeck & Ruprecht, 2006), 243–264.

24 Chisholm, 'The Navy in the Atom Age', 7–13, 32.

25 Peter Bacon Hales, 'The Atomic Sublime', *American Studies*, vol. 32, no. 1 (1991): 11.

26 Cf. James Farrell, 'The Crossroads of Bikini', *Journal of American Culture*, vol. 10, no. 2 (1987): 55–66.

27 Jane Caputi, *Gossips, Gorgons and Crones: The Fates of the Earth* (Santa Fe, NM: Bear & Company Publishing, 1993), 30.

28 Weart, *Nuclear Fear*, 402–403.

29 Victor Turner, *The Forest of Symbols: Aspects of Ndembu Ritual* (Ithaca, NY: Cornell University Press, 1967), 30.

30 Cf. Michael Light, *100 Suns* (New York: Knopf, 2003); Peggy Rosenthal, 'The Nuclear Mushroom Cloud as Cultural Image', *American Literary History*, vol. 3, no. 1 (1991): 63–92; Szasz, *Atomic Comics*, 48–49, 56–58, 69, 77, 81–82, 122, 124, 127; Bryan Taylor, '"Our Bruised Arms Hung Up as Monuments": Nuclear Iconography in Post-Cold War Culture', *Critical Studies in Media Communication*, vol. 20, no. 1 (2003): 1–34; Bryan Taylor, 'Nuclear Pictures and Metapictures', *American Literary History*, vol. 9, no. 3 (1997): 567–597; Constandina Titus, 'The Mushroom Cloud as Kitsch', in *Atomic Culture: How*

We Learned to Stop Worrying and Love the Bomb, ed. Scott Zeman and Michael Amundson (Boulder, CO: University of Colorado Press, 2004), 101–123.

31 James Chadwick, 'Initial Report by Sir James Chadwick', 16 July 1945, The National Archives, Kew, Richmond, Surrey, United Kingdom (hereafter TNA), CAB126/250.

32 Otto Frisch, 'Eye-Witness Report of Nuclear Explosion, July 16, 1945', n.d., TNA, CAB126/250.

33 James Tuck to James Chadwick, 17 July 1945, TNA, CAB126/250.

34 William Laurence, *Dawn over Zero: The Story of the Atomic Bomb* (New York: Knopf, 1946), 46.

35 Stephen Petersen, 'Explosive Propositions: Artists React to the Atomic Age', *Science in Context*, vol. 17, no. 4 (2004): 583.

36 Anthony Michaelis, 'Film Studies and Armament', *Quarterly of Film, Radio and Television*, vol. 6, no. 3 (1952): 239.

37 Ferenc Szasz, *Larger than Life: New Mexico in the Twentieth Century* (Albuquerque, NM: University of New Mexico Press, 2006), 175–178.

38 Chisholm, 'The Navy in the Atom Age', 7–13.

39 'Atom Parade: As Britain Explodes Her First Atomic Bomb, We Review Names that Make Atomic History', *Picture Post*, 11 October 1952, 10–11.

40 'A Hydrogen Atomic Bomb: Controversy in U.S.' *The Times*, 13 January 1950, 4; Derek Wragge Morley, 'Can Man Survive the Hydrogen Bomb?' *Picture Post*, 18 February 1950, 31–37.

41 Morley, 'Can Man Survive the Hydrogen Bomb?' 31. Morley referred to Hans Thirring's *Die Geschichte der Atombombe* (Vienna: Neues Österreich, 1946), and *Atomkrieg und Weltpolitik* (Vienna: Danubia, 1948).

42 Morley, 'Can Man Survive the Hydrogen Bomb?' 35–36.

43 Ibid., 31, 33–34.

44 Harold Hinton, 'Urey Fears Entry of Bombs in Ships', *New York Times*, 9 February 1950, 1; 'Search for Smuggled A-Bombs', *Bulletin of the Atomic Scientists*, vol. 6, nos 8–9 (1950): 288.

45 Cf. Mick Broderick and Robert Jacobs, 'Nuke York, New York: Nuclear Holocaust in the American Imagination from Hiroshima to 9/11', *The Asia-Pacific Journal*, vol. 10, issue 11, no. 6 (2012), available at: <http://japanfocus.org/-Mick-Broderick/3726>, accessed 28 January 2013.

46 Paul Saint-Amour, 'Bombing and the Symptom: Traumatic Earliness and the Nuclear Uncanny', *diacritics*, vol. 30, no. 4 (2000): 61.

47 Morley, 'Can Man Survive the Hydrogen Bomb?' 32–33.

48 The British Mission to Japan, *The Effects of the Atomic Bombs at Hiroshima and Nagasaki* (London: HMSO, 1946).

49 'To Make You Think: The Effect of an Atomic Bomb on a British City', *Illustrated London News*, 13 July 1946, 44.

50 Atomic Scientists' Association, *Atom Train: Guide to the Travelling Exhibition on Atomic Energy* (London: Atomic Scientists' Association, 1947), 22. See also

Christoph Laucht, 'Atoms for the People: The Atomic Scientists' Association, the British State and Nuclear Education in the Atom Train Exhibition, 1947–1948', in *British Nuclear Culture*, ed. Jonathan Hogg and Christoph Laucht, special issue of the *British Journal for the History of Science*, vol. 45, no. 4 (December 2012): 591–608.

51 Morley, 'Can Man Survive the Hydrogen Bomb?' 32–33.

52 Ibid., 32–33.

53 'A Possible Destroyer of the World Illustrated Diagrammatically: How the Heavy Hydrogen Bomb Works', *Illustrated London News*, 4 February 1950, 176–177. See also later: 'The Hydrogen Bomb: A Drawing Illustrating Its Destructive Power, Forms of Protection, and Characteristics of Hydrogen Bomb Explosions', *Illustrated London News*, 5 July 1958, 24–25.

54 Morley, 'Can Man Survive the Hydrogen Bomb?' 36–37; Charles Pellegrino, *The Last Train from Hiroshima: The Survivors Look Back* (New York: Henry Holt, 2010), 328.

55 William Laurence, 'Ending of All Life by Hydrogen Bomb Held a Possibility', *New York Times*, 27 February 1950, 1, 7.

56 Cf. P.D. Smith, *Doomsday Men: The Real Dr Strangelove and the Dream of the Superweapon* (London: Lane, 2007).

57 Robert A. Divine, *Blowing on the Wind: The Nuclear Test Ban Debate 1954–1960* (New York: Oxford University Press, 1978), 3–35.

58 Tim Raison, 'Who Is Right in the Row about Civil Defence?' *Picture Post*, 16 October 1954, 20–21; Grant, *After the Bomb*, 78–79.

59 Raison, 'Who Is Right in the Row about Civil Defence?' 21.

60 Home Office, *Civil Defence Manual of Basic Training, Volume II: Atomic Warfare, Pamphlet No. 6* (London: HMSO, 1950), 19–41; The British Mission to Japan, *The Effects of the Atomic Bombs at Hiroshima and Nagasaki*, unpaginated picture insert. For later civil defence pamphlets, see Home Office, *Nuclear Weapons* (London: HMSO, 1956); and Home Office, *The Hydrogen Bomb* (London: HMSO, 1957).

61 Jeff Hughes, 'The Strath Report: Britain Confronts the H-Bomb, 1954–1955', *History and Technology*, vol. 19, no. 3 (2003): 263.

62 Melissa Smith, '"What To Do If It Happens": Planners, Pamphlets and Propaganda in the Age of the H-Bomb', *Endeavour*, vol. 33, no. 2 (June 2009): 62.

63 Raison, 'Who Is Right in the Row about Civil Defence?' 20–21, 53.

64 Cf. Guy Oakes, *The Imaginary War: Civil Defense and American Cold War Culture* (New York: Oxford University Press, 1994).

65 David Monteyne, *Fallout Shelter: Designing for Civil Defense in the Cold War* (Minneapolis, MN: University of Minnesota Press, 2011), 21–33.

66 John Stobbs, 'Civil Defence: Do We Need It? Because We Haven't Got It', *Picture Post*, 4 October 1952, 28.

67 Tracy Davis, *Stages of Emergency: Cold War Nuclear Civil Defense* (Durham, NC: Duke University Press, 2007), 1.

68 Jack Winocour, 'The Ten Minutes That Shook America', *Picture Post*, 2 July 1955, 30.

69 Oakes, *The Imaginary War*, 84–96.

70 Winocour, 'The Ten Minutes That Shook America', 30–31.

71 Henry Stimson, 'The Decision to Use the Atomic Bomb', *Harper's Magazine*, vol. 194, no. 1161 (February 1947): 97–107; Samuel Walker, *Prompt and Utter Destruction: Truman and the Use of Atomic Bombs against Japan*, rev. edn (1997; Chapel Hill, NC: University of North Carolina Press, 2004), 99–100.

72 Barton Bernstein, 'Compelling Japan's Surrender Without the A-Bomb, Soviet Entry, or Invasion: Reconsidering the US Bombing Survey's Early Surrender Conclusions', *Journal of Strategic Studies*, vol. 18, no. 2 (1995): 102.

73 Winocour, 'The Ten Minutes That Shook America', 30–32.

74 Cf. Davis, *Stages of Emergency*, 2.

75 Winocour, 'The Ten Minutes That Shook America', 32.

76 Cf. Laura McEnaney, *Civil Defense Begins at Home: Militarization Meets Everyday Life in the Fifties* (Princeton, NJ: Princeton University Press, 2000).

77 Oakes, *The Imaginary War*, 5–6.

78 Divine, *Blowing on the Wind*, 186–187.

79 J.B.S. Haldane, 'H-Bomb Dust: Is It a Danger to the Living or Unborn?' *Picture Post*, 23 April 1955, 11–12.

80 'Science Must Warn Mankind', *Picture Post*, 23 April 1955, 13.

81 James Chapman, '*Quatermass* and the Origins of British Television sf', in *British Science Fiction Television: A Hitchhiker's Guide*, ed. John Cook and Peter Wright (London: I.B. Tauris, 2006), 29; Peter Hutchings, *Hammer and Beyond: The British Horror Film* (Manchester: Manchester University Press, 1993), 41–42.

82 Catherine Jolivette, 'Science, Art and Landscape in the Nuclear Age', *Art History*, vol. 35, no. 2 (April 2012): 261–262.

83 Brian Horrocks, 'The Lesson of Korea – Could This Happen in Western Germany?' *Picture Post*, 12 August 1950, 39.

84 Liddell Hart, 'Six Crucial Questions of the Day Including – Will There Be War?' *Picture Post*, 24 July 1948, 12.

85 Liddell Hart, 'Is Our Atom Strategy Suicidal?' *Picture Post*, 19 February 1955, 22–24.

86 Trevor Philpott, 'Guided Robots Go to War: All Over the World Scientists Are Pushing Guided Missile Research to the Limit', *Picture Post*, 22 November 1952, 15–19.

87 James Chapman, *British Comic Books: A Cultural History* (London: Reaktion Books, 2011), 45–75.

88 'The Red Moon Mystery', *Eagle*, vol. 2, no. 51 (28 March 1952); 'The Red Moon Mystery', *Eagle*, vol. 3, no. 1 (10 April 1952), reprinted in Frank Thompson, *Dan Dare: Pilot of the Future. The Red Moon Mystery* (London: Titan Books, 2004), unpaginated. I thank James Chapman for pointing me towards this source.

89 Woodrow Wyatt, 'Guided Missile Race: What Britain Should Do', *Picture Post*, 10 March 1956, 11–14.

90 'The Shape of Wars to Come', *Picture Post*, 26 November 1955, 54–55.

91 'After the Atom Bomb: An Astonishing Rebirth', *Picture Post*, 24 August 1946, 7.

92 See, for example, 'Atom Parade', 10–11; Chisholm, 'The Navy in the Atom Age', 13; Morley, 'Can Man Survive the Hydrogen Bomb?' 36–37.

93 Sylvain Mangeot, 'Ten Years after Hiroshima: Japan's Gentle Protest', *Picture Post*, 19 March 1955, 40, 42.

94 Trevor Philpott and John Chillingworth, 'A "Picture Post" Team Reports from Hiroshima Ten Years After', *Picture Post*, 6 August 1955, 10–11, 13–14.

95 Scott Zeman, '"Taking Hell's Measurements": *Popular Science* and *Popular Mechanics* Magazines and the Atomic Bomb from Hiroshima to Bikini', *Journal of Popular Culture*, vol. 41, no. 4 (2008): 697.

96 Philpott and Chillingworth, 'A "Picture Post" Team Reports from Hiroshima Ten Years After', 10–11.

97 Ibid., 10–11, 13–14.

98 Martha Bartter, 'Nuclear Holocaust as Urban Renewal', *Science-Fiction Studies*, vol. 13, no. 2 (1986): 148–158.

99 Philpott and Chillingworth, 'A "Picture Post" Team Reports from Hiroshima Ten Years After', 10–11, 13.

100 Ibid., 13–14.

101 Rod Farmer, '*Life*'s Atom Bomb: The Image of Nuclear War in *Life* Magazine, 1945–1955', *New England Journal of History*, vol. 52, no. 3 (1995): 33.

102 Hulton, 'The Last Picture Post', 3; Stuart Laing, *Representations of Working-Class Life 1957–1964* (Basingstoke: Macmillan, 1986), 141–142.

4

Representations of Atomic Power at the Festival of Britain

Catherine Jolivette

In 1951 the Nobel Prize in Physics was awarded jointly to nuclear scientists John Cockcroft and Ernest Walton. The laureates were honoured for their research on the transmutations of atomic nuclei, having successfully conducted the first controlled artificial disintegration of elements at the Cavendish Laboratory in Cambridge in 1932 under the direction of the pioneering physicist Ernest Rutherford. In delivering the presentation speech, Professor Ivar Waller, member of the Nobel Committee for Physics, announced that the work of the British scientists had 'introduced a totally new epoch in nuclear research'.[1]

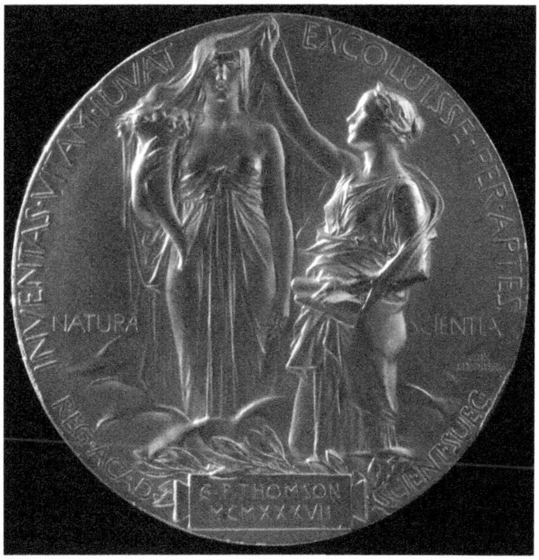

4.1 The *Nobel Medal for Physics*, 1937, photo © Science Museum/ Science & Society Picture Library. All rights reserved.

At the time of the award Cockcroft (who had received a knighthood in 1948) was the Director at the Atomic Energy Research Establishment at Harwell in Berkshire,[2] while Walton was a Fellow at Trinity College, Dublin. During Cockcroft's acceptance speech at the award banquet in Stockholm, Sweden, as he examined his medal (Figure 4.1), the physicist declared:

> On one side of the Nobel Medal you can see the genie of Science lifting the veil which obscures the Goddess of Nature. I was fortunate in working in the Cavendish Laboratory in the 1930s when the time was ripe for a further lifting of the veil, since new ideas in theoretical physics combined with the development of technology in physics made possible the many great discoveries of that 'annus mirabilis' of 1932 ... Since that time many people have wished perhaps that the genie and with him the scientist could be put back in the bottle and have blamed science for the troubles of our time. I feel myself that the overwhelming evil and danger comes not from science but from political ideas which reject the freedom of the human spirit and the values and rights of individual human beings. In

these difficult times science can be one of the strongest shields of our Western Civilization. The honour which you have rendered to scientists today will fortify us at a time when more than ever we need understanding.[3]

Cockcroft's statements – about the promise and peril of nuclear physics, about public apprehension regarding science in general, and about political responsibility during a period of international anxiety – reflect some of the complex attitudes shared in the early 1950s toward the power of the atom. On the one hand, technological developments offered exciting advances in medicine, engineering, manufacturing and energy; on the other, the destructive force of the nuclear fission atomic bomb had already been proven in Japan; meanwhile participation in the Korean War had further weakened Britain's fragile post-war economic situation and, with the escalation of Cold War tensions, the hard-won peace seemed increasingly under threat.[4]

The Themes of the Festival

In the same year that Cockcroft and Walton received the Nobel Prize in Physics, their groundbreaking achievements, along with those of numerous other notable Britons, were honoured in the Festival of Britain: a summer-long, nationwide calendar of events and exhibitions that ran from May until September of 1951.[5] The celebration was planned as a display of civic optimism that would mark both the centenary of the Great Exhibition of 1851 and the beginning of a new era of increased productivity and security for Britain. The aims of the exhibition extended far beyond those of a national trade fair. It was also an occasion for the redefinition of British identity, through a series of exhibitions that aspired to present 'an act of national autobiography'[6] through the re-telling of the national past.

Discovery was a central theme of the Festival, with special emphasis placed on 'the pre-eminent achievements of British men and women in mapping and charting the globe, in exploring the heavens, and in investigating the structure and nature of the universe'.[7] In telling the story of British nuclear physics at the Festival of Britain, visual displays played a key role. In this chapter, exhibits and works of art from various Festival sites are examined in terms of the discourses that they engaged concerning both the potential – for better or for worse – of nuclear energy, and the impact of the atom on both the artistic output and popular imagination of the era.

Within the programme of Festival events there were multiple exhibitions that showcased nuclear science: in London these included the *Exhibition of Science* at the Science Museum in South Kensington, a display on nuclear physics as part of the *Physical World* section of the *Dome of Discovery* at the *South Bank Exhibition*, and the *Growth and Form* exhibition at the Institute of Contemporary Arts (ICA); while in Scotland, Glasgow's Kelvin Hall played host to the *Exhibition of Industrial Power.*

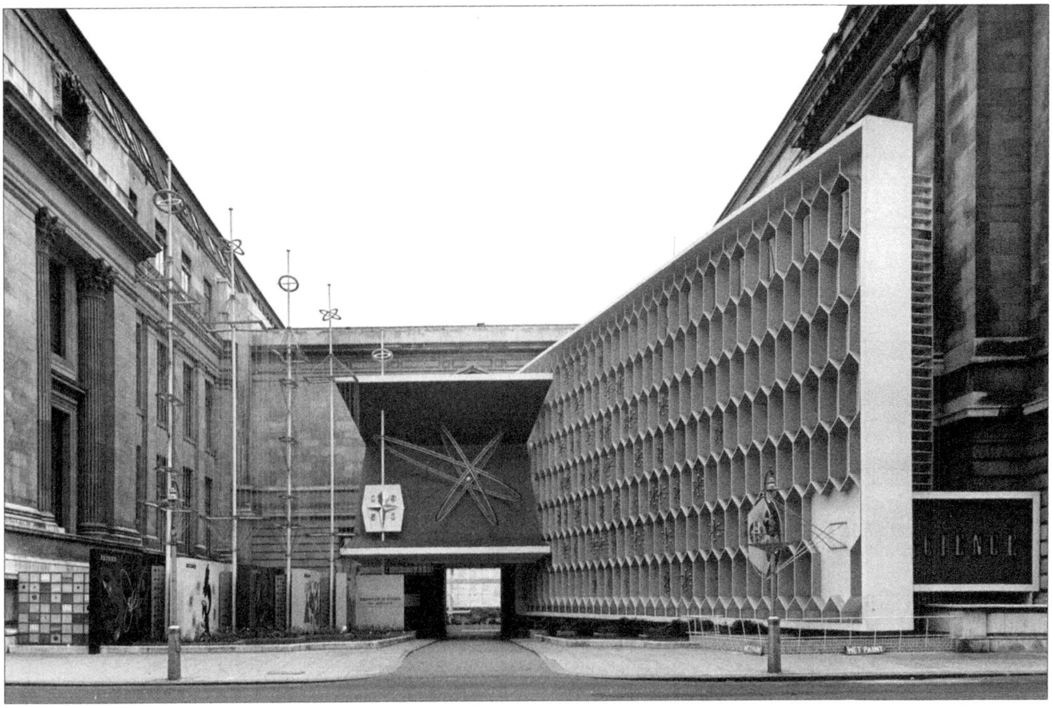

4.2 Left: Brian Peake, Screen based on carbon atoms; right: Gordon Andrews, Screen based on atomic energy, *Exhibition of Science*, Science Museum, London, Festival of Britain, 1951, photo © The National Archives, UK.

Atomic motifs featured strongly in the Festival displays both inside and outside of the exhibition halls. Architectural examples included a three-dimensional atomic structure model that was transformed into Edward Millis' multi-coloured decorative *Abacus* that partitioned the *South Bank Exhibition* from the cityscape beyond the traffic of Waterloo Road. Outside the Science Museum in South Kensington, Brian Peake's screen, comprising hexagonal aluminium units based on the composition of carbon atoms, announced the theme of the first five exhibition rooms inside,[8] while, above the entrance, Gordon Andrews' sculptural form communicated the peaceful application of atomic energy (Figure 4.2).

The *Exhibition of Science* at South Kensington

The theme of the *Exhibition of Science* at South Kensington was matter itself, understood through the structure and nature of the atom. Here, the visitor was encouraged to imagine themselves, like *Alice in Wonderland*, shrinking to atomic size in order to see what is invisible to the naked eye. In the guidebook introduction, 'The Story the Exhibition Tells', scientist Jacob Bronowski explained: 'You come into the exhibition through five rooms which take you, step by step, into the heart of matter. Going through these rooms you seem to shrink like Alice in Wonderland, and the things round you seem to grow larger and larger.'[9]

Sophie Forgan has suggested that the invocation of Alice and her 'Wonderland' engendered 'memories of childhood innocence and a world of fantasy' while providing a 'welcome counterbalance to images often found in the press, of bombs, death and destruction, and fears about uncontrolled chain reactions'.[10] Forgan has also noted that, since 1945, government-sponsored exhibitions at the Science Museum had tended toward the use of the term 'atom' as opposed to 'nuclear' and that such language was significant in terms of public presentation.[11] 'Atoms,' she remarks, were 'small and relatively inoffensive, and had interesting properties; in contrast, nuclear physics promised the release of enormous power and energy.'[12]

As the *Exhibition of Science* caption writer Bronowski insisted, science was not to be presented as something 'remote or frightening, a magic and a mystery', but instead the exhibition was to make 'you feel at home with the knowledge of science, and to make you take pride in it, because it shows science as it is – fascinating, yes, but real and downright'.[13] Inside the Science Museum, then, it was the 'atomic' rather than the 'nuclear' that was brought into focus through a three-part story that told the 'physical and chemical nature of matter', the 'structure of living things', and concluded in a section titled 'stop press' that described some of the most recent topics of scientific research.[14]

For the first section, in the main hall, designer Ronald Dickens created a giant periodic table illustrating the over 90 different elements, each made up of different types of atoms. Geologic side displays looked at the structures of rocks and crystals, and the chemical behaviours of the elements were explored through presentations of alloys and carbon compounds.[15] In the second section, in keeping with the positive presentation of science at the Festival, rather than highlighting the destructive capabilities of nuclear power, emphasis was placed on the natural processes of life through illustrations of all kinds of microscopic arrangements, from the formation of proteins to the genetic make-up of fruit flies.

Crystallography and the Festival Pattern Group

Not only were these atomic images displayed in didactic presentations but since 1949 the Festival Pattern Group (under the aegis of the Council of Industrial Design) had been developing designs based on the atomic patterns of substances including insulin, haemoglobin and boric acid, which appeared everywhere from furniture, to wallpaper, to tableware and textiles.

Lesley Jackson has traced the story of the Festival Pattern Group in her book, *From Atoms to Patterns: Crystal Structure Designs from the 1951 Festival of Britain*.[16] Jackson charts the group's progress, from the initial proposal in 1946 – by crystallographer Dr Helen Megaw of Birkbeck College, London, to Marcus Brumwell of the Design Research Unit, that designers might be intrigued by the forms that scientists saw daily under the microscope – to the displays at the Festival five years later.[17] Within the exhibition hall at the Science Museum, for example, John Line's wallpaper, based on the form of the mineral afwillite, was used as a backdrop to the exhibits; while Brian Peake designed a 600-feet long ball-and-spoke light fitting for the exhibition space that was based on the atomic structure of carbon (and elegantly translated by the General Electric Company).[18] Fabric designs based on crystal structure diagrams were also used in the foyer of the cinema housed in the basement of the museum.[19] As the *Exhibition of Science* guidebook explained: 'Many of these diagrams have an attractive decorative quality which has impressed crystallographers enough for them to suggest they should be used as design motifs for industrial products. A group of leading manufacturers has accepted this proposal, and examples of their work provide some of the furnishings of this exhibition.'[20]

Other ways in which these designs were used were demonstrated in exhibits by Festival Pattern Group manufacturers at the *Dome of Discovery* and in the design scheme for the Regatta Restaurant at the *South Bank Exhibition*. The Regatta Restaurant showcased crystallography forms in the decoration of its curtains, upholstery, menus and dinnerware, while a display in the foyer was dedicated to samples from all of the Festival Pattern Group participants.[21] The restaurant also featured a mural on a crystallographic theme by artist John Tunnard that, at 20 feet tall and 63½ feet long, spanned the full length of the vestibule wall (Figure 4.3).[22]

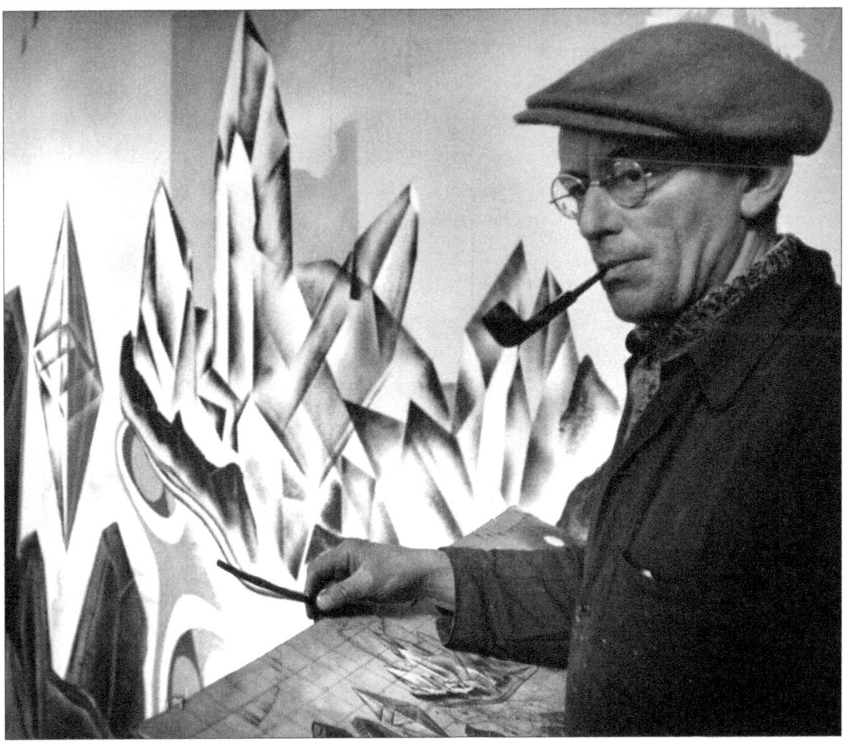

4.3 John Tunnard works on his mural for the Regatta Restaurant at the Festival of Britain, 1951, photo © The National Archives, UK/ Science & Society Picture Library. All rights reserved.

To aid with his commission, Tunnard, like the members of the Pattern Group, was sent copies of crystal structure diagrams which he translated into studies for the mural. The design of Tunnard's finished painting encompassed many different forms of crystals, suggesting, as Simon Martin has noted, a crystallographic journey from the earth's core to outer space.[23] Magma is shown forced up through volcanic eruptions; rocks and minerals form; frost flowers emerge and snowflakes float against the skies as the eye travels onwards and out of the earth's atmosphere, past the planets and into the infinite darkness of the universe.[24]The final scheme pleased the restaurant's designer, Misha Black, who described it as having 'crystals all over the place'.[25]

Growth and Form at the Institute of Contemporary Arts

A fascination with atomic and microscopic imagery also informed the *Growth and Form* show that was held at the ICA to coincide with the Festival summer. The single-room exhibition included large-scale models and diagrams of molecular structures, together with photographs of organic growth that were hung suspended from wires, projected onto walls, ceilings and floors, and refracted by mirrors, to create a complete walk-through environment for the viewer (Figure 4.4). Here too, crystallography was invoked in a 'wall-sized blow-up of an x-ray diffraction pattern of a crystal' appropriated from the research of crystallographer Kathleen Longsdale.[26]

The *Growth and Form* exhibition developed out of a proposal presented by Richard Hamilton and Nigel Henderson to Roland Penrose in 1949.[27] It was originally suggested that the show should be sponsored by the ICA as a contribution to the Festival of Britain, an idea that was vetoed by Herbert Read on the grounds that it did not fit the Festival's theme of a century of British achievement,[28] and so the exhibition, although housed at the ICA, was independently run by the artists.[29]

The exhibition reflected Hamilton's interest in scientific and technological imagery and was named after the title of Scottish mathematical biologist D'Arcy Wentworth Thompson's book, *On Growth and Form*,[30] a text that had circulated widely in art schools since its publication in 1917 and had influenced artists such as Naum Gabo and Barbara Hepworth. Like the Festival Pattern Group, the *Growth and Form* exhibition planners also clearly saw their own project as a further source of inspiration to designers, stating in their proposal that: 'Modern science has made available a new world of visual experience which has yet to make its full impact on all the branches of design. The most obvious benefits of the exhibition would be in the influence it may have on design trends.'[31]

As Isabelle Moffat has noted, Hamilton's drive to provide a clear, didactic, visual presentation of scientific advancement might be located within the same intellectual and philosophical milieu as the Festival-sponsored *Exhibition of*

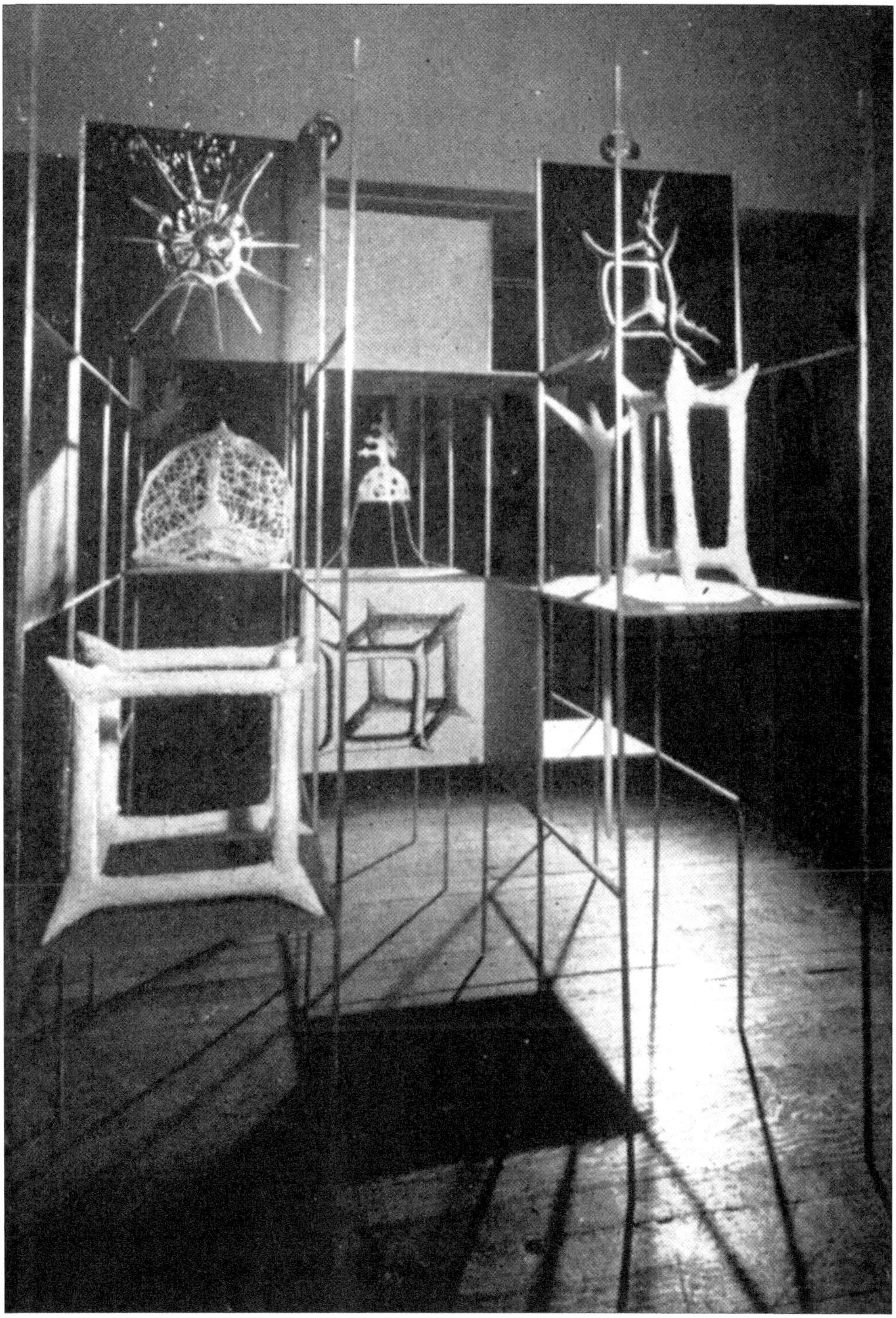

4.4 *Growth and Form* Exhibition, installation view, Institute of Contemporary Arts, London, 1951, photo © Architectural Press Archive/RIBA Library Photographs Collection.

Science in South Kensington.[32] True to the 'scientific spirit of his times', Moffat describes Hamilton's proposals to the ICA exhibition committee as having 'the earnestness of a dutiful student', and states that he 'took seriously the role of the artist and of the designer in facilitating the education of the consumer/spectator'.[33]

Yet Moffat has also observed that, while both exhibitions utilized shifts in scale, at the Science Museum the viewer's perception of shrinking to the size of an atom was progressive, whereas at the *Growth and Form* show the enlargements demonstrated simultaneous representations of wildly different dimensions.[34] Moreover, while *Growth and Form* was hung without didactic labels, the *Exhibition of Science* had a prescriptive route and a didactic commentary. The notion that there was a right way to tell the story was, Moffat argues, an assumption held by the Festival in general.[35] For the Festival planners, she asserts, the 'right' way to tell the story of modern British science was to disassociate it from Cold War fears of 'The Bomb'.[36] Unlike the spatial dislocations and polysemous visual readings encouraged by Hamilton's multi-layered single-room experience at the ICA, the Science Museum exhibition, like those at many other Festival venues, offered a much more linear narrative.[37]

The London *South Bank Exhibition: The Dome of Discovery*

The centrepiece of the Festival of Britain was the London *South Bank Exhibition*, a sequence of 'pavilions' located on the embankment of the River Thames between County Hall and the recently completed new Waterloo Bridge.[38] Here too, there was a carefully guided narrative ('the way to go around'), as mapped out in the exhibition guidebook (Figure 4.5).[39] At this South Bank site the campus was divided by Hungerford Railway Bridge into two narrative circuits, designed to lead the visitor through the twin themes of the 'Land' and the 'People' of Britain, traced throughout their interwoven histories to the present day.[40] Heralded by the futuristic architecture of the *Dome of Discovery* and the *Skylon*, the Festival also showcased the latest advances in technology and design, and included pavilions dedicated to telecommunications, nuclear energy and the exploration of outer space.

The Upstream theme of 'The Land' ended within the 'Physical World' segment of the *Dome of Discovery*. This exhibition section led the visitor through ground-floor displays that explained the history of British physics and chemistry. As the guidebook noted, 'whichever of these sequences you take first you can follow out the second later; both, however, come together again (as the sciences do) in the subject of the atom and its nucleus which is displayed in the upper gallery'.[41]

Inside the *Dome of Discovery* the emphasis was very much on the story of the British contribution to science, through the efforts of men such as Robert Boyle, Sir Isaac Newton, Michael Faraday, Lord Kelvin, Ernest Rutherford

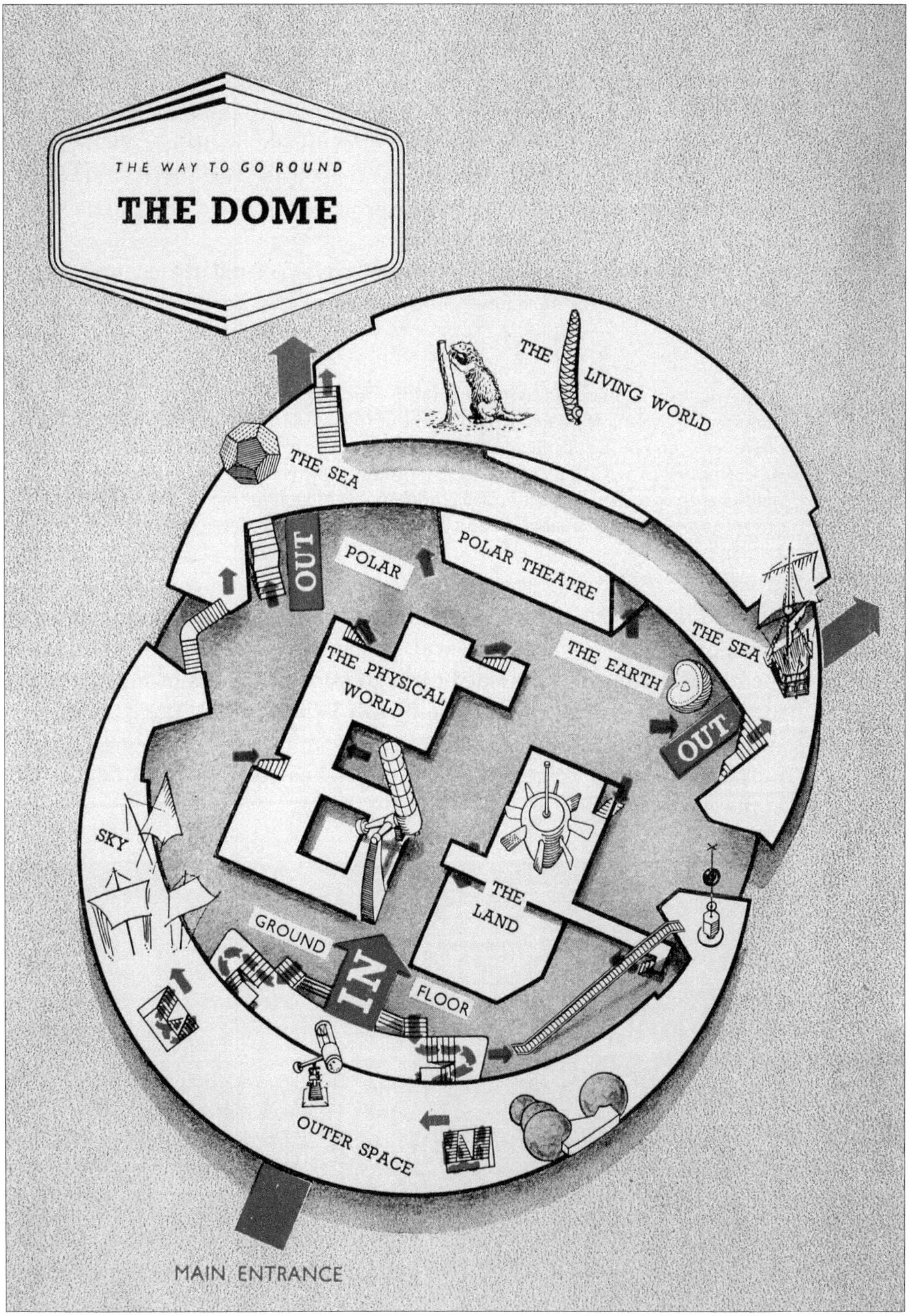

4.5 'The Way To Go Around: The Dome', from Ian Cox, *The South Bank Exhibition: A Guide to the Story It Tells* (London: HMSO, 1951), photo © Science Museum/Science & Society Picture Library. All rights reserved.

and J.J. Thompson.[42] As elsewhere in the Festival, the narrative was attentive not just to the history of British achievement but to how these revolutionary discoveries had led to the improvement of present-day life. Scientific theory was shown to lead to pragmatic results: chemical experiments, the guidebook explained, had led to the production of synthetic medicines, clothing and plastics; Kelvin's mathematical insight helped to create refrigerators; and J.J. Thompson's discovery of the electron had enabled the production of cathode ray tubes for television sets.[43]

The exhibition treatment of nuclear physics similarly conveyed these present-day benefits of science; the catalogue was emphatic on this point, stating that:

> the displays of this subject, which form the conclusion of the 'Physical World' section, do not embark on speculation about future marvels; they do show, however, some of the applications that are being developed at the present time. Important among them are the by-products of research which have given medicine the very important new technique of radioactive tracers. With these an atom can, as it were, be labelled and observed in its movements through the human body.[44]

As with the *Exhibition of Science* at South Kensington, the floor plan of the 'Physical World' followed a logical, positivist progression. Exhibits within the raised platform display dedicated to nuclear physics included a model of the nuclear reactor, BEPO, the British Experimental Atomic Pile at the Atomic Energy Research Establishment near Harwell (Figure 4.6), and a mural by Laurence Scarfe based on studies he had made from 1950 to 1951 at the site.[45]

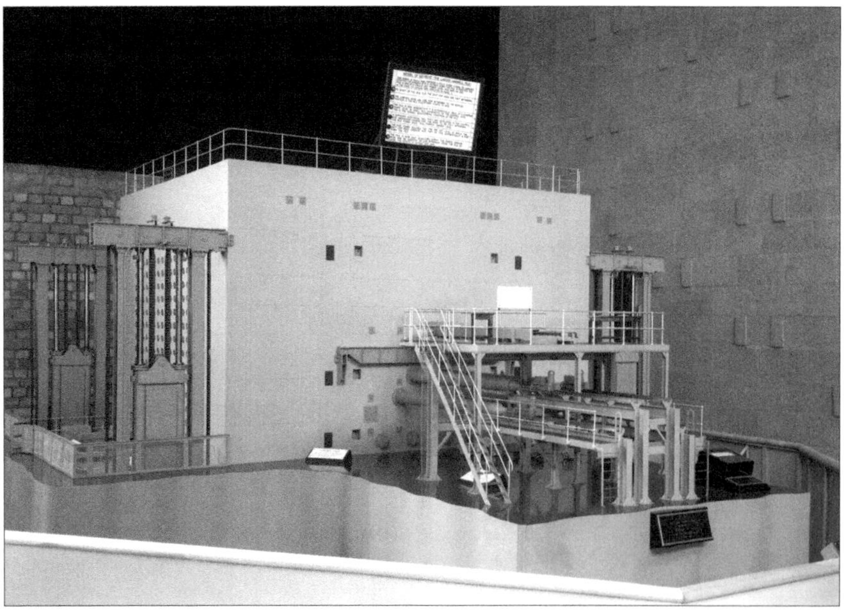

4.6 BEPO, the Harwell atomic pile, 1948, photo © Science Museum/Science & Society Picture Library. All rights reserved.

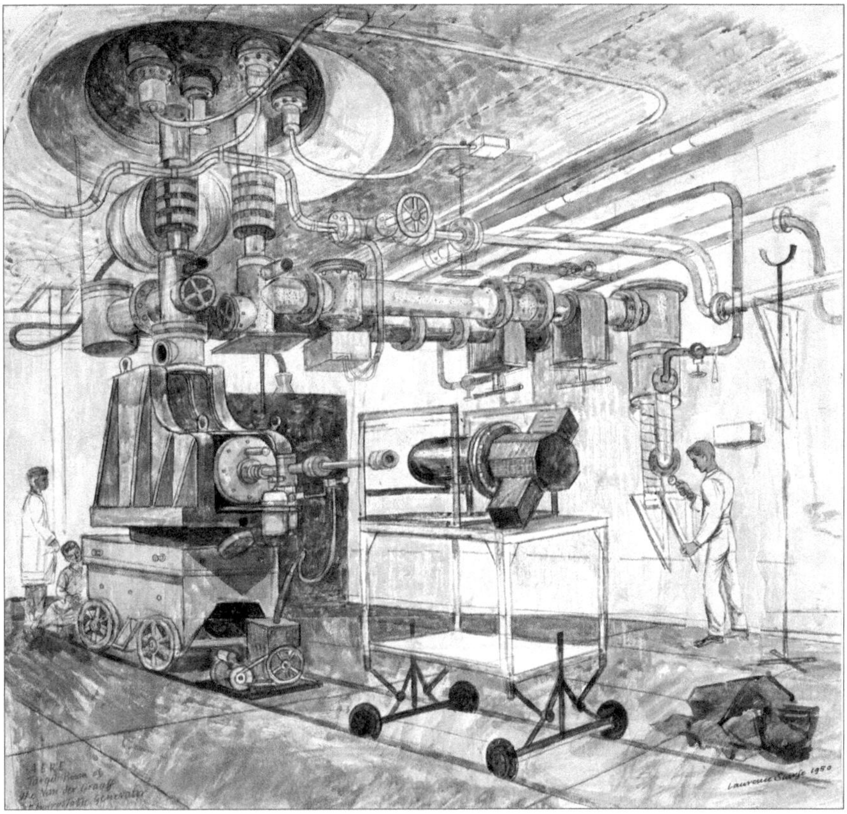

4.7 Laurence Scarfe, 'Target room of the Van der Graaf Electrostatic Generator', August 1950, one of six watercolour studies of the nuclear physics lab at Harwell for the atomic mural in the *Dome of Discovery* at the *South Bank Exhibition*, London, Festival of Britain, 1951, photo © Science Museum/Science and Society Picture Library.

Scarfe's *Atomic Energy* mural was derived from a series of six watercolours he had painted of interior views of Harwell that included renderings of huge machinery such as the 'Van der Graaf Electrostatic Generator' (Figure 4.7) and the 'Cyclotron Pit', as well as smaller-scale processes such as the depiction of a radiochemical experiment in progress on a workbench in the isotope production lab.[46] Scarfe's drawings also showed details of the ventilation plant for the radiochemical laboratories, with close-up views of the ducting system.

Sophie Forgan has explored how the Festival of Britain was instrumental in shaping post-war public images of science,[47] just as other government-sponsored initiatives also sought to clearly, calmly and rationally communicate to the public the activities of British nuclear research. Scarfe's detailed and cleanly drawn illustrations of the nuclear physics lab at the British Atomic Research Establishment at Harwell echo photographs in books such as *Harwell*, published in 1952 by HMSO under the aegis of the Ministry of Supply and the Central Office of Information.[48] As Forgan has observed, these photographs represent a research centre engaged with 'hugely complex processes and machinery' while at the same time embodying a sense of 'order and control'.[49] Scarfe's mural and the South Bank BEPO model inside the *Dome of Discovery* allowed the visitor a behind-the-scenes view of this secret, and somewhat daunting, world.

A balance of didacticism and wonder seems to have been a goal of the 'Physical World' theme conveners, Arthur Garratt and Jan Read, but with the emphasis on the instructive. The field of crystallography (in which the British were at the forefront of advances) was included in this section of the *Dome*, with scientific explanations of crystal structure diagrams complemented by a display of products by Festival Pattern Group manufacturers (Barlow & Jones, British Celanese, Chance Brothers, Dunlop, ICI Leathercloth, John Line, Old Bleach Linen Company, Vanners & Fennell and Vernon Industries).[50] The catalogue, however, made a distinction between the displays inside the *Dome* and those at the *Exhibition of Science*:

> Pure science – discovery for its own sake – has huge rewards to offer the human mind: whole new territories of beauty and order, fantastic in their intricacy. This you can see in the Exhibition of Science in South Kensington. But these explorations have produced something else of vast importance – the basis for most, if not all, of the great material achievements of the modern world. It is these discoveries that form most of the displays in the Physical World.[51]

Discovery, as Forgan has noted,[52] was the theme of the South Bank's exhibition halls and, as the introduction to the South Bank guidebook pointed out, 'certain important aspects of science which do not properly fall within the terms of reference of the South Bank exhibition' were displayed at South Kensington.[53] Similarly, the possibilities of crystallography for fine art and applied design were given fuller expression in the fixtures and furnishings of the *Exhibition of Science*, and reached their apotheosis in the Regatta Restaurant.

In none of the Festival displays were there any discussions of the militaristic application of nuclear physics. As the South Bank guide exhorted, the exhibition told 'the story of British contributions to world civilization in the arts of peace'.[54] In January 1947, Prime Minister Clement Attlee's resolution to go ahead with the research and manufacture of British nuclear weapons had been shrouded with such secrecy that the majority of his party would not even hear that a nuclear deterrent was in development until over a year later.[55] The public was even more in the dark; thus, in the official narratives of the Festival, power was the theme instead.

The Glasgow *Exhibition of Industrial Power*

Although London's *South Bank Exhibition* is the best known Festival vehicle, the celebration was a national event, with satellite exhibitions throughout the country; and while in London, architects and designers celebrated microscopic atomic imagery, it was Glasgow that hosted the *Exhibition of Industrial Power*. Outside of the *South Bank Exhibition*, the *Exhibition of Industrial Power* was the largest of all the official Festival of Britain exhibitions. Held at the Kelvin Hall (itself named after pioneering physicist,

Lord Kelvin), the theme of the exhibition was 'Heavy Engineering' and the narrative sequence comprised a series of halls, each showing a different aspect of power and technology, from coal and steel to shipbuilding and hydro-electric power.

Aiming to tell 'the history of power, from the mill-wheel to the atom', the exhibition was aimed not at 'the engineer and the trade buyer' but at the general public: 'a show not only for the technician but for the ordinary holidaymaker in search of pleasure and excitement' and billed as 'by far the most spectacular exhibition ever held in Scotland'.[56] With an entrance fee of three shillings for adults and one-and-sixpence for children, the show promised the experience of a lifetime with 'the best of Britain's heavy engineering equipment' shown against 'the most striking background ever devised in this country'.[57] A description of the setting could be found on an advertising flyer promoting the upcoming exhibition:

> A waterfall forms one entire wall of one hall. A million-volt machine sends flashes of man-made lightning to a domed roof representing the night sky. A 250-foot ship, three decks high, runs the length of another hall. After passing through a full-scale pre-historic coal swamp, the visitor descends in real pit cages to a modern mine.[58]

Within architect Basil Spence's scheme the atomic energy display was located in the *Hall of the Future* (Figure 4.8). The visitor approached this circular room via a stairway and glass tunnel, entering the 'present' to look down upon a series of pits below the floor level in which 'the great pioneers of the past', James Watt, Richard Trevithick, Michael Faraday, Charles Parsons and Ernest Rutherford, were seen at work.[59] Looking up, the visitor would see 'a shining cone rising from the floor, its tip pulsating and throwing off great crackling flashes of lightning to a night sky which curves above it in a twinkling hemisphere – the limitless future'.[60] Between these, on floor level, 'suspended as it were between yesterday and tomorrow',[61] was a display on current atomic research; the 'plain man's guide to nuclear energy as it is understood to-day', as the exhibition organiser Alastair Borthwick described it.[62]

The lightning machine of section designer Victor Prius' display was used to explain the theory of nuclear fission. Around the machine a series of panels explained the processes of releasing atomic energy (either by splitting the heavier atoms as in the uranium pile, or by combining lighter ones as in the creation of energy by the sun).[63] Here, the official guidebook read: 'scientists have placed a new source of power in the hands of engineers, the basic power of the sun. If it is used for peaceful ends, anything is possible.'[64]

This was the theme of Scotsman William Crosbie's large-scale mural that screened the exit from the *Hall of the Future*. Crosbie's painting (Figure 4.9) was intended as a grand finale to the most spectacular room of the whole exhibition. Within the darkened hall the luminous mural, ingeniously

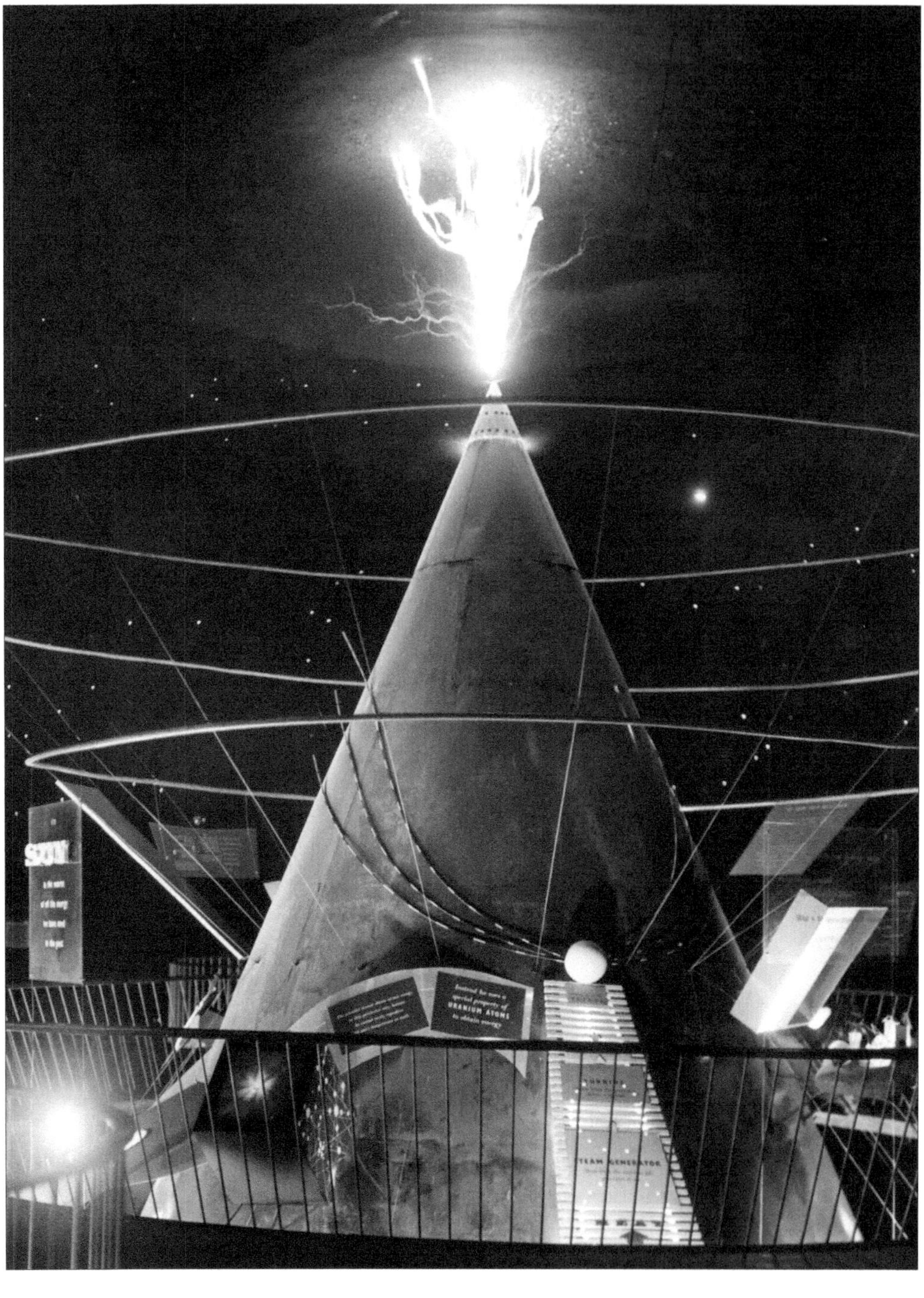

4.8 Scheme for the atomic energy display in the *Hall of the Future* at the *Exhibition of Industrial Power*, Glasgow, Festival of Britain, 1951, photo

painted in fluorescent paint, was lit by a series of infra-red lamps. 'I have watched this mural growing up for some time now,' wrote Borthwick in his anticipation of the exhibition, 'and can say with my hand on my heart that I have never seen anything like it. It glows all over with a weird intensity; when it is in position, floating in darkness beyond the lightning machine, it will end the exhibition with a flourish.'[65]

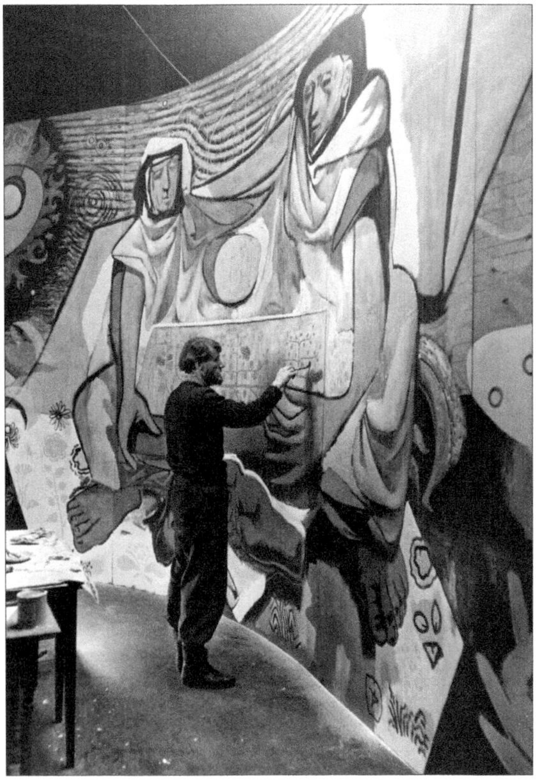

4.9 William Crosbie at work on his atomic mural for the *Hall of the Future* at the *Exhibition of Industrial Power*, Glasgow, Festival of Britain, 1951, photo © RCAHMS (Sir Basil Spence Archive), licensor: www.rcahms.gov.uk.

The mural featured an immense tableau representing humans as the guardians of the earth, deriving sustenance from the land and sea while surrounded by the elements of earth, air, fire and water. At the centre of this visionary scene, a pair of barefoot, timelessly classicizing figures (referred to in the artist's sketches as 'The Gardeners') cradle a miniature landscape tableau in which organisms thrive beneath a luminous nuclear sun. A preparatory sketch for the mural lists 'chlorofil, coal and plankton',[66] as created by the warming rays, while the finished painting also included depictions of fish, plant life and atomic diagrams alongside the human figures who pose as caretakers of this new source of energy. The setting for the mural is some undetermined future time in which the scenery is natural, without overt reference to architecture or modern scientific equipment and agricultural methods. Sketches for unexecuted schemes show figures fishing with primitive nets, and either naked or robed in simple garments that transcend the specifics of time and place. The mood is elemental, primordial even, in contrast to popular science-fiction depictions of the future. 'What atomic energy will mean to us is still largely a matter for conjecture',[67] the guidebook hesitated – just as the scriptwriters for the *Dome of Discovery* had been unwilling to speculate too far beyond the present day. In keeping with the message of the rest of the Festival, the vision of the future presented in Crosbie's mural was peaceful, the mood sober, yet benevolent.

A recurrent theme of the Glasgow exhibition, and indeed of the broader Festival itself, was continuity. Discovery was presented as a consolidation and expansion of previous national achievements, rather than as a rupture with the past. Unsurprisingly then, the story of 'power' also unfolded in evolutionary terms. On entering the exhibition, the visitor came first to the *Hall of Power*, a great oval chamber dominated by Thomas Whalen's sculptural coal cliff mural (Figure 4.10). At 105 feet long, and the largest

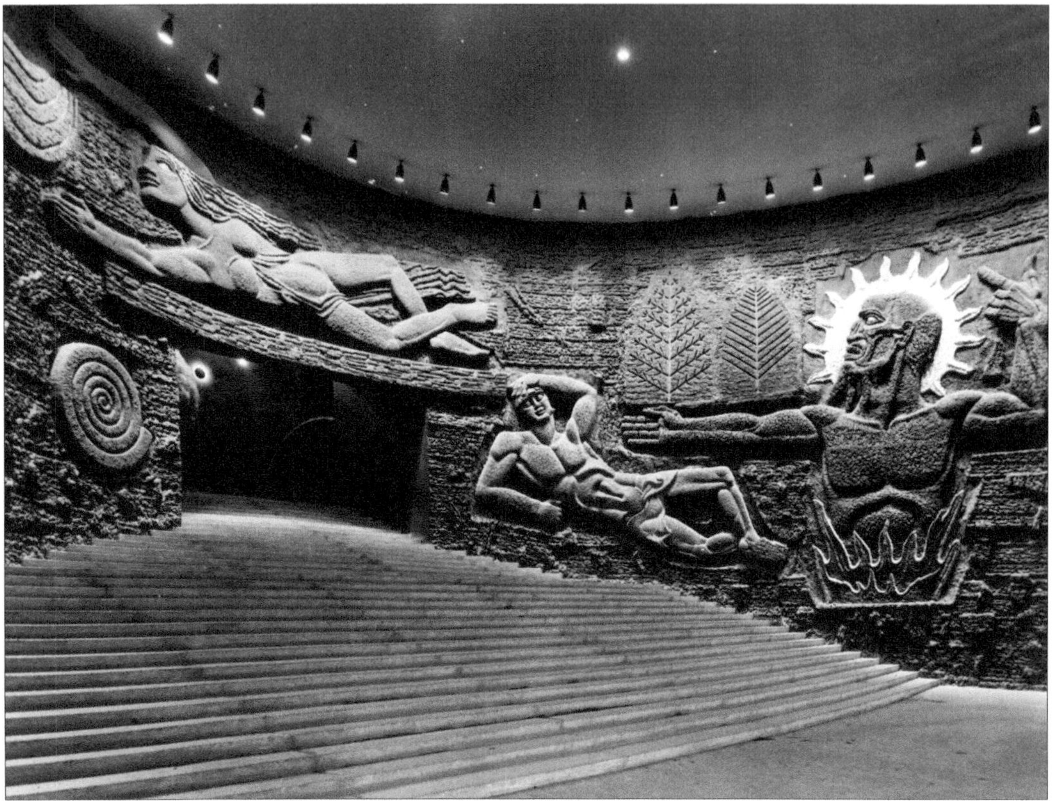

4.10 Thomas Whalen, Coal cliff sculptural mural for the *Hall of Power*, designed by Basil Spence, at the *Exhibition of Industrial Power*, Glasgow, Festival of Britain, 1951, photo © The National Archives, UK/ Science & Society Picture Library. All rights reserved.

sculpture at that time to have been created in Scotland, Whalen's relief was conceived as a 'giant visual chronicle', the figure of the 'God of Nature' extending his right arm to awaken the recumbent figure of 'Man', to show him 'the latent possibilities of coal' and to 'urge him to win this fuel and make it his servant'.[68] Beneath the stylized figure of 'Air' was the entrance to the 'Coal Sequence' and, to the left, the figure of 'Man' was repeated, 'awakened and active, digging in the coal seams'.[69] Whalen's figures shared the same blocky forms as 'The Gardeners' in Crosbie's mural, and both were reminiscent of works by a young Jacob Epstein or Henry Moore. It was a visual vocabulary that paralleled the formation of fossils with the evolution of humankind and the spirit of discovery, linking (as Borthwick stated) 'labour and production and progress with prehistory and the great natural forces'.[70]

This theme was to be found elsewhere at the Festival in the South Bank *Land of Britain* pavilion, where Graham Sutherland's *The Origins of the Land* mural portrayed geological and organic forms simultaneously coming into being (Colour plate 6). Jacquetta Hawkes, archaeological advisor to the Festival, also spoke of such a concept in her book *A Land*, published during the Festival year. 'The image I have tried to evoke,' she wrote in her introduction, 'is of an entity, the land of Britain, in which past and present, nature, man and art all appear in one piece.'[71] All three images – Whalen's relief and both Sutherland and Crosbie's murals – constructed a mythological account of Britain's past, present and future, relying on an iconographic and stylistic presentation that fused the enduring and the contemporary in a way that underscored the reverence for the national past and the embrace of modernization that the Festival of Britain promoted.

The presence of such works of art might have found criticism from J.M. Richards, writer of 'The Exhibition Buildings' for *Architectural Review*, for at the South Bank he noted a tendency to 'over-installation' in buildings such as the *Dome of Discovery* that showcased products as well as ideas.[72] 'The special fault here,' he wrote, 'is that the scientific displays are adorned with a profusion of murals, peepshows and pictorial symbols whose significance (except as decoration) is not made clear enough to enable them to convey any message to the lay public and which are a mere distraction to the serious student.'[73]

Perhaps this is why, in the *Land of Britain*, the ambiguity of Sutherland's iconography resulted in mixed reviews from art critics, with Toni del Renzio for *Art News and Review* applauding the mural as 'perhaps the most impressive failure at the South Bank'.[74] Other reviewers questioned the juxtapositioning of art and heavy industry at the Glasgow exhibition, offering that people might possibly 'find it absurd for a Mercury Rectifier and a Turbine to stand on carpets in front of such excellent murals'.[75] But, as the Festival guidebook explained, the function of the artworks was to 'set the mood' for each exhibition space and to illustrate the fundamental concept of each display sequence.[76] The murals differed from the other perhaps more didactic features of the exhibitions in that they were able to express more abstract or emotive concepts, 'the kind of reality that can best be translated into images by a painter'.[77]

In the case of the Crosbie mural, the painting addressed unspoken fears behind the upbeat scientific displays. For while Britain's contributions to the development of the atomic bomb were absent from the *Hall of the Future*'s narratives of nuclear research, the *Exhibition of Industrial Power* catalogue cautioned: 'the use which has been made of these discoveries and the work which is being based on them today will determine whether we are entering an age of un-dreamed plenty and comfort, or whether we are working out our complete extinction.'[78]

Such anxieties underlay even the optimistic prognostications of the Festival, for (as the Archbishop of Canterbury put it in his opening speech), when the Festival of Britain was first planned, it was intended to mark 'the end of a tormented half century and the beginning of one in which the arts of peace could flourish'. 'Things have not worked out like that,' the Archbishop regretted, claiming that 'we are still in a world dominated by fear and threatened by catastrophe.'[79]

Commentators on the *Exhibition of Industrial Power* recognized the shadow of the Bomb as the subtext to Crosbie's painting, even if the artist had sought out a hopeful portrayal. In a review sub-titled 'Power Uncontrolled', a reporter for the *Glasgow Herald* described the protagonists of the atomic mural as:

an Adam and Eve very obviously after the Fall, lugubriously contemplating the fruits of the earth and the creations of the sun-shine with lurid fluorescent colour; a taste no doubt of painting in futurity. They are a sad, wide-eyed, wondering pair, and well they may be: they are looking at a lesson in nuclear physics.[80]

The writer went on to detail the rest of the *Hall of the Future*, characterizing the large black cone of the lightning machine as 'an immense and sinister Fool's cap', the electricity 'filling the air with frightful, unrelenting crackling, with pallid light, with the smell of ozone; and minds with fissionary nightmares'.[81] Other reporters similarly remarked on the 'day of judgment fireworks and universal alarm',[82] and how 'the long, brilliant blue spark leaps from the top of the cone to the ceiling and wavers there and crackles thunderously and all the children run away screaming. Then they come back for more.'[83] This final room of the exhibition, the *Glasgow Herald* concluded, was 'dreadfully eloquent. Mr Spence and his colleagues have not thought it necessary to provide here any sort of written moral: the lesson of power, terrible and uncontrolled, is clear enough to see.'[84]

When Princess Elizabeth visited the *Exhibition of Industrial Power* on 28 May 1951, in her opening speech she reiterated the importance of turning the forces of nature to serve human welfare. Speaking of the discoveries of the past, she exclaimed:

These devices stand for great victories of genius, but in this Exhibition we are going to be shown new inventions that leave them far behind. We shall see power harnessed to the needs of industry with greater effect than it has ever been harnessed before. We shall even have some revelation of the peaceful use that awaits the tremendous forces locked in the tiny world of the atom.[85]

Five years later, when, as Queen, she performed the opening ceremony at the world's first nuclear power station at Calder Hall in Cumberland,[86] she acknowledged the potential for both progress and ruin that came with the new technology:

Today, as power from Calder Hall begins to flow into the national grid, all of us here know that we are present at the making of history. For many years now, we have been aware that atomic scientists, by a series of brilliant discoveries have brought us to the threshold of a new age. We have also known that on that threshold mankind has reached a point of crisis. Today we are in a sense seeing a solution of that crisis, as this new power which has proved itself to be such a terrifying weapon of destruction, is harnessed for the first time for the common good of our community.[87]

Conclusion

The extensive documentation that survives from the Festival of Britain provides an opportunity to examine several popular representations of British nuclear science within the context of a series of well-planned, government-sanctioned events. The Festival also offers a moment at which to assess how

Britain projected its national image during a pivotal period of early Cold War history. By the Festival summer of 1951 the country was on the atomic threshold.[88] The imminent provision of electricity from nuclear sources was presented as a societal good, with atomic technology offering further benefits in medicine and agriculture. Discussion of the military applications of nuclear science was omitted from Festival exhibition narratives, at the same time as British atomic weapons were secretly under construction. The official displays at the *Exhibition of Science*, the *Dome of Discovery* and the *Exhibition of Industrial Power* communicated a fascination with atomic forms that was also demonstrated by the work of the Festival Pattern Group, and the guidebook and didactic texts extolled the specifically British contributions to the history of scientific discovery.

While the carefully arranged sequences of the Festival exhibits told a unified national story, a 'corporate reaffirmation of faith in the nation's future',[89] the commissioned works of art, such as the atomic-themed murals, spoke to a more complex and subtle range of uncertainties and aspirations. Having witnessed the destructive power of the atomic bomb at Hiroshima and Nagasaki, by the early 1950s, fear was growing as to the potential of the even greater force of the hydrogen bomb,[90] and Britain itself was on the brink of change. November 1951 would see the return of Winston Churchill and the Conservative government; in early 1952 there would be a new monarch; and within a year of the close of the Festival, the nation would become the world's third nuclear power.[91] Paintings such as William Crosbie's Glasgow mural acknowledged both the possibilities and the misgivings ushered in by the atomic age, while projecting a guarded hope for the future.

Notes

1 Professor Ivar Waller, 'Nobel Prize in Physics 1951, Presentation Speech', in *Nobel Lectures, Physics 1942–1962* (Amsterdam: Elsevier Publishing Company, 1964), <http://www.nobelprize.org/nobel_prizes/physics/laureates/1951/press.html>.

2 Harwell was in Berkshire until the English county boundaries were changed in 1974. It is now in Oxfordshire. For the purposes of this chapter, the original county names have also been retained (for example, Cumberland instead of Cumbria).

3 'John Cockcroft – Banquet Speech', in *Les Prix Nobel en 1951*, ed. Arne Holmberg (Stockholm: Nobel Foundation, 1952), <http://www.nobelprize.org/nobel_prizes/physics/laureates/1951/cockcroft-speech.html>.

4 See Robert Hewison, *Culture and Consensus: England, Art and Politics Since 1940* (London: Methuen, 1995), 61.

5 For the best overviews of the art and architecture at the *South Bank Exhibition*, see: Mary Banham and Bevis Hillier, eds, *A Tonic to the Nation: The Festival of Britain, 1951* (London: Thames and Hudson, 1976); James Hamilton, *25 for 51: Paintings for the Festival of Britain* (Sheffield: Mappin Art Gallery, 1978); Elain

Harwood and Alan Powers, eds, 'Festival of Britain', Special issue, *Twentieth Century Society Journal*, no. 5 (2001); and Becky E. Conekin, *'The Autobiography of a Nation': The 1951 Festival of Britain* (Manchester: Manchester University Press, 2003).

6 *Official Book of the Festival of Britain, 1951* (London: HMSO, 1951), 68.

7 Ian Cox, *The South Bank Exhibition: A Guide to the Story It Tells* (London: HMSO, 1951), 9.

8 See Brian Peake, 'The Science Exhibition', in Banham and Hillier, *A Tonic to the Nation*, 146.

9 *Exhibition of Science South Kensington: A Guide to the Story It Tells* (London: HMSO, 1951), 7, The National Archives UK (TNA), WORK 25/230/E1/A1/3.

10 Sophie Forgan, 'Atoms in Wonderland', *History and Technology*, vol. 19, no. 3 (2003): 181.

11 Ibid., 179.

12 Ibid., 179.

13 Jacob Bronowski, 'The Story the Exhibition Tells', introduction to the guide for the *Exhibition of Science* at the Science Museum, as excerpted in Banham and Hillier, *A Tonic to the Nation*, 144.

14 Ibid.

15 Ibid.

16 See Lesley Jackson, *From Atoms to Patterns: Crystal Structure Designs from the 1951 Festival of Britain* (Yeovil: Richard Dennis, 2008).

17 Ibid., 5.

18 Ibid., 23.

19 See Peake, 'The Science Exhibition', 146.

20 'Crystal Structure as a Motif for Design', in *Exhibition of Science South Kensington: A Guide to the Story It Tells*, 40.

21 See Jackson, *From Atoms to Patterns*, 19.

22 For dimensions, see Alan Peat and Brian A. Whitton, *John Tunnard: His Life and Work* (Aldershot: Scolar Press, 1997), 96.

23 See Simon Martin, *John Tunnard: Inner Space to Outer Space* (Chichester: Pallant House Gallery, 2010), 83.

24 Ibid., 83–84.

25 Letter from Misha Black to Mark Hartland Thomas, 1 September 1950, Design Archives, University of Brighton, Design Council Archives (DCA), Festival Pattern Group Files, DCA 5348-2, quoted in Jackson, *From Atoms to Patterns*, 23.

26 See David Mellor, 'The Pleasures and Sorrows of Modernity: Vision, Space and the Social Body of Richard Hamilton', in *Richard Hamilton* (London: Tate Gallery, 1992), cited in Conekin, *'The Autobiography of a Nation'*, 62.

27 British surrealist Roland Penrose was a founding member of the ICA, which opened in 1948.

28 See Anne Massey, *The Independent Group: Modernism and Mass Culture in Britain, 1945–1959* (Manchester: Manchester University Press, 1995), 42–43.

29 Hamilton and Henderson were nonetheless supported by a distinguished exhibition committee, including Read, Penrose and scientists Dr Jacob Bronowski, Dr Alex Comfort, Dr Joseph Needham, Professor C.H. Waddington and Lancelot Law Whyte.

30 D'Arcy Wentworth Thompson, *On Growth and Form* (1917; reprint, Cambridge: Cambridge University Press, 1992).

31 Institute of Contemporary Arts, 'Proposals for an exhibition: "Growth and Form"', Tate Gallery Archives (TGA), TGA 955.1.12.26, 3/36 [n/d].

32 See Isabelle Moffat, '"A Horror of Abstract Thought": Postwar Britain and Hamilton's 1951 "Growth and Form" Exhibition', *October*, no. 94, The Independent Group (Autumn 2000): 89–112.

33 Ibid., 99.

34 Ibid., 110.

35 Ibid., 108.

36 Ibid., 104.

37 Ibid., 110.

38 Cox, *The South Bank Exhibition*, 7.

39 Ibid., 4.

40 Ibid., 8.

41 Ibid., 61.

42 Ibid., 62.

43 Ibid., 62.

44 Ibid., 62.

45 *1951 Exhibition London Catalogue of Exhibits* (London: HMSO, 1951), 105–110, cited in Forgan, 'Atoms in Wonderland', 181.

46 Image descriptions from the Science and Society Picture Library, <http://www.scienceandsociety.co.uk>, image numbers 10424058–10424062.

47 Sophie Forgan, 'Festivals of Science', *The British Journal for the History of Science*, vol. 31, no. 2 (June 1998): 117–240.

48 Ibid., 184.

49 Ibid., 185.

50 See Jackson, *From Atoms to Patterns*, 23.

51 Cox, *The South Bank Exhibition*, 61.

52 Forgan, 'Festivals of Science', 234.

53 Cox, *The South Bank Exhibition*, 6.

54 Ibid., 8.

55 See Niklaus Thomas-Symonds, *Attlee: A Life in Politics* (London: I.B. Tauris, 2009), 95.

56 Advertising flyer for the *Exhibition of Industrial Power*, Kelvin Hall, Glasgow, 28 May–18 August 1951, TNA, WORK 25/230/E1/A3/3.

57 Ibid.

58 Ibid.

59 From 'The Hall of The Future', in the guide catalogue to the *Exhibition of Industrial Power*, Kelvin Hall, Glasgow, Festival of Britain (London: HMSO, 1951), 42, TNA, WORK 25/230/E1/A1/5.

60 Ibid.

61 Ibid.

62 Alastair Borthwick, 'The Conquest of Power', *Scotland's Magazine* (1951): 28, Royal Commission on the Ancient and Historical Monuments of Scotland (RCAHMS), Bib. no: S7237 MS2329/SCT/25/1/14.

63 Catalogue to the *Exhibition of Industrial Power*, 45.

64 Ibid.

65 Borthwick, 'The Conquest of Power', 28.

66 Atomic mural, unexecuted scheme, 1949, from the Basil Spence Archive, Royal Commission on the Ancient and Historical Monuments of Scotland, Item DP012308.

67 Catalogue to the *Exhibition of Industrial Power*, 45.

68 From 'The Hall of The Future', in the guide catalogue to the *Exhibition of Industrial Power*, 8.

69 Ibid.

70 Ibid.

71 Jacquetta Hawkes, *A Land* (London: Cresset Press, 1951; reprint, Harmondsworth, Middlesex: Penguin Books Ltd, 1959), 9.

72 J.M. Richards, 'The Exhibition Buildings', *Architectural Review*, vol. 110, no. 656 (August 1951): 124.

73 Ibid.

74 Toni del Renzio, 'Review of Graham Sutherland Exhibition at the Redfern Gallery', *Art News and Review*, vol. 4, no. 23 (1952): 4.

75 Patrick O'Donovan, 'Festival Journey – 6, Power and No Glory', review of the *Exhibition of Industrial Power*, RCAHMS, Bib. no: 67661, MS2329/SCT/25/2/90.

76 Cox, *The South Bank Exhibition*, 12.

77 Ibid.

78 Banham and Hillier, *A Tonic to the Nation*, 154.

79 Archbishop of Canterbury, speech, 17 July 1950, quoted in the Festival of Britain press release, *Coming Events in Britain*, with an introduction by General Lord Ismay, TNA, WORK 25/250 E3/B32.

80 'Glasgow's Festival "Segment" a Piece of Wizardry: Kelvin Hall Gets Final Touches', *Glasgow Herald*, 25 May 1951, RCAHMS, Bib. no: 67611, MS2329/SCT/25/2/36.

81 Ibid.

82 'Functional Sculpture', *Architect's Journal*, 7 June 1951, RCAHMS, Bib. no: 67627, MS2329/SCT/25/2/62.

83 O'Donovan, 'Festival Journey – 6, Power and No Glory'.

84 'Glasgow's Festival "Segment" a Piece of Wizardry'.

85 'Princess Opens Glasgow Exhibition: Story of Man's Achievement "One Long Tale of Marvels"', *Edinburgh Evening News*, 28 May 1951, 5.

86 The first connection to the grid was on 27 August 1956, and the plant was officially opened by Queen Elizabeth II on 17 October 1956. See <http://www.sellafieldsites.com/calderhall/content.php?pageID=11>.

87 Ibid.

88 See Alfred Goldberg, 'The Atomic Origins of the British Nuclear Deterrent', *International Affairs*, vol. 40, no. 3 (July 1964): 424.

89 Cox, *The South Bank Exhibition*, 6.

90 See Goldberg, 'The Atomic Origins of the British Nuclear Deterrent', 423.

91 In a strange twist of fate, the command ship for 'Operation Hurricane' was HMS *Campania*, the same Royal Navy aircraft carrier that had served as a 'Sea Travelling Exhibition' ship for the Festival. See <http://www.royalnavyresearcharchive.org.uk/ESCORT/CAMPANIA.htm>.

1 Peter Rose Pulham, *Ville Entière*, c. 1949–1951, oil on canvas, 130 × 95 cm, private collection © Estate of Peter Rose Pulham.

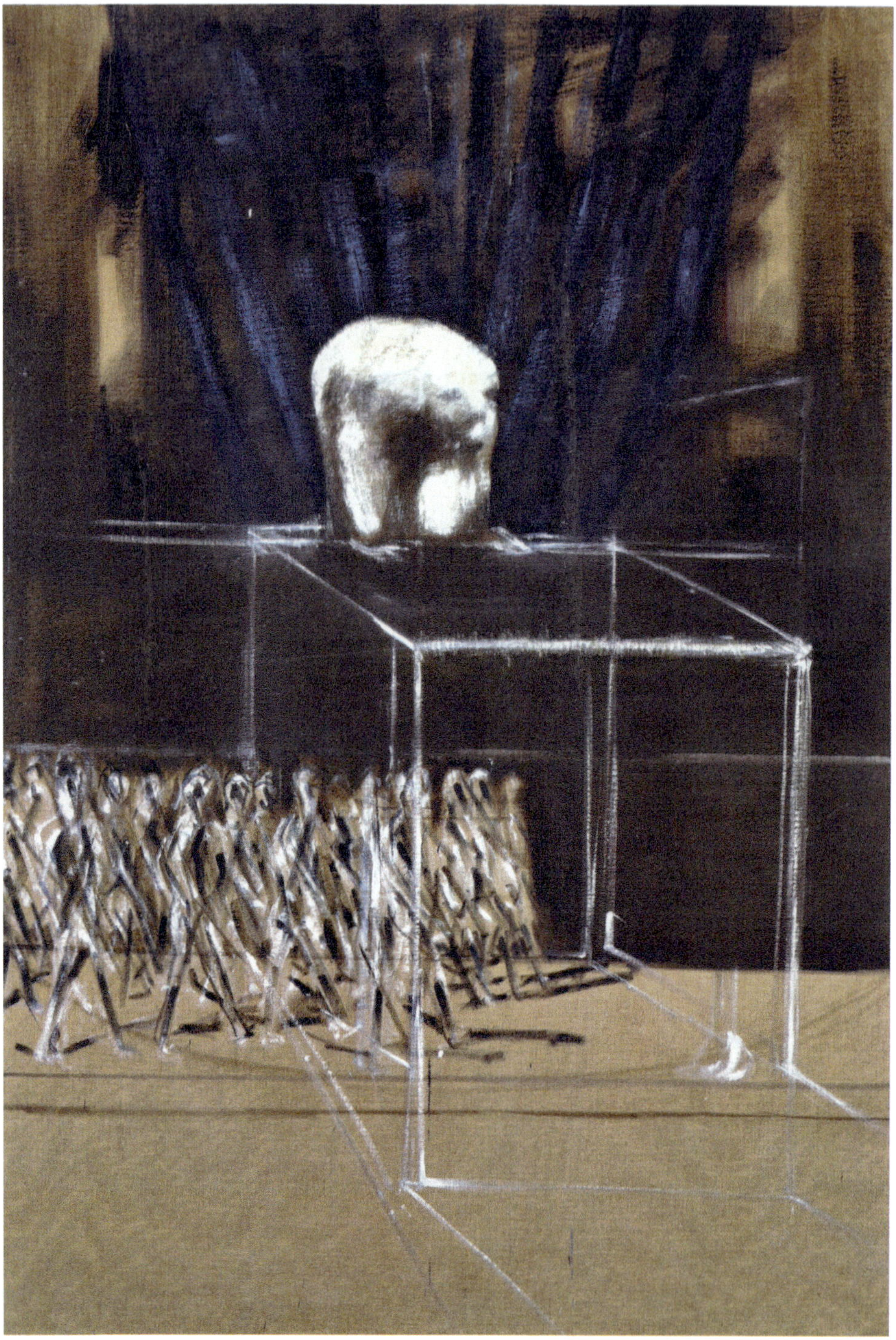

2 Francis Bacon, *Untitled (Marching Figures)*, 1952, oil on canvas, 198 × 137 cm, by kind permission of the Estate of Francis Bacon.

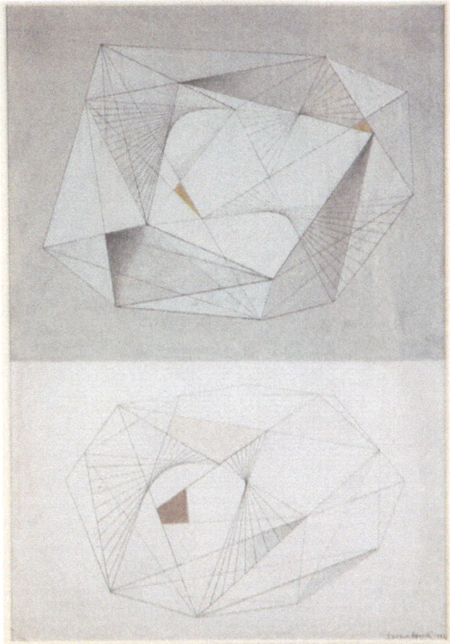

3 Barbara Hepworth (1903–1975), *Forms (Brown, Grey and White)*, 1941, gouache and pencil/paper, 50.3 × 35.3 cm, coll. Hepworth Estate, photo © The Hepworth, Wakefield.

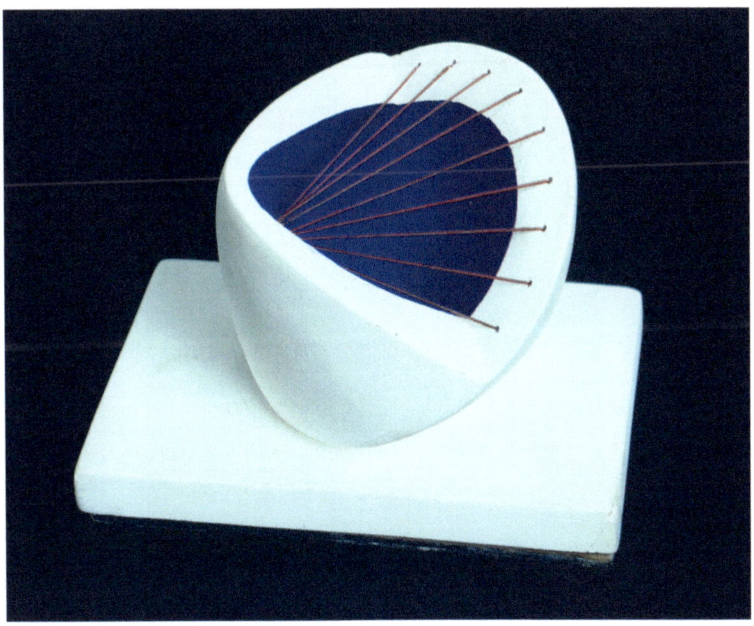

4 Barbara Hepworth (1903–1975), *Sculpture with Colour (Deep Blue and Red)*, 1940, painted plaster and string on plaster base, 10.5 × 14.9 × 10.5 cm, coll. Tate Gallery, photo © Bowness, Hepworth Estate.

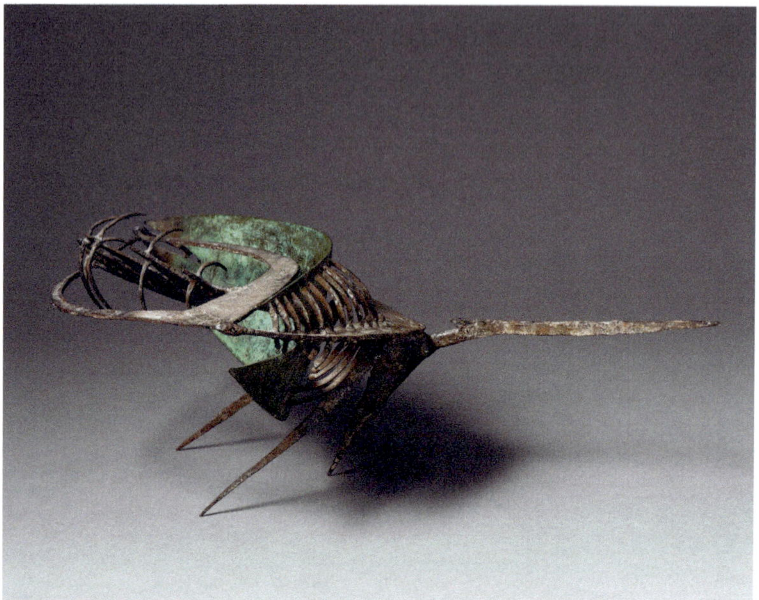

5 Lynn Chadwick (1914–2003), *Balanced Sculpture*, 1952, welded iron, h. 49 cm, exhibited British Pavilion, Venice Biennale, 1952, private collection.

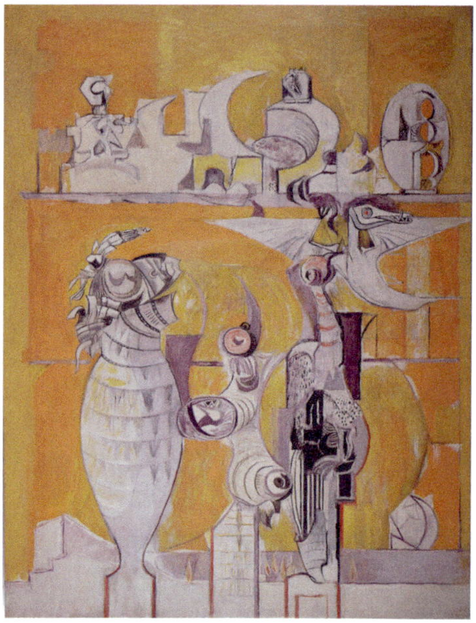

6 Graham Sutherland, *The Origins of the Land*, 1950–1951, oil on canvas, 425.5 × 327.7 cm, Tate Gallery, London © Tate Images, © Tate, London 2013/Estate of Graham Sutherland.

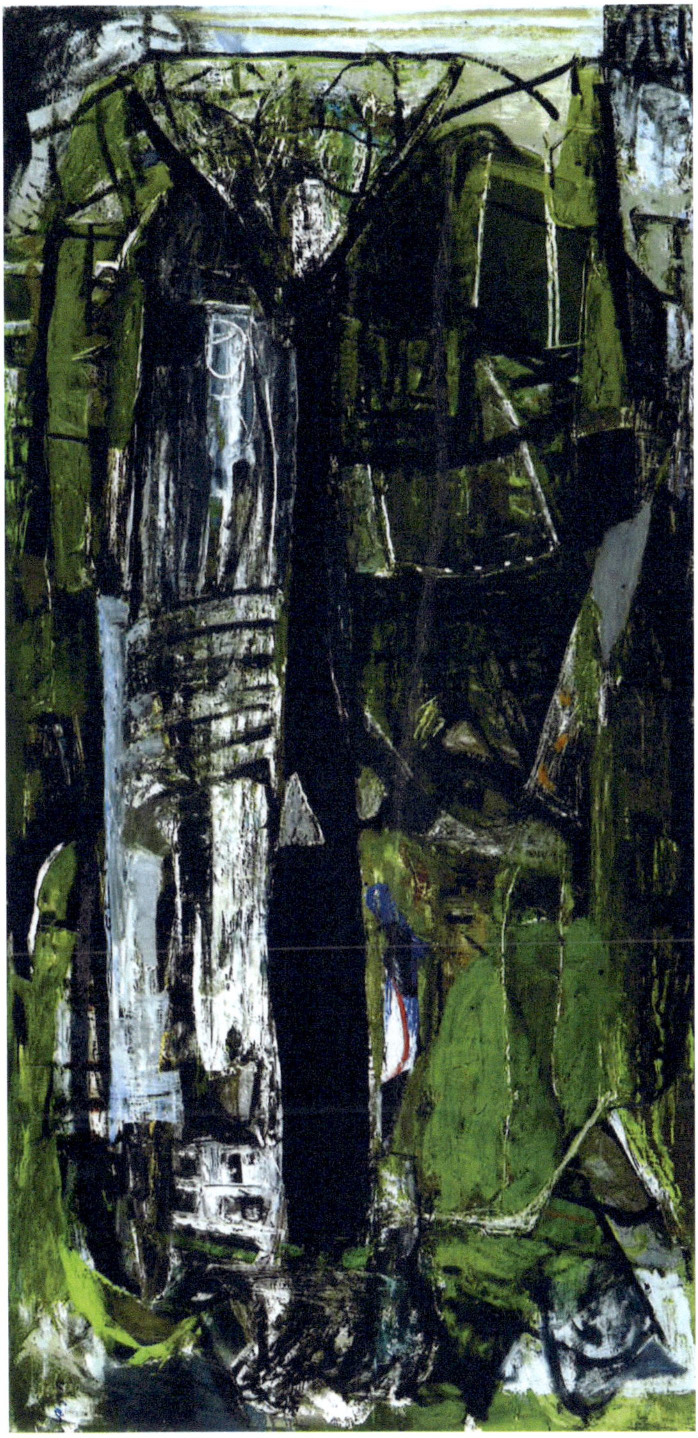

7 Peter Lanyon, *St Just*, 1953, oil on canvas, 243.8 × 121.9 cm, private collection © Estate of Peter Lanyon.

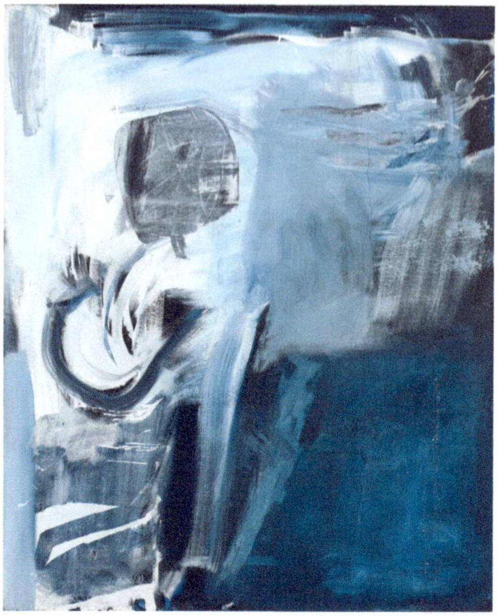

8 Peter Lanyon, *Thermal*, 1960, oil on canvas, 182.9 × 152.4 cm, Tate Gallery, London © Tate, London 2013/Estate of Peter Lanyon.

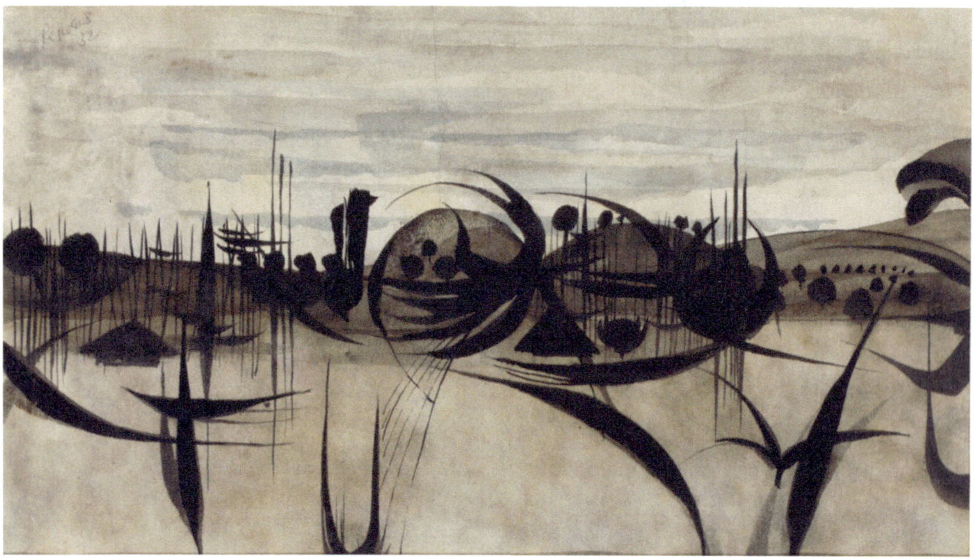

9 Alan Reynolds, *Hopfields*, 1952, ink and watercolour, 19.5 × 35 cm, private collection © Sotheby's 2000/the artist.

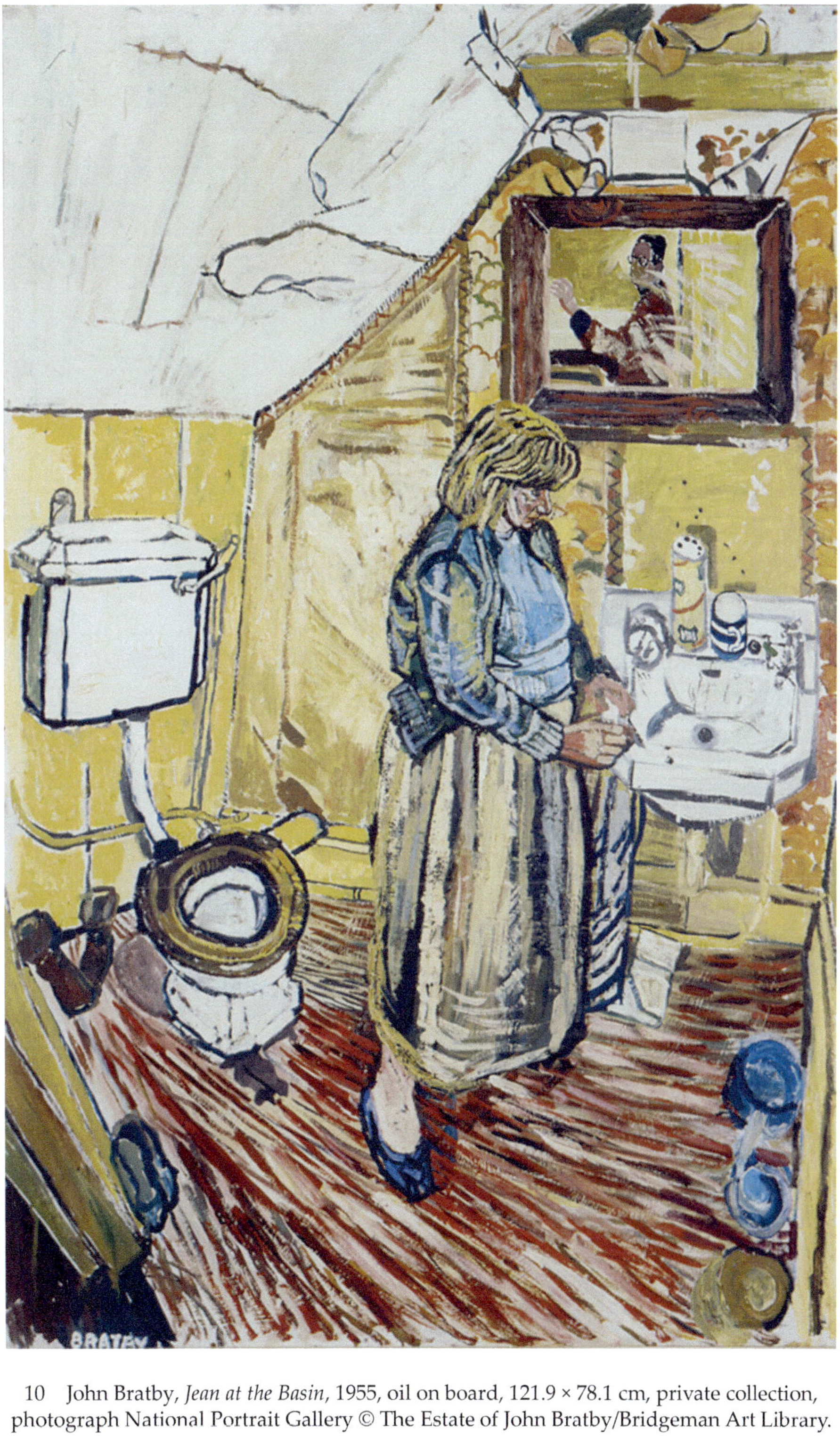

10 John Bratby, *Jean at the Basin*, 1955, oil on board, 121.9 × 78.1 cm, private collection, photograph National Portrait Gallery © The Estate of John Bratby/Bridgeman Art Library.

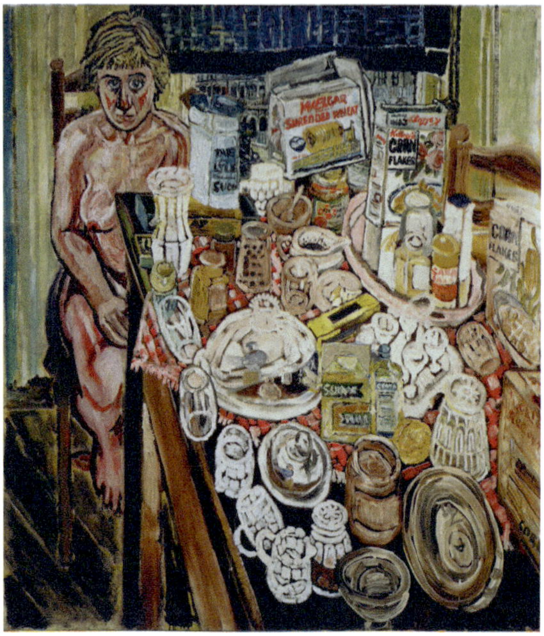

11 John Bratby, *Jean and Still Life in front of a Window*, 1954, 122 × 108 cm, Southampton City Art Gallery © The Estate of John Bratby/Bridgeman Art Library.

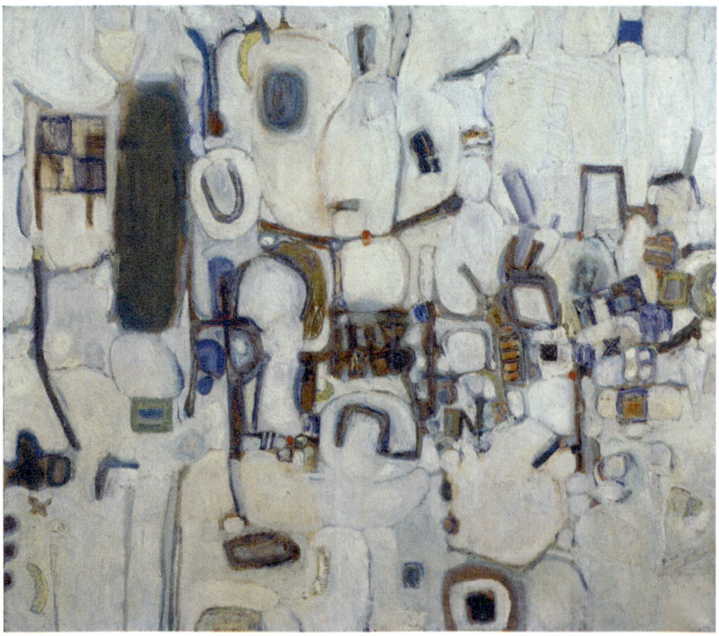

12 Prunella Clough, *Bypass I*, 1960, oil on canvas, 96.5 × 112 cm, The Courtauld Gallery, London © Estate of Prunella Clough 2013. All rights reserved DACS.

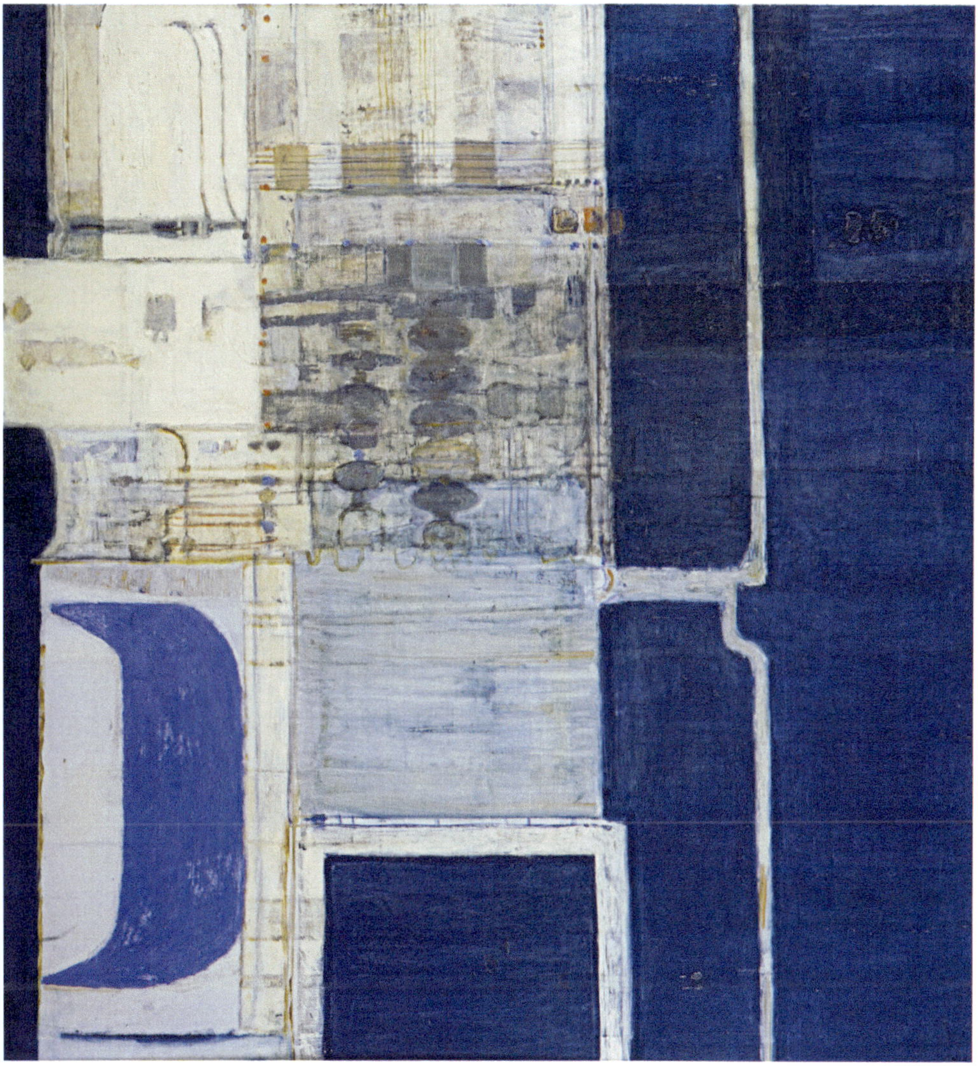

13 Prunella Clough, *Electrical Landscape*, 1960, oil on canvas, 162 × 152.8 cm, National Galleries of Scotland © Estate of Prunella Clough 2013. All rights reserved DACS.

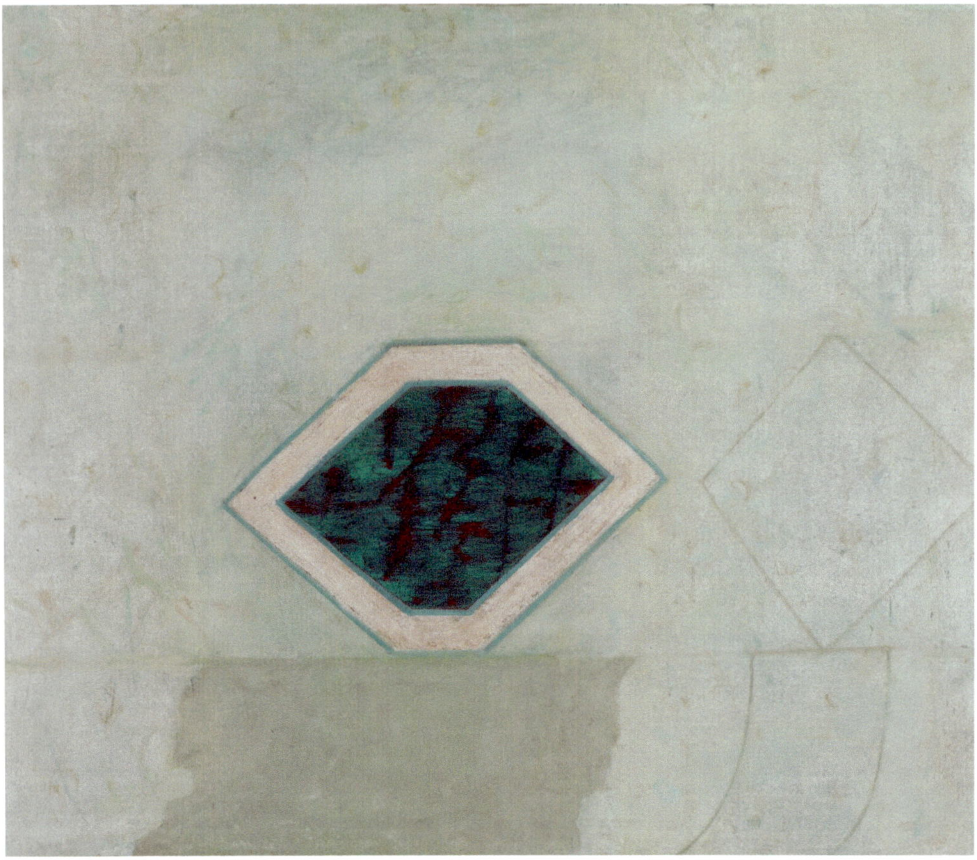

14 Prunella Clough, *Untitled*, 1965, oil on canvas, 72.2 × 85 cm, Arts Council Collection, Southbank Centre, London © Estate of Prunella Clough 2013. All rights reserved DACS.

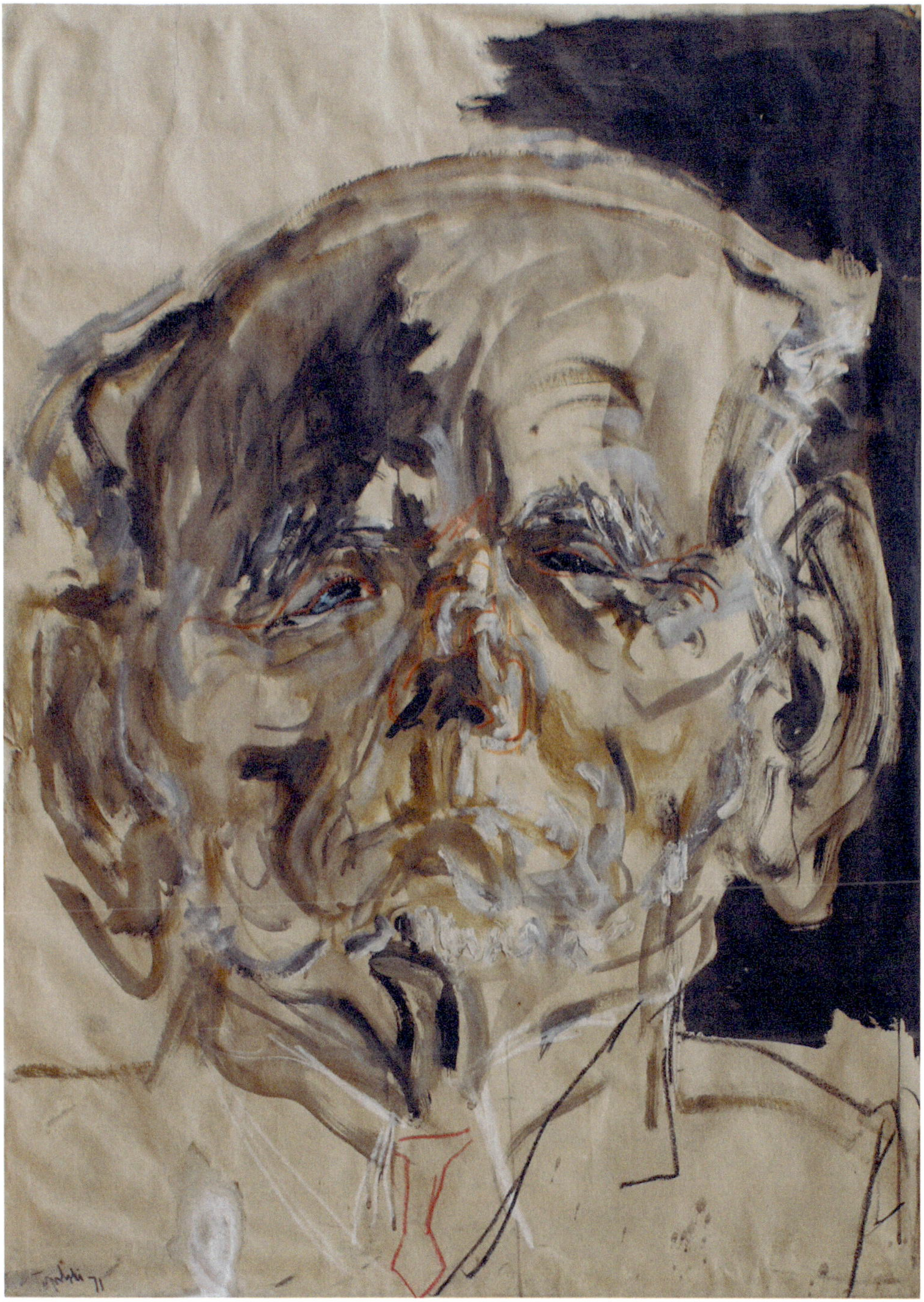

15 Feliks Topolski, *Stefan Borgrajewicz*, 1971, ink, conté crayon and acrylic paint on paper, 126.7 × 93.4 cm, Estate of Feliks Topolski © By kind permission of the Estate of Feliks Topolski.

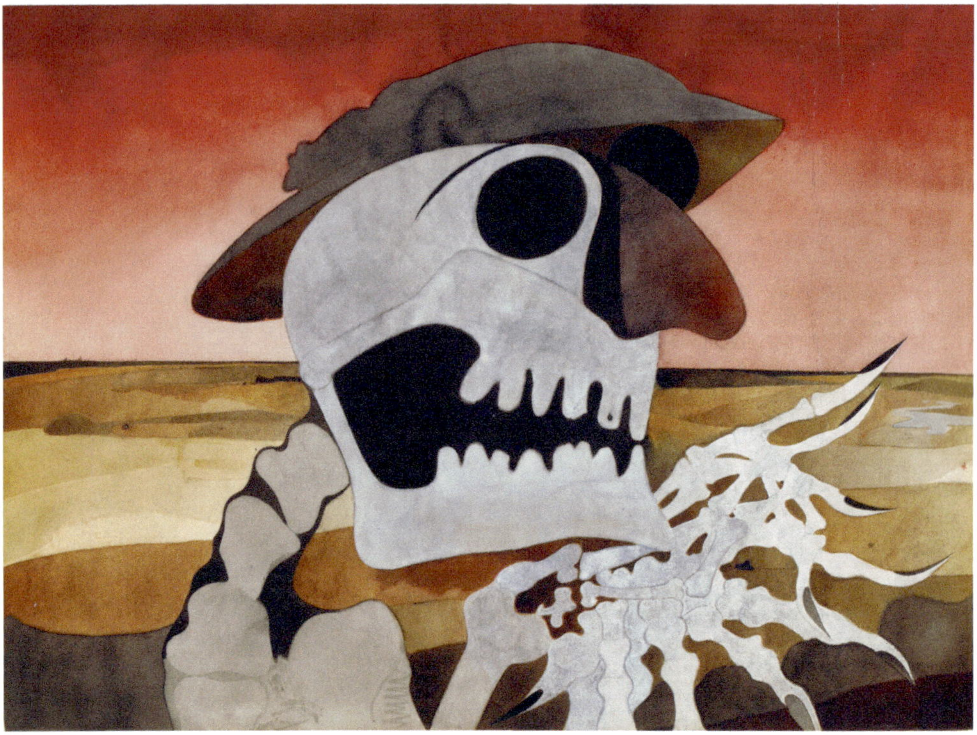

16 Edward Burra, *Skull in a Landscape*, 1946, watercolour on paper, 55.2 × 76.2 cm, Imperial War Museum, London © Estate of Edward Burra/Lefevre Fine Art.

5

The *Genius Loci* of Cold War Britain: The Metamorphic Landscapes of Graham Sutherland, Peter Lanyon and Alan Reynolds

Fiona Gaskin

Introduction

The expression of insecurity, anxiety and despair can be found in a number of art forms created in Britain during the early Cold War period prior to the Cuban missile crisis. The Geometry of Fear movement in sculpture and the competition for the Monument to the Unknown Political Prisoner produced arguably the most haunting examples of Cold War imagery in British art. However, perhaps less well known and analysed was an equivalent tendency in landscape painting. In this tendency, the mutants and iron waifs of sculpture were echoed in troubling 'personages' constructed from elements of the natural world. Alternatively, landscapes were imbued with unsettling and violent anthropomorphic or metamorphic qualities that read as metaphors of the nuclear threat.

Metamorphosis had its origins in the *objet trouvé* of Surrealism (in the concern of Surrealism with the poetically revealed or discovered in everyday and discarded objects)[1] and in the metonymic symbolism and preoccupation with the *genius loci* of both Surrealism and Neo-Romanticism.[2] It is an artistic tendency that revivifies in the late 1940s and early 1950s by which time the motivations for it are different. During this period, it can be seen as part of a wider tendency towards allusion and the denial of fixed meanings that manifested itself in both abstract and figurative painting as a form of anti-rationalism after the Second World War. Rationalism – including Constructivism – was associated with the thought processes that had given rise to, or at least had not prevented, the Second World War and the Holocaust, and anti-rationalism emerged in art as a form of revulsion for it.

This chapter looks at the re-emergence of the metamorphic tendency in landscape painting in the period 1945–1963 and at how the concept of *genius loci* was adapted to become a metaphor of the Cold War in the work of three artists: Graham Sutherland, Peter Lanyon and Alan Reynolds. These artists produced very different imagery but are linked through their use of the medium of landscape painting to express their experience of place.

It should be acknowledged from the outset that Lanyon was not alone amongst artists of the St Ives School to paint metamorphic works – in fact the theme of the body as landscape was being explored by a number of his contemporaries, notably John Wells, William Scott and Roger Hilton; and in sculpture, Barbara Hepworth went through a period of referencing the body and landscape. However, amongst painters, Lanyon tackled the theme over the longest period and, even when he ceased to use it, his most gestural and Expressionist works could be read as records of the actions of his own body in landscape. Sutherland was preoccupied with metamorphosis before and during the war, but after the war it became a dominant theme for a large part of the period under discussion. Of all three artists, he was the most obviously associated with this tendency. Finally, Alan Reynolds' work has not been explored in this context before. However, as we shall see, his extreme popularity was due in a large part to the archetypal nature of his landscapes. For a period from 1952 to 1956, the archetypes could be read as embodiments of the dread and fear created by the nuclear threat, placing him within this tendency.

Graham Sutherland

Sutherland's preoccupation with the concept of *genius loci* – literally, the spirit of a place – had strong links with the Surrealist idea of the *objet trouvé* under which the significance of objects is dependent on their capacity to represent a wish fulfilment or to concretize a psychic state in objective, external form.[3] However, while his interest in the *objet trouvé* may have originated with Surrealism in the 1930s, it was to endure throughout his *oeuvre,* assuming a new significance in the early Cold War period. Unlike most Surrealists, Sutherland looked for such objects in nature – in rocks and tree roots and other vegetation, as well as in man-made objects like walls and hedges – and then painted them in their settings, as landscapes. He sought out forms in the natural world that were ambiguous or hybrid, as well as inanimate objects into which could be read human or animal characteristics. Frequently, these 'personages'[4] also had a metonymic relationship with the landscape, for example, by representing its contours in miniature.[5]

Sutherland responded to natural objects 'as if in response to some internal need of the nerves'.[6] Where he did so, his practice was to make notes in the form of a series of sketches of the object from different viewpoints.[7] He was searching for an angle that best disguised the object (both as rock, tree root and personage). Disguise was an important part of the success of his work: the viewer's engagement would be achieved first through the initial lack of a clear resemblance to a known subject, then by participation in the shock of

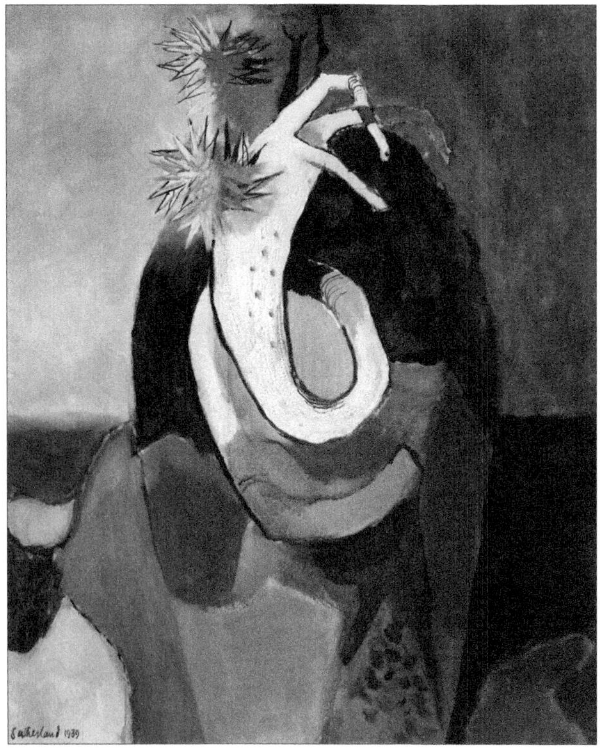

5.1 Graham Sutherland, *Gorse on a Sea Wall*, 1939, oil on canvas, Ulster Museum © National Museums Northern Ireland, © National Museums Northern Ireland/ Estate of Graham Sutherland.

recognition, and finally in the further unsettling realization of the possibility of more than one reading. The sketches were subsequently worked up in the studio – a laborious process in which the original encounter was 'redefined, paraphrased and changed into something new'.[8]

Two distinctive phases are discernible in Sutherland's obsession with personages. In the first phase, spanning the 1930s and 1940s, the personages are relatively naturalistic. For example, works like *Gorse on a Sea Wall* (1939; Figure 5.1) and *Green Tree Form: Interior of Woods* (1940) are recognizably organic forms, although invested with a life of their own: the gorse is like a vicious claw ready to strike, while the roots resemble a primeval fish – a forgotten throwback to an earlier age. Sutherland was attracted to such subjects by their strangeness and 'element of disquiet'.[9] In later works, like *Palm Palisade* (1947), the organic forms (and figure) are less readable, being more angular and abstract. By the time of the second phase, which includes his mural *The Origins of the Land* for the Festival of Britain in 1951 (Colour plate 6) and his series of *Standing Forms* in the early 1950s, these personages are entirely fictitious constructs of his imagination, although suggesting parallels with recognizable forms. For example, *Standing Form against a Hedge* (1950; Figure 5.2) suggests body parts in its legs, torso, neck and eye, but the personage is essentially mechanistic and sculptural.[10] Similarly, in his painting *Three Forms in a Garden* (1951), the standing forms seem to be composed of a combination of metallic parts and roots or tuber, the central figure being part robot and part abstraction of stooked corn. Both the mechanistic and vegetable elements of Sutherland's personages indicate an important difference from the personages that inhabit the works of his close friend Francis Bacon, such as in *Three Figures at the Base of the Crucifixion* (1945). The starting point of Bacon's figures is the human flesh – 'the reek of human blood smiles out at me', as Bacon was wont to quote Aeschylus – whereas for Sutherland it was the natural world and later also machinery.[11]

Sutherland described his standing forms as 'figures once removed' and derived inspiration for them from seeing garden sculptures in the shade at an old house in St-Jean-Cap-Ferrat.[12] More generally, as we have seen, they resulted from an animist approach to nature. Their alien character is

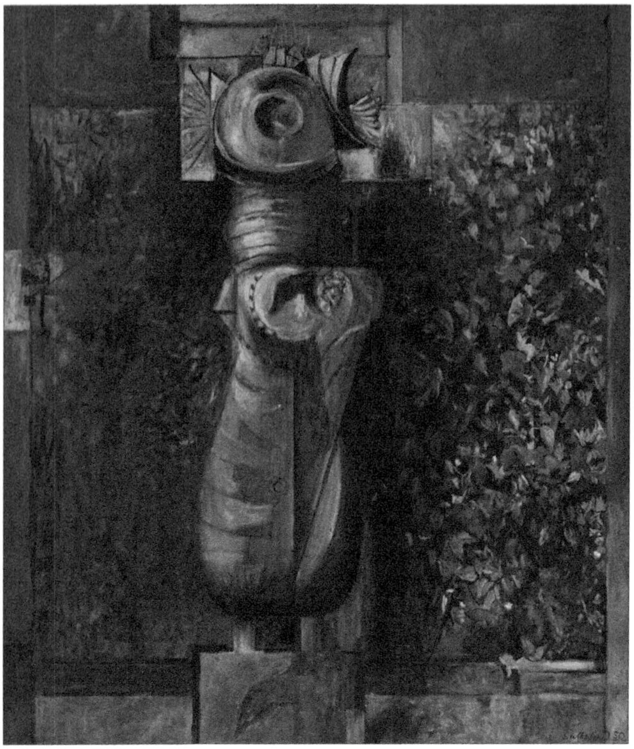

5.2 Graham Sutherland, *Standing Form Against a Hedge,* 1950, oil on canvas, 134.6 × 116.8 cm, Arts Council Collection, Southbank Centre © Arts Council Collection/ Estate of Graham Sutherland.

accentuated by the precision with which they are painted and by the claustrophobic, unreadable spaces they inhabit. The spaces – or places – are also constructs from elements of nature. So, for example, the backdrop to *Standing Form against a Hedge* seems at first glance to be readable but on a closer examination is curious: the hedge is only in leaf and flower on one side of the personage; on the other it resembles a metallic panel, suggesting that it is either dead or in winter. A sense of depth is denied, forcing the viewer to confront the personage. The dramatic backdrops, looking not unlike theatrical 'flats', provide settings of heightened intensity for the personages which, placed on plinths, take on the aura of statuary or objects in a still life.

In understanding Sutherland's purpose in depicting these personages, it is instructive to compare them with his handling of human figures both in the same canvases and in his portraits. While most of his metamorphic pictures are 'peopleless' (although traces of acculturation or man-made structures are usually present), some contain schematic figures, which are sometimes gendered. (Certain of the *Standing Forms* appear waisted and therefore vaguely female, but otherwise his personages are not gendered.)[13] These figures are stilted and mannequin-like by comparison with the expressive power of his personages and suggest that Sutherland was less at ease with them as subject matter. However, one of the more compelling of his figures is that appearing in *Personal,* an illustration to David Gascoyne's poems of 1937–1942. The figure, which is naked and melancholy, its arms raised in mute appeal, is a vivid expression of human precariousness in a dangerous and inclement world. Its title invites it to be read as a reference to the artist and his view of the artist's condition.[14]

A sense of existential anguish is also to be found in the works of his later personages, which are marked by a darkening of mood. Although the metamorphic forms of Sutherland's earlier works are accompanied by a sense of highly charged unease, in part due to their animism and in part to the portending, tenebrous tones interspersed with splashes of lurid, sulphurous colour, their meaning is unclear beyond inspiring a sense of foreboding. Paradoxically, although the personages of his later works are entirely imaginary and bizarre, their meaning is unmistakable. They appear as totemic or fetishistic images designed to inspire terror – as 'emotional paraphrases of reality' that represent 'the implications of the outer chaos of twentieth-century civilisation'.[15] And whereas in the animist works of his first phase Sutherland is transforming an observation from nature into a state of mind, creating a sense of *genius loci* in the process, in these later works he is objectifying a state of mind through constructs derived from nature. The *genius loci* has become a metaphor not only for the artist's own experience of place but of the human condition and, in particular, the place of man in the universe.

Perhaps the most potent of these images was the enormous mural for *The Origins of the Land*, which was positioned at the entrance to the *Land of Britain* pavilion at the Festival of Britain.[16] Again, it has a blank, unreadable background, although the warm colours are suggestive of the life-giving force of the sun. The canvas is arranged in layers, perhaps evoking geological strata, with the flames of the lowest level representing the molten core of the earth. The personages are part pterodactyl and part root forms. Their grey, stone-like colours resemble fossils of prototype life forms, and the rock-like shapes lying around hint at the process of erosion and fossilization beneath the earth's surface. The subterranean personages are fearsome in their strangeness, the beady eye in profile on the winged, prehistoric bird and the sightless eyes of the root forms heightening the sense of terror. Yet the overwhelming emotion engendered in the viewer is that of being repelled – the imagery is essentially terrifying in its repugnancy.[17] The mural seems a curiously uncomfortable and tense evocation of, or commentary on, national identity at what was a nostalgic and nationalistic celebration of being British. It was greeted with incomprehension and a sense of lost opportunity at the iconography. The critic Tony del Renzio described it as 'perhaps the most impressive failure of the South Bank',[18] although many also understood it as a metaphor for the 'now prevailing cosmic anxiety'[19] or, at the very least, suggesting a terrain under threat.[20]

Although their aesthetic was not generally admired, Sutherland's personages were nevertheless promoted by the 'establishment', notably the British Council, in the same way that the Geometry of Fear sculpture was heavily promoted abroad despite adverse popular criticism at home.[21] Sutherland was the artist chosen to represent Britain at the Venice Biennale in 1952 (the same year as the Geometry of Fear sculptors) and, following his success at Venice that had resulted in widespread demands for exhibitions of his work, at the São Paulo Bienal in 1955. As Margaret Garlake reports, the British Council's representative in São Paulo felt that Sutherland might have won the prize for a foreign painter instead of Alberto Magnelli had there been a British juror, but no money had been available to send one.[22] However, because of the international acclaim given to Lynn Chadwick and Reg Butler at the Biennales, as well as more generally to Henry Moore and Barbara Hepworth, the British Council's efforts to promote British artists became directed particularly at sculpture as the field in which they excelled.[23] To some extent, the emphasis on sculpture detracted from its efforts to promote painting. Certainly, landscape painting was not to make an appearance at Venice or São Paulo in any significant way until the inclusion of Peter Lanyon at the São Paulo Bienal in 1961.[24]

The stimulus for Sutherland's paintings of personages was the outside world, with the paintings themselves containing constructs of elements of the natural environment. Such paintings are essentially about place – place as experienced by the artist and then, at second hand, by the viewer – but they can only really be considered as landscape paintings if we accept that place is a component or metonymic of landscape. Indeed, the paintings deny any sense of perspective or vista to a wider area. In this they are representative of post-war landscape paintings generally, which became increasingly concerned with the artist's unique relationship with a place or locality. This process had already been foreshadowed in the Neo-Romantic preoccupation with *genius loci* in which place was the dominant interpretative framework for landscape painting. Sutherland's particular contribution was to convert place into an organic presence through a process of metamorphosis. The result is a kind of 'mindscape', rather than a representative landscape, that records Sutherland's own experience of place, overlaid with his perception of its ideological symbolism. For Sutherland, this symbolism is concerned with the nuclear threat – an entirely new form of threat which, when unleashed, could not be controlled and would wreak devastation on both mankind and the natural world. As a consequence, place is no longer enshrined in *genius loci* but has become a metaphor for the unique experience of the artist.

Peter Lanyon

There is perhaps no artist more associated with this process than Peter Lanyon who was the pre-eminent British landscape painter of the period of the Cold War under discussion. Lanyon sought to render in paint his sensory experience of the West Cornish landscape of his birth. He saw that landscape both as a palimpsest of ancestral memories of living on and toiling the land, and of his own experience of existing within it.

In the late 1940s, Lanyon, like Sutherland, was also experimenting with landscapes that involved subterranean elements or underground strata. However, whereas Sutherland's interest in strata derived in part from depicting mines as an Official War Artist, Lanyon's stimuli were the influences of the spiral forms of Naum Gabo, and the preoccupation with internal and external forms in the sculptures of Barbara Hepworth.[25] (Both artists encouraged Lanyon as he started out as a professional artist.) Early paintings like *The Yellow Runner* (1946), a painting replete with symbolism about Lanyon's homecoming from war and his settling down to marriage and family life in Cornwall, used the device of depicting interior and exterior forms simultaneously: the stockade of horses is contained within a womb-like structure that is at the same time represented as underground and on the surface. As Garlake explains, as well as suggesting the need for security, the painting implies 'a process of metamorphosis or interchange between categories: body and earth or animal and human'.[26] Similarly, Lanyon's *Generation Series* of eight paintings, which also explores the themes of renewal after returning from war and his wife's first pregnancy, is heavily indebted to Gabo and Hepworth in its interplay of interior and exterior forms, and to Hepworth in its metamorphic references to a female form and to a mother and child.

Two paintings made in 1949 signal the development of Lanyon's own personal idiom of depicting actual landscapes simultaneously from multiple viewpoints to render the complexity of the experience of place. *West Penwith* and *Portreath* both depict the surface details and morphology of actual places while allowing the viewer to see into the landscape by representing it at the same time from above and in section, as well as frontally. The cavities and subterranean passages revealed in the images may be understood as a depiction of the 'bones' of the landscape. They symbolize not only the mining history of the area but the ancestral bones of Lanyon's forebears and other inhabitants of West Cornwall who worked the land and sea.[27] There are also explicit references to the body within the landscape contained within each painting. Lanyon claimed that he saw landscape as gendered – in his personal mythology, the land was female and the sea was male – and in these works there are references to gender and sexuality in the penetration of the sea into the harbour and in the intermingling of the sea with the land.[28] There are also figurative elements within the landscapes themselves: *West Penwith* may be read as a reclining female figure, the contours of her body reflected in the undulations of the landscape, while *Portreath* appears to contain a woman's head in profile looking towards the harbour. The allusiveness of landscape to the body – particularly to the female form – was a subject being explored contemporaneously in the sculpture of Moore and Hepworth. These early works of Lanyon share the same mood of solemn fecundity and monumentality found in their sculptures of the period.

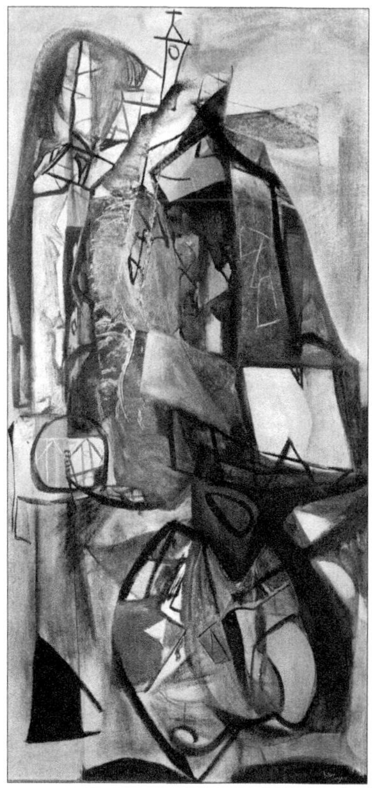

5.3 Peter Lanyon, *Porthleven*, 1951, oil on board, 244.5 × 121.9 cm, Tate Gallery London © Tate, London 2013/ Estate of Peter Lanyon.

Lanyon's entry for the Festival of Britain exhibition *60 Paintings for '51*, *Porthleven* (Figure 5.3), a painting of the Cornish harbour of Porthleven, continued to explore the idea of landscape seen from multiple viewpoints and its relationship with the body. Yet the mood is very different from his earlier works for two important reasons. First, whereas in *West Penwith* and *Portreath* the artist occupies the same space as the viewer and is external to the image (that is to say, landscape is depicted as 'other' or separate from self), in *Porthleven*, the artist's presence, although not represented, is implied within the painting itself. Indeed, the artist's body is the fulcrum of the image – he experiences and views the landscape from all its directions and angles – and renders that experience on canvas in a highly gestural expression of the key elements. As a result, the painting reads as a record of the artist's personal experience of place, an impression accentuated by the large, 'human proportions' of the canvas.[29] The second reason is the spontaneity of the execution of the paint. Both early paintings have been worked and reworked extensively and display Lanyon's characteristic style of layering the paint and then scraping and rubbing it away – a process that reinforces the idea of strata. By contrast, the paint in the final version of *Porthleven* was applied rapidly and appears fresh and liquid in its immediacy.[30] The spontaneity serves to reinforce the expressiveness of the image and the sense of the artist's presence within it. The painterly strokes are markers of the artist's existence and their visceral quality is suggestive of the existentialist idea of the artist battling against the absurdity of existence by conferring meaning through his actions.[31]

Porthleven is a harbour that consists of three basins and has a distinctive clock tower at its headland. However, when depicted in the painting from different viewpoints simultaneously, it appears semi-abstract with the schematic clock tower being the most recognizable element. In some ways more readable than the landscape itself are two rudimentary but life-sized metamorphic figures made out of the elements of the landscape that are contained within it in addition to the implied presence of the artist. Most clearly visible on the right is a woman dressed in a shawl and skirt, her rectangular face with mouth and eye in profile just below the clock tower; she is looking up at another hooded figure – perhaps a man – whose body lies along the left-hand edge of the canvas. The clock tower may be read as a lantern that he is holding above her. Lanyon claimed that the figures were accidental (seen after the event) and represented archetypes of a fisherman and his waiting wife.[32] According to Chris Stephens, the accidental figures may have appeared to Lanyon at an early stage in the development of the painting and he then nurtured them.[33]

The most important metamorphic work of Lanyon's early period was his painting of *St Just* (1953; Colour plate 7), a richly allusive painting of a Cornish mining town made notorious by a mining disaster in 1919 when 31 men were killed by defective machinery at a nearby mine. Like Lanyon's earlier works, underground strata is suggested by the Y-shaped mining shaft that divides the canvas down its centre and is seen at once frontally, in section and from above. The shaft has obvious metamorphic significance and may be viewed as a spread-eagled person or, aptly referencing the religious place name, as a crucifixion with a crown of thorns or barbed wire at its apex. Flanking it are two other figures: on the right, a kneeling woman with head bowed and arms raised either in supplication to the crucified saint or in grief at the loss of life; on the left, a standing man whose bared rib cage is perhaps a deliberate reference to Bacon's association of a carcass with the crucifixion in *Painting* (1946), or to Sutherland's *The Crucifixion* (1946) with its pronounced, emaciated rib cage. In Sutherland's painting the rib cage was directly inspired by photographs he had seen of starving bodies in the concentration camps of Buchenwald, Belsen and Auschwitz. In Lanyon's case, even if there is no intended allusion to the other artists' treatment of the theme, there is at the very least a suggestion that the miners were 'so much meat' or 'fodder' for the exploitative mine owners.[34] A dagger is visible to the right of the mineshaft. It appears to have a streak of blood on it – the only red of the painting. It represents the violence of the accident with its unnecessary deaths, as well as the danger lurking beneath the surface of the mine.

Margaret Garlake summarizes the painting's theme as being both an elegy for the tragic accident and a passionate denouncement of man's inhumanity to man.[35] (The political message of the painting reflects Lanyon's desire to identify with the common working man, which at other times led him to defend the mining industry against attempts to preserve the countryside as a timeless landscape.) Undoubtedly, the painting is primarily about a historically specific place and event. However, it may also be interpreted as a treatment through landscape painting of the iconography of the crucifixion. The religious undertones are clearly deliberate – Lanyon describes the painting in a letter to Ivon Hitchens, saying he had constructed and executed Christ in the painting and that the result is like 'the residue of a NAPALM bomb. There is no sign of him but he is everywhere.'[36] Indeed, he referred to it as his crucifixion painting before he gave it a title.[37]

As noted above, the iconography of the crucifixion was being explored contemporaneously by other artists, notably by Bacon and Sutherland. The crucifixion represents the issue of coming to terms with man's terrible past, as revealed through his recent actions, and the possibility or not of redemption. In the hands of Bacon and Sutherland, it is variously a metaphor for the hopelessness and horror of the human condition (revealed in the killing and concentration camps of the Second World War and in the subsequent development of nuclear weapons). In Bacon's works, the possibility of salvation seems denied – the three figures at the base of the cross are terrifying mutants that have invaded the mythology of the crucifixion and are re-writing its ending. In Sutherland's works, the cruelty of his series of *Thorn Heads* contains a savage beauty that may yet suggest the possibility of salvation. Lanyon cannot have been unaware of the parallels with their work in his treatment of a crucifixion theme.[38] It is unlikely, however, that he wished to reference their work in his as he was by then too intent on developing his own idiom. More likely, he is tackling a theme that was 'in the air' generally and giving it his own interpretation. The fact that Lanyon was not himself religious supports this interpretation. Similarly, his allusion to a napalm bomb (used in the Korean War) suggests that he saw his work as having a wider stimulus and relevance, perhaps as an expression of despair at the brutality and futility of war.

As Andrew Causey has observed, Lanyon's metamorphic works represent the collapse of accepted genres: figure as harbour, figure as landscape, landscape as figure. The compression is the equivalent of Merlau Ponty's phenomenology in which external reality is not seen as objective or permanent but as particular to the presence of a person in one place at one time.[39] In effect, the artist 'sees' or knows the world with his senses as much as through his consciousness, and the subject of what he sees is not an unchanging object but the correlate of the artist's body and its sensorimotor experience. Lanyon was to give this specificity a new dimension when he took up gliding in 1959 and started recording the experience of landscape as seen from the air. When this happened, the phenomenological experience of his being in a particular place and time was to become the predominant theme of his art. However, even before this, Lanyon went to great lengths to 'feel' the landscape by immersing himself in it, encountering it in all weathers, lying on cliff edges, touching it, looking at it upside down and trying to catch a view of the landscape behind him 'unawares'.[40]

Before taking up gliding, Lanyon had been depicting landscape from an aerial or bird's-eye viewpoint, flattening the fields and water basins as if pinned out beneath him. For example, in *Silent Coast* (1957), Lanyon experienced the landscape from a great height looking down and produced a work that resembles the flat stained colour fields of a Rothko. Once he took to the air, however, his depictions of the aerial viewpoint also became more multi-directional as he sought to render the totality of the sensory experience of both seeing and being above the landscape. He observed that 'for the first time I was able ... to experience my country from outside returning to land rather than emerging from inside'.[41] For example, in paintings like *High Wind* (1958), *Offshore* (1959) or *Thermal* (1960; Colour plate 8), landscape, sea and cloud are all intermingled as his glider spirals above them. His gliding paintings are, if anything, more obviously sensory than his early works with exquisitely diaphanous layers of paint suggesting the misty weather conditions, sea spray and clouds. Importantly, these paintings also seek to depict movement – as, for example, in *Soaring Flight* (1960), which depicts the lift of the glider as it rises in a spiral on the warm air current of a thermal, or in *Cross Country* (1960) where the black arcs and loops record the traces of the glider through the air – and the space/time continuum implicit in his earlier works through their multi-directional viewpoints has become explicit in the representation of the artist's journey through the landscape.

The theme of Lanyon alone in his glider confronting the uncertain weather conditions inevitably places the artist's body at the centre of the paintings. This extreme awareness of self translates to an extreme awareness of the act of painting. Perhaps unsurprisingly then, the gliding paintings are the most abstract works that Lanyon produced. In these later works, abstraction is no longer being used simply as a method, like Lanyon's earlier use of Constructivism, to arrive at an image. It has become an end in itself. As a consequence, Lanyon positions himself alongside the movements of action painting and American Abstract Expressionism, although he has arrived at their methods independently and would always deny any influence or association. However, the resultant works are in fact never truly abstract. They continue to use representative colours and contain figurative elements that are simply depicted in a gestural way.

As well as depicting movement, the gliding paintings are also often suggestive of danger or vertigo, which Lanyon defined as 'a sort of Angst applied to landscape'.[42] As Lanyon explains, his aim was:

> to explore the region of vertigo and of all possible edges where equilibrium is upset and I am made responsible by my own efforts for my own survival. Without the urgency of the cliff-face or of the air which I meet alone, I am impotent. I think this is why I paint the weather and the high places and places where solids and fluids meet. The junction of sea and cliff, wind and cliff, the human body and places all contribute.[43]

As we have seen, vertigo – or vertiginous viewpoints – has been an underlying theme throughout his paintings, and danger has been suggested in the hidden, subterranean caverns or more directly in the imagery of the dagger in *St Just*. However, perhaps inevitably given the risks he was now taking to experience landscape from the air, the suggestion of danger moves to the foreground. In *Soaring Flight* (1960), the danger is suggested by the use of the colour red to depict the path of the glider. Similarly, in *Solo Flight* (1960), the path of the glider is also picked out in red (likewise, the dagger in *St Just* is streaked in red).

The theme of danger has, of course, strong Cold War connotations. The threatening personages of the early 1950s concealed in landscape or constructed out of it have, in Lanyon's hands, been replaced with representations of the experience of danger by the artist. However, the representations of danger serve a similar metaphorical purpose as the personages by suggesting the 'prevailing cosmic anxiety'.[44] In fact, they are more readily comprehended by the viewer, who lives the experience with the artist. Moreover, in the gliding paintings, the artist has assumed a heroic quality as he swoops and dives – like Saint-Exupéry's Little Prince exploring the universe – alone against the elements, fearless in his quest to experience the sublime. Like Lanyon, the viewer is able to believe momentarily in his own ability to transcend the forces of evil through his self-expressive actions.

In its primitive theme and deeply subjective handling, there is no more telling metaphor of the freedom of the individual in the democratic, capitalist West (as embodied in the archetype of the heroic, risk-taking artist) against the suffocating conformity of the totalitarian, Communist East. Garlake's summary underscores this point most eloquently when she says:

> The modern artist was reinvented by post-war writers as an idealised, solitary figure, untouched by external forces, ideology or material needs: a privileged innocent required only to fulfil a personal vision. This absurdly over-simplified model replaced

the older idea of the Bohemian, in order to represent the artist as a metaphor for a free society. Alone in his glider, answerable to no-one and performing great loops and arcs through the empty sky, Lanyon may be seen as the archetypal modern artist.[45]

Whereas Sutherland's paintings were essentially about objectifying a state of mind by embodying it in landscape, Lanyon's works are, as we have seen, increasingly about objectifying the sensory experience of place. After early works like *Long Sea Surf* (1958) and *Low Tide* (1959), which both depict figures within the image, the personages or metamorphic elements disappear from the gliding paintings, although the gendering and sexualizing of sea and land remains, and the body's presence is simply implied. This is because *the painting has become the artist.* The conflation of artist, object and place has become complete, and the viewer sees and experiences the landscape through the artist's eyes.[46]

In fact, Lanyon was to reintroduce figures into his works in the 1960s, as, for example, in the highly autobiographical *Two Place* (1962) in which the body and landscape become interchangeable colour fields. However, the figures are no longer accidental or concealed. Moreover, they are flesh-toned, nude and sensual (see also *Beach Girl* (1961)). By the time of *Clevedon Night* (1964), one of his last works, the nude is exuberantly erotic in her thin, pure colour and languorous pose, and much more closely aligned to pop art (or the flesh tones of De Kooning) than to his early Cold War personages. In his last works, which are also much more brightly coloured, Lanyon is reflecting, or perhaps deliberately aligning himself with, the new mood of optimism emerging in the 1960s after the Cuban missile crisis – an optimism due to a growing acceptance of the balance of power and the ability to see beyond it, and to rising consumer prosperity.

Alan Reynolds

Alan Reynolds, the third artist to be considered here, occupies an important and perhaps now underrated position in the context and history of British landscape painting in the 1950s due to his total embrace of abstract art by the early 1960s. During the early to mid-1950s, however, Reynolds was a highly popular landscape artist whose work was seemingly more accessible than Lanyon's.

Like many artists of his generation, he studied art after serving as a soldier in the Second World War. While still a student, he exhibited at the Artists' International Association in 1950 and in two group shows at the Redfern Gallery in 1952 before being accorded a solo exhibition by the Redfern Gallery in 1953. The popularity of his work was immediate: he was soon being fêted as the golden boy of Neo-Romanticism, featured as a student in *Vogue*, and enjoying commercial success and critical acclaim.[47] Reynolds' style of painting landscape was highly distinctive, with what many considered a recognizably 'English' pastoral quality. In fact, his landscapes were painted in the studio without preparatory sketches and, according to Reynolds, were true inscapes, although he did also make botanical studies from nature and, as is clear from the subject matter of hop poles and orchards, drew inspiration from the Kent countryside where he lived. He worked straight onto the canvas or, more usually, board. Nevertheless, the products of his imagination were rooted initially in the Romantic and Neo-Romantic tradition of Samuel Palmer, Paul Nash and Graham Sutherland; early landscapes of sunlit hills as, for example, *In Summer* (1953) – a point that did not go unremarked at the time.[48]

From the outset, however, Reynolds was also strongly influenced by German art, which he first encountered as a soldier stationed near Hanover. Like many of his contemporaries, he was drawn particularly to the work of Paul Klee and the latter's impulse towards abstraction and pattern.[49] The influence of Klee on Reynolds cannot be overstated. Although Reynolds did not see the important exhibition of Klee's work held at the National Gallery in 1945, he did fall under the influence of one of the two paintings illustrated in the catalogue for the exhibition, which was a small oil called *Nocturnal Festival* (1921). The composition of this painting fascinated Reynolds and played a key role in his artistic development. Essentially, the picture breaks down into three horizontals: a foreground of field/garden that is closest to the viewer, a middle ground of roof/mountain that is more distant, and a far ground of cosmos/sky. The division is dynamic rather than rigid in the sense that the same object can occupy more than one plane. Reynolds was to use this composition consistently in his landscape paintings. For example, the horizontal lines demarcating the three planes are clearly visible despite the vertical accents that cut across them in 'typical' watercolours from 1952, the year in which he first came to public and critical attention – see, for example, *Hopfields* (1952; Colour plate 9) or *Brown Landscape* (1952). Similarly, they can be seen very clearly in later works such as *Winter Legend* (1954), *Winter Saga* (1955) and *Sunrise* (1956–1957).

Like Klee (although this resemblance was coincidental) and Lanyon, Reynolds collected botanical samples – seed pods, flowers, leaves, roots, and so on – which he stored in cellophane in his studio and, as well as making ink and pen wash studies, pressed and gummed into albums.[50] The experience of collecting intermeshed with his phenomenological experience of place that, in Reynolds' case, was otherwise more aesthetic than sensory. Reynolds frequently made his botanical studies the central image of a landscape: for example, by adopting dramatic viewpoints (which often have the viewer looking up from ground level towards the plants), or by placing plants drawn with botanical exactitude in close-up in the foreground (as if magnified or seen under a microscope) while the other natural elements are stylized or simplified. G.S. Whittet described this practice thus in his account of his interview with Reynolds in 1958 for the 'Painter's Purpose' column in *The Studio*:

> The use of these landscape plants is evident in the wind-blown 'stars' of Goat's Beard, looming large in the air in the foreground sky of some of the wide landscape panels. These intricate and yet formally decorative silhouettes are there for a purpose. Not only do they supply a spatial datum point, setting up a tension between the major form and the foreground plane but they supply the shock of impact such as we feel in the country 'when a bee comes near one's face and it seems larger than it is'.[51]

Reynolds was ultimately to find this approach distracting and it disappears from his paintings by the mid-1950s. By then, he was schematizing all natural motifs. He had always done this with trees, which were usually round or leaf-like on thin stems, but in his later works all the natural elements are imaginary. As John Hodin said, nature is typified rather than abstracted.[52] Again, this was a trajectory that Klee had followed in his search for the 'essence' of natural forms.

Klee's most important influence, however, was to liberate Reynolds from the need to paint landscapes in a naturalistic way. Reynolds had found in Klee the theoretical support to move on from his early interest in botanical drawing to his highly personal idiom of synthesized landscapes. In this, he was strongly influenced by Klee's 'untreadable space and semi-geometrical signs for natural phenomena'.[53] David Sylvester described Klee's treatment of space as uncentred or 'afocal': 'The evident order of a Klee is only in its parts, as with a landscape, a forest, a hedge, a crowd of people. To find an overall order, you must look for it yourself by taking hints from the given sensation … You must commune with the picture and its order will become manifest – not in space but in space-time.'[54] Klee's afocalism was highly influential on British artists generally and was particularly evident in the grids and loose frameworks of Tachisme. Lanyon's paintings are also usually afocal, as he attempts to record his experience of place over time while travelling through it. Reynolds' spaces are initially as complex as Klee's. For example, it is impossible to progress through the landscape of *Hopfields,* which succeeds in being both near and far at the same time and un-joined up in its elements, and yet maintains the appearance of landscape. However, by the mid-1950s, Reynolds' treatment of space becomes more readable.

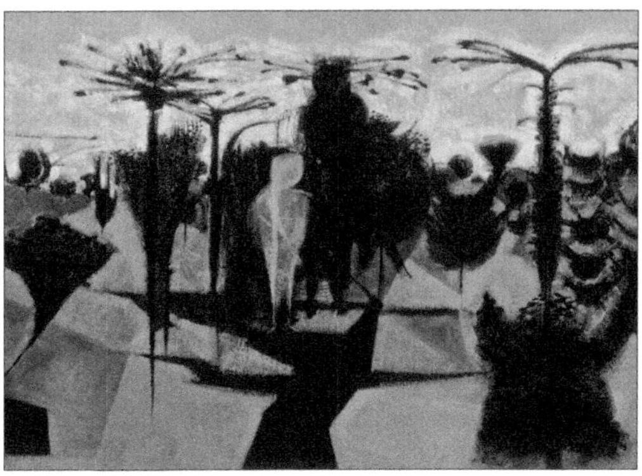

5.4 Alan Reynolds, *Keeper of the Dark Copse II*, 1951–1952, oil on board, 79.2 × 106.7 cm © Tate, London 2013/the artist.

Like Sutherland and Lanyon, Reynolds' paintings often contained metamorphic and anthropomorphic undertones as, for example, in *Keeper of the Dark Copse II* (1951–1952; Figure 5.4).[55] The painting has a shadowy presence in the centre. In conversation with the artist, Reynolds claimed that this was accidental, although he says that he noted it and allowed the presence to remain.[56] The trees are also 'triffid-like', as they frequently are.[57] Indeed, in many works, trees and hop poles have a totemic quality. They can be regarded as poetic

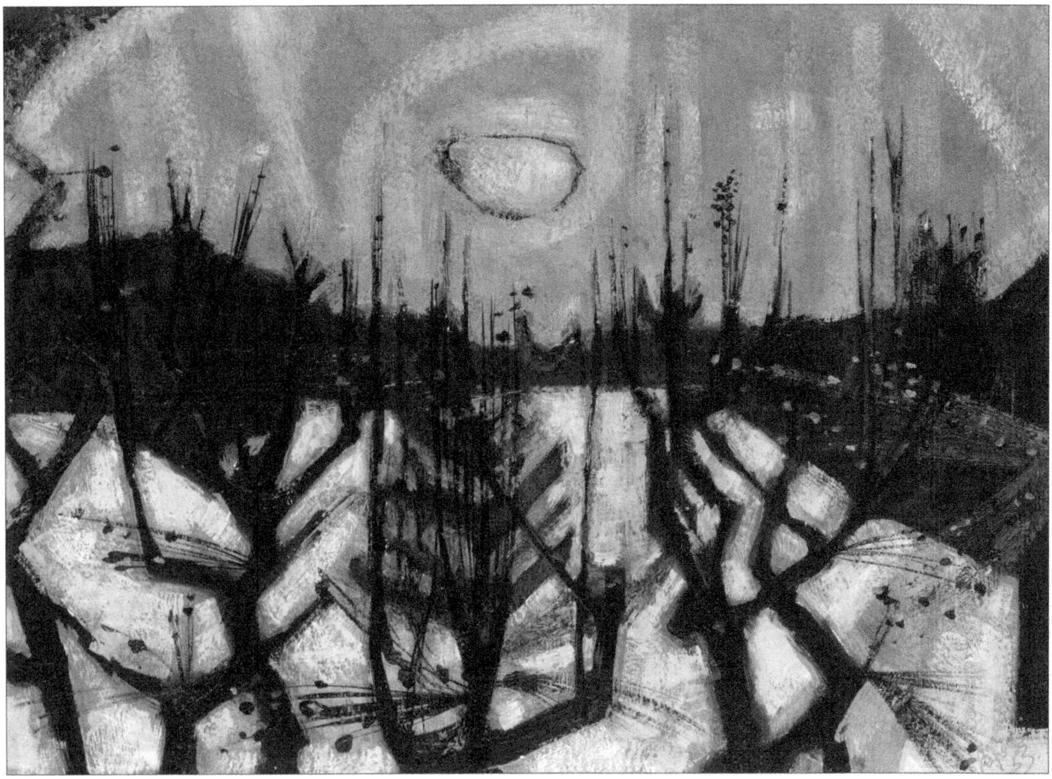

5.5 Alan Reynolds, *Bleak Orchard*, 1953, oil on board, 37 × 31.5 cm © Sotheby's 2007/the artist.

in the sense used by Sutherland when he says: 'I mean poetry in another and greater sense – poetry which enlightens the significance and nature of things. Poetry also which makes things hold more than their original meaning, and which defines their essence.'[58] In the idealism of youth, Reynolds certainly identified with the idea of the artist as a poet or visionary, but he never saw it as an end in itself, saying: 'Poetry is never absent from Nature, but alone it cannot constitute a work of art. It must be reconciled with the elements of design and composition.'[59]

Reynolds was preoccupied by metamorphosis, both in his constant treatment of the changing seasons and in the metamorphic nature of many of his motifs for trees and plants that display characteristics of 'personages' or transmogrification, and transform his gardens and landscapes into a 'phantasmal park'.[60] Initially, as in *Keeper of the Dark Copse II*, the purpose of the metamorphosis seems essentially poetic – the wish to create heightened meaning or *genius loci*. (Reynolds did not, however, see himself as a romantic artist in the manner of Sutherland, believing that his emphasis on form rendered his works classical.)[61] Yet, as the Cold War gathered momentum, the imagery becomes increasingly violent and menacing, as, for example, in *Bleak Orchard* (1953; Figure 5.5) or *Landscape with Hedge* (1955). They may now be regarded as metaphors of Cold War 'danger'. However, according to the artist, the viewer would be mistaken if he thought that the paintings were metaphors of the danger or menace posed to Britain – or to British national identity – by the nuclear threat as symbolized by its landscape.[62] Firstly, Reynolds claims that he never sought consciously to make any sort of political statement through his art, although he acknowledged that the prevailing mood of angst, reflected so visibly in the Geometry of Fear sculpture of the same period, was hard to escape.[63] Secondly, like so many artists of his generation, Reynolds was of a left-wing disposition. Although not a Communist, he saw Russia as the victim in East–West relations – a view that prevailed in artistic circles until the crushing of the Hungarian Uprising in 1956. Therefore, insofar as these paintings can be read metaphorically, they are about the English landscape being used to signify the threat posed to the human condition by nuclear capability and, in particular, by the aggression of Britain and America.

Very few of Reynolds' admirers saw these paintings in those terms, however, and he enjoyed extraordinary popularity both in Britain and internationally. Perhaps the most significant coup for Reynolds was to share the first prize for best painting (along with American artist John Hultberg and Italian artist Gianni Dova) at the *Young Painters* show organized by the Congress for Cultural Freedom in 1955.[64] Reynolds' prize was followed by a highly successful solo show, *The Four Seasons*, in March 1956 at the Redfern Gallery.[65] Rex de C. Nan Kivell, a director of the Redfern Gallery, described the public reaction in a letter: 'Swarms of people are still pouring through the Gallery, it is extraordinary that he has fascinated so many people, including the really hard core of the tougher intelligentsia.'[66]

The typified, often archetypal nature of Reynolds' motifs contributed to his landscapes having a very pronounced element of pattern.[67] Other factors played a part in this, notably, the rhythmical arrangement of the elements on the canvas and the repetition of motifs. In fact, many of his paintings can be read as abstract patterns rather than as landscapes, and increasingly the viewer requires an underlying familiarity with Reynolds' personal language of forms to be able to interpret the later works as landscapes.

At the height of his popularity and financial success, Reynolds gave up representation but not, initially, allusion, producing a series of works based on loose grids with colours evocative of landscape. As with the public's negative reaction to the highly abstract *Autumn Landscape* (1951) by William Gear that won the *60 Paintings for '51* competition, so Reynolds' gradual embrace of abstraction and his subsequent conversion to concretion were met with dismay and incomprehension by a public and establishment that largely turned against him, feeling cheated out of a visual language by which they could first deny and later rationalize the uncertainties and anxieties of their times.[68] The sense of disillusion was particularly bitter, not simply because Reynolds had abandoned representation in favour of abstraction – something which the British public mistrusted – but because he had 'betrayed' an image that the British held of themselves. That image was their national identity as enshrined in the untouchable 'shibboleth' of the pastoral English landscape.

Conclusion

The three artists discussed here are linked by their varying use of metamorphosis and metaphor to reinvest the materiality of landscape and, in particular, the concept of *genius loci* with contemporary ideology concerned primarily with attitudes towards the Cold War. Sutherland provides the common thread between Lanyon and Reynolds. His personal mythology of organic and mechanistic personages to symbolize his feelings about the Cold War, nuclear annihilation and the Holocaust started the tendency that the other artists develop. In other ways, however, Lanyon and Reynolds could not be more different in their approach to landscape painting – Lanyon is experiential and sensory and Reynolds is conceptual and synaesthetic.

Reynolds produced images that were seemingly accessible and represented an idyll that audiences longed for. In terms of the political context of the early Cold War, his work started as the perfect instrument by which to affirm the continuity of the British way of life after the destruction wrought by the Second World War, and was seized on and promoted as such. Despite the schematic nature of many of his early landscapes, the images were understood simply as depictions of a 'Garden of Eden' in the English landscape tradition – a consoling reading at a time when Britain's status as a world power was increasingly precarious.[69] Gradually, the troubling, metamorphic qualities that were never far from the surface of his early landscapes became more insistent, and the pleasing images of sunlit hills gave way to threatening phalanxes of spikey orchards and thorn trees that read as metaphors for the Cold War itself. By now unable to ignore the reduced power of Britain on the global stage that was soon to become manifest in the flashpoints of Suez and the Hungarian Uprising, the public seemed to assimilate this change without difficulty and continued to flock to his shows and buy his works.

By contrast, Lanyon's imagery was always too abstract and difficult to be popular in Britain, although it was more readily accepted in America because of its parallels with action painting and Abstract Expressionism.[70] As we have seen, Lanyon's use of the metamorphosis undergoes a shift. Initially, the men and women buried in the landscape imagery tell of the human histories embedded or etched in an ancient landscape where the experience of place cannot be separated from traces of acculturation and human dwelling. Even the land and sea are gendered. In later works, the human element is provided by the artist's being, rather than by the presence of hidden personages. The subject of the painting has ceased to be about layers of meaning and has been 'simplified' to a single theme – the subjectivity of place. As a consequence, in Lanyon's hands, *genius loci* is no longer an immanent mood that is external to the artist's own mood as engendered by his experience of place. And what we see is a record in paint of the artist's sensory experience of a place in which the personages have been stripped away but the artist's presence is inevitably implied.

In terms of Cold War politics, the work of all three artists can be read as metaphors of the danger posed by the nuclear threat. But, whereas in the works of Sutherland and Reynolds the metaphor is figurative and the imagery seeks to unsettle, in Lanyon's later works the danger is simply implied as part of the expression of the artist's vertiginous experience of place from a glider. As a consequence, the metaphorical value of these later paintings lies not so much in their imagery as in their gestural expressiveness and extreme individualism. It was for these qualities that Lanyon's gliding paintings were to become the supreme metaphor for the freedom of the artist and therefore the ultimate propaganda tool in the cultural war between East and West – a fact also reflected in Lanyon's selection to represent Britain at the São Paulo Bienal in 1961 at the height of the Cold War.

Notes

1 As, for example, in the pictures Paul Nash took between 1931 and 1935. See *A Private World: Photographs by Paul Nash*, exh. cat. (London: Fisher Fine Art, 1978); and Andrew Causey, *Paul Nash's Photographs: Document and Image*, exh. cat. (London: Tate Gallery, 1973). Nash wrote about the animism and animatism in found objects (such as the weathered log he submitted to the *International Surrealist Exhibition* in London in 1936 as *Found Object Interpreted, Vegetable Kingdom*, which subsequently anthropomorphized into *Marsh Personage*, 1939) in 'The Life of the Inanimate Object', *Country Life*, 1 May 1937, 496 – see Andrew Causey, *Paul Nash* (Oxford: Oxford University Press, 1980), 479; Malcolme York, *The Spirit of the Place: Nine Neo-Romantic Artists and Their Times* (London: Tauris Parke Paperbacks, 2001), 61; and *Paul Nash Writings on Art*, selected with an introduction by Andrew Causey (Oxford: Oxford University Press, 2000), 137. See also more generally, J.H. Matthews, *The Surrealist Mind* (Selinsgrove: Susquehanna University Press; London and Toronto: Associated University Presses, 1991); and Briony Fer, 'Surrealism, Myth and Psychoanalysis', in *Realism, Rationalism, Surrealism: Art between the Wars*, ed. Briony Fer, David Batchelor and Paul Wood (New Haven and London: Yale University Press, 1993), chapter 3, 223–224.

2 Nash attempts to define the English *genius loci* in an essay in *Unit One*, ed. Herbert Read (London: Cassell & Co., 1934), 80.

3 See 'Graham Sutherland: The Damned', in David Mellor, *A Paradise Lost: The Neo-Romantic Imagination in Britain 1935–55*, exh. cat. (London: Barbican Art Galleries, Lund Humphries, 1987), 72–76; Virginia Button, *The Aesthetic of Decline: English Neo-Romanticism c. 1935–1956*, PhD thesis, Courtauld Institute of Art, University of London, 1991, 96–101, on Sutherland's preoccupation with *genius loci*; and Martin Hammer, *Graham Sutherland: Landscapes, War Scenes, Portraits 1924–1950*, exh. cat. (London: Dulwich Picture Gallery and Scala Publishers Ltd, 2005), 16, on Sutherland and the Surrealist preoccupation with the found object.

4 A word Sutherland perhaps borrowed from John Nash (see Yorke, *The Spirit of the Place*, 118). Paul Nash's first recorded use of the word 'object-personage' is in a letter to Henry Moore, 15 September 1933, quoted by Causey in *Paul Nash*, 253–254. Sutherland defines personages thus: 'Shadows had a presence;

certain conformations of rock seemed to go beyond just being a rock, they were emanations of some kind of personality', in 'Landscape and Figures', *The Listener*, 26 July 1962, 132.

5 'Sometimes through sheer laziness, I would lie on the warm shore until my eye, becoming riveted to some sea-eroded rocks, would notice that they were precisely reproducing, in miniature, the forms of the upland hills', Graham Sutherland, letter to Colin Anderson, published as 'A Welsh Sketchbook', *Horizon*, vol. v, no. 28 (April 1942): 225–236, republished in Graham Sutherland, *Correspondences: Selected Writings on Art*, ed. Julian Andrews (Picton and Geneva: The Graham and Kathleen Sutherland Foundation, 1982), 52.

6 Graham Sutherland, 'Thoughts on Painting', *The Listener*, 6 September 1951, 376.

7 He described the process thus: 'I may go for a walk. There is everything around me. I look: certain things that one sees, almost by accident, seem to strike me more than others. Out of the thousand things I see – *one* juxtaposition of forms – above all others seem to have a meaning. I don't always understand what I am doing – or what I am likely to do … I make notes following the first "frisson" of an encounter. Gradually an idea emerges: there is the knowledge that, almost in spite of myself – something has arrived.' From a letter by the artist to G., 11 March 1972, reproduced in *Sutherland in Wales*, the catalogue for the collection at Picton Castle, Haverfordwest, Dyfed, 1976.

8 Ibid.

9 Graham Sutherland, letter to Colin Anderson, 'A Welsh Sketchbook', republished in Sutherland, *Correspondences*, 51. He added: 'These and other things delighted me. The twisted gorse on the cliff edge, such as suggested the picture "Gorse on Sea Wall" – twigs, like snakes, lying on the path, the bare rock, worn, and showing through the path … the deep green valleys and the rounded hills and the whole structure, simple and complex', ibid., 52. Geoffrey Grigson argues that the image is in fact of Sutherland himself: it was 'not gorse. The caption is indispensable. It is Sutherland', in Grigson, 'Authentic and False in the New "Romanticism"', *Horizon*, vol. xvii, no. 99 (March 1947): 207 and 211.

10 Sutherland's interest in machines had been stimulated by his early training as an engineer and later by his encounter with devastated machinery as a war artist.

11 Aeschylus, *The Eumenides*, I. 252. The line as Bacon quotes it is translated in W. Stanford, *Aeschylus in His Style* (Dublin, 1942), quoted in fn. 38 of 'A Web of Images' by Dawn Ades, in *Francis Bacon*, exh. cat. (London: Tate Gallery, Thames & Hudson, 1985), 8.

12 See John Hayes, 'Graham Sutherland', 18, quoting Sutherland in 'An English Stone Landmark', in *The Painter's Object*, ed. Myfanwy Evans (London: Gerald Howe, 1937).

13 There was a strong equation at this time in the public's mind between the contours of landscape and the female form, derived in part from the sculpture of Moore and Hepworth, and so the non-sexualization of the personages is an interesting contradiction of this tendency and perhaps enhances their power to disturb.

14 Mellor, 'Graham Sutherland: The Damned', 72–73.

15 Sutherland, 'Thoughts on Painting', 376.

16 The mural was 16 × 10 feet. At the Festival of Britain, British industry was showcased in a series of themed pavilions that also included: *Natural Resources, Power and Production, Transport,* and *Sea and Ships.* A series of painted murals on related subjects accompanied a range of exhibits chosen to illustrate the themes.

17 Roger Berthoud recalls that Bernard Denvir, writing for the *Daily Herald* on the 1953 Tate retrospective for Sutherland (at which *The Origins of the Land* was exhibited), described Sutherland's work of the early 1950s as 'probably the most repellent pictures ever to be seen inside the Tate'. But, as Berthoud goes on to observe, this was the age of Belsen and the Bomb: 'when men wanted to know what it was like to be living in 1953, they would look at his paintings.' See Roger Berthoud, *Graham Sutherland: A Biography* (London: Faber & Faber, 1982), 162, cited in Catherine Jolivette, *Landscape, Art and Identity in 1950s Britain* (Aldershot and Burlington, VT: Ashgate, 2009), 33.

18 Toni del Renzio, 'Review of Graham Sutherland Exhibition at the Redfern Gallery', *Art News and Review,* vol. 4, no. 23 (1952): 4.

19 Herbert Read, 'Introduction', *An Exhibition of Paintings and Drawings of Sutherland,* exh. cat. (London: Arts Council and Tate Gallery, 1953), unpaginated.

20 At the Festival of Britain, the theme of 'the people' sat alongside the individual themes of the pavilions. In the *Land of Britain* pavilion, in particular, the twin themes of identity of the 'land' and 'people' were inextricably entwined. A metaphoric landscape shown in such a disturbing way inevitably suggested that both the geography of Britain and national selfhood were in peril. See further: Jolivette, *Landscape, Art and Identity in 1950s Britain,* chapter 1.

21 See Francis Watson reviewing Sutherland in 'Recent Paintings by Graham Sutherland', *The Listener,* 8 February 1951, 226. Sutherland was also promoted by the Tate Gallery in London, which held a retrospective in 1953 (see note 19 above). Various permutations of an exhibition called *Young British Sculptors* that included the Geometry of Fear sculptors toured Europe, North and South America, Australia and New Zealand.

22 Margaret Garlake, *The Relationship between Institutional Patronage and Abstract Art in Britain c. 1945–1956,* PhD thesis, Courtauld Institute of Art, University of London, 1987, 299 and fn. 145.

23 Chadwick won the Grand Prix for Sculpture at the 1956 Venice Biennale and was made *hors concours* at the 1961 Bienal at São Paulo. Moore won the International Prize for Sculpture at Venice in 1948 and the Best Foreign Sculptor prize at São Paulo in 1953, while Hepworth won the Grand Prix for Sculpture at São Paulo in 1959. The 1954 Venice Biennale was used by the British Council to display the prize-winning maquette for the international sculpture competition organized by the ICA on the theme of the 'Unknown Political Prisoner' that had been won by Reg Butler in 1953.

24 Peter Lanyon was never shown at the Venice Biennale – a surprising omission perhaps attributable to his untimely death.

25 Lanyon may also have been influenced by Sutherland's vertiginous depictions of tin mines in Cornwall made in 1942. Sutherland was recording their temporary re-use during the war.

26 Margaret Garlake, *Peter Lanyon,* St Ives Artists series (London: Tate Gallery Publishing, 1998), 25.

27 According to Causey, Lanyon's curiosity about the old mines and workings went back to his childhood. He had descended the old shafts in the 1930s and went down Geevor Mine with John Wells and Sven Berlin in 1946. See Andrew Causey, *Peter Lanyon: Modernism and the Land* (London: Reaktion Books, 2006), 50.

28 Peter Lanyon talking to Lionel Miskin about his early life, transcribed audio tapes, Tate Gallery Archive, TAV 215AB 1962, 20–21.

29 In fact, the painting is on Masonite board as Lanyon had destroyed the canvas either from over-working it or by deliberately slashing it out of frustration (there are varying accounts of the cause). See Garlake, *Peter Lanyon*, 36.

30 Although Lanyon worked on the painting for over a year, making a large number of drawings and constructions, the final work was painted much more rapidly. According to an undated letter from 1951 from Lanyon to William Scott in the William Scott Archive, the second version was painted in a matter of a few hours: 'I rang the Arts Council this morning and I find that they only want the pictures on March 1st … I took 4 hours over mine / having worked for 10 months on preparation. It looks like it too.'

31 See Fiona Gaskin, 'British Tachisme in the Post-War Period, 1946–1957', in *Artists and Patrons in Post-War Britain*, ed. Margaret Garlake (Aldershot: Ashgate, 2001), 36–40.

32 Peter Lanyon, 'Recorded Talk' for the British Council, 1963 (refers to conversations between Lanyon and Alan Bowness adapted as lecture notes for Lanyon's use).

33 Details of suggested figures can be discerned in the photographs of the first stages of *Porthleven* as it proceeded and in the reference materials used by Lanyon and placed next to the painting in the photographs that included the linocut of *The Returned Seaman* with its two figures of a seaman and waiting wife occupying similar positions in the image to those in *Porthleven*. See *Peter Lanyon*, exh. cat. (St Ives: Tate St Ives, 2010–2011), 40. See also Causey, *Peter Lanyon: Modernism and the Land*, 93.

34 See Letter to Rosalie Mander, 17 March 1952, where he refers to miners as 'so much meat' for the 'man-engine', quoted in Causey, *Peter Lanyon: Modernism and the Land*, 130.

35 Ibid., 40.

36 See Garlake, *Peter Lanyon*, 38–39. Lanyon also referred to *Bojewyan Farms* as a predella to *St Just*, ibid., 40.

37 Letter to Ivon Hitchens, 10 January 1952, quoted by Garlake, ibid., 39. Lanyon says something similar in a letter to Roland Bowden, 16 December 1952, Tate Gallery Archives, TGA 942.13, cited in Andrew Causey, *Peter Lanyon* (Henley-on-Thames: Aiden Ellis Publishing, 1971), 135: 'I have just finished my crucifixion and Christ is nowhere … The crucifixion (which is un-named yet) is all gossip and mocking and napalm burning WITHOUT the Francis Bacon illustrative obsession.'

38 See Chris Stephens, *Peter Lanyon: At the Edge of Landscape* (London: 21 Publishing, 2000), 102.

39 Ibid., 100.

40 'Offshore in Progress', *Arstcribe*, no. 34 (March 1982): 58–61.

41 Lanyon, 'Recorded Talk' The crucifixion represents the issue of coming to terms with man's terrible past, as revealed through his recent actions, and the possibility or not of redemption.

42 Peter Lanyon, 'A Sense of Place', *Painter and Sculptor*, vol. 5, no 2 (Autumn 1962): 3–7.

43 Ibid.

44 Sutherland, 'Thoughts on Painting', 378.

45 Garlake, *Peter Lanyon*, 55–56.

46 There are parallels with sculpture and, in particular, with Barbara Hepworth's description of herself as the object of her sculptures: 'For a few years I became the object. I was the figure in the landscape and every sculpture contained to a greater or lesser degree the ever changing forms and contours embodying my own response to a given position in that landscape', from her autobiography, *Barbara Hepworth: Carvings and Drawings* (London: Lund Humphries, 1952), unpaginated.

47 *Vogue* (May 1953): 118. *Vogue* repeats an observation by G.S. Whittet in his 'London Commentary', *The Studio*, vol. 148, no. 738 (September 1954): 92: 'Since March 1952 Alan Reynolds has scored one of the most phenomenal successes of recent years in four exhibitions at the Redfern Gallery.' A more sceptical view of his success was found in the Perspex column in *Apollo* (April 1953): 112: 'Another semi-abstractionist who is rapidly – perhaps too rapidly – making a name for himself is Alan Reynolds, who has a show at the Redfern. He is very young, still at the Royal College of Art, but he is being bought by the Big Museums, and greatly admired by the advanced critics ... He is successful, I feel, with the hop-pole and farm-building Kentish countryside; but having evolved a singularly ugly formula for trees and being defeated by natural forms, which he renders in amorphous smudges of dark tone, he fails with normal landscape. He is in danger of parodying his own style. Mr Reynolds could afford to have much more to do with nature, and much less with intellectual concerns like "the Relativity of All to the Rectangle".'

48 See, for example, Robert Melville, 'Exhibitions', *Architectural Review* (September 1952): 202–203.

49 Paul Klee enjoyed a revival of interest and appreciation after the Second World War. An exhibition of 137 pictures by Klee opened in December 1945 at the National Gallery in London. In 1948, a major Klee exhibition was held at the Musée National d'Art Moderne in Paris, which was marked by a special edition of *Cahiers d'art* (published in 1946) consisting of 65 pages of essays and plates devoted to the artist. Klee's work could also be seen fairly regularly at the Hanover Gallery and was the subject of a solo show at the Mayor Gallery in 1950, at the ICA in 1953 (the catalogue is dated 1954) and at the Hanover Gallery in 1956. There was also a useful catalogue to the Paul Klee exhibition at the Leicester Galleries, 1941, with an introduction by Herbert Read. For the influence of Klee's legacy, see James Hyman, *The Battle for Realism: Figurative Art in Britain During the Cold War, 1945–1960* (New Haven and London: Yale University Press, 2001), 21.

50 See Neville Wallis, 'A Critic's Choice', *The Studio*, vol. 154, no. 775 (October 1957): 118; and G.S. Whittet, 'Painter's Purpose', *The Studio*, vol. 155, no. 781 (April 1958): 114–115. See also Lanyon's letter to William Scott, 16 February 1951, William Scott Archive, London.

51 Whittet, 'Painter's Purpose', 114.

52 *Alan Reynolds, Towards a Retrospective: From Early Landscapes to Abstraction*, exh. cat. (London: Agnews & Sons, 1982).

53 Robert Melville, 'Alan Reynolds', *The Studio*, vol. 150, no. 751 (October 1955): 113.

54 David Sylvester, 'The Auguries of Experience', *Tiger's Eye*, vol. 1, no. 6 (15 December 1948): 48–51.

55 'Chimneys and trees are invested with a totemic symbolism and the vague human figures that appear occasionally have the shadowy quality of a character in an epic poem', G.S. Whittet, 'London Commentary', *The Studio*, vol. 145, no. 722 (May 1953): 171.

56 Interview between the author and Alan Reynolds, 18 July 2006.

57 The comparison with triffids is apt because John Wyndham's *The Day of the Triffids* was published in 1951 (London: Penguin) and the idea of invasion by an intelligent, alien species quickly became a staple of science-fiction novels and films. On a similar theme, there is a vein of painting of threatening insects produced towards the end of the decade by artists such as John Tunnard and Paul Feiler – for example, see the latter's metamorphic image of a landscape/insect, *Porthledden III* (1960).

58 Sutherland, 'Thoughts on Painting', 376.

59 Quoted in the foreword to his first solo show at the Redfern Gallery, London, 1953.

60 Robert Melville, introduction, *Alan Reynolds*, exh. cat. (London: Redfern Gallery, 1954).

61 Indeed, Reynolds believed that Melville's identification of his early works with Neo-Romanticism damaged his reputation (interview between the author and Alan Reynolds, 18 July 2006). It is perhaps noteworthy that in Hodin's Redfern monograph on Reynolds of 1962, the artist's statements make no mention of the influence of Palmer, and any references to Neo-Romanticism are dropped.

62 Interview between the author and Alan Reynolds, 18 July 2006. The expressiveness was noted, amongst other things, by Robert Melville: 'The persistence with which he stretches the downs across his pictures and hangs the sun or moon above them suggests that there is a primitive pictograph for a landscape at the back of his mind – a horizontal line with a circle above it – which operates as compulsively as a magic sign. This pictograph may well stand for our situation in the universe, and if so, it is obvious that it has to be manipulated with the utmost circumspection if disaster is to be avoided. His work is a fascinating manifestation of a conservatism that goes deeper than that of his most enthusiastic admirers. His landscapes are the observation platforms of original man, whose hopes and fears are inextricably involved with the rising and setting of the sun,' *Architectural Review* (May 1958): 348–349.

63 Interviews between the author and Alan Reynolds, 18 July 2006 and 2 April 2009.

64 John Hultberg, who had studied with Mark Rothko and Clyfford Still, was accorded his own show at the ICA in October 1956.

65 In the Perspex column of *Apollo* (April 1956): 104, it was described as 'the most exciting exhibition held in London during March'.

66 Letter to Humphrey Brooke, Secretary to the Royal Academy of Arts, dated 21 March 1956, RAA/SEC/15.

67 However, Reynolds was not interested in Jungian archetypes and they had no relevance to what he was painting (interview between the author and Alan Reynolds, 18 July 2006).

68 Reynolds ultimately left the Redfern Gallery in 1976 because the gallery was unable to sell his reliefs.

69 Robert Melville, foreword, *Alan Reynolds*, exh. cat. (London: Redfern Gallery, 1954).

70 Although it should be noted that Reynolds had two successful shows in America, at Durlacher Bros in 1954 and again in 1958 – the appeal of the English idyll was not confined to Britain.

6

Cold War at Home: John Bratby, the Self and the Nuclear Threat

Gregory Salter

John Bratby rose to prominence in the British art world in the 1950s as one of the so-called 'Kitchen Sink Painters' with a series of paintings depicting himself and his family in their home over this period. This chapter focuses on the connections between Bratby's art and discourses surrounding the family in post-war Britain alongside the anxieties of the Cold War and the nuclear bomb. These connections become clear when Bratby's art is placed in the context of contemporary psychoanalysis. Specifically, the theories of Melanie Klein and D.W. Winnicott (particularly influential at this time) present the family as simultaneously vital to an individual's development – and by extension the development of society and even democracy – but also ridden with anxieties, fear and conflict. Bratby's painted world of cramped domesticity and fraught, often violent, relationships with his family lends itself to an interpretation built on contemporary psychoanalysis. With its enclosed spaces and piles of food, it also presents the home as a kind of bunker or shelter from the threats of the outside, while the artist even imagines the effects of radiation on himself in a self-portrait. With parallel examples drawn from contemporary literature, this chapter draws out the connections between anxieties about the family and nuclear war to present a new way of considering Bratby's art.

In the Mirror and on Canvas: The Self

In *Jean at the Basin* (1955; Colour plate 10) John Bratby paints his wife Jean Cooke in the windowless, run-down bathroom of the house they shared with Jean's father in Greenwich. The couple had married shortly after meeting while they were both art students in early 1953, and had been living here since then.[1] Jean stands in the centre of the room, her head bowed slightly. She could already be pregnant with the Bratbys' first son David at this point, judging by her rounded belly and the way in which she leans on the bathroom sink for support as she poses for her husband. Around her, Bratby takes great care to pick out the cracks running across the ceiling, the peeling wallpaper, and small details such as the mug and cleaning powder that sit on the sink. He is present too: he paints his reflection in the mirror above the sink, his face turned to take the scene in and his hand and paintbrush poised, ready to paint. At the bottom of the canvas, due to the painting's warped perspective, the viewer can make out the bathroom doorframe and the wall, suggesting that the canvas has been placed in the doorway, effectively blocking the exit. The whole work, through its furious streaks of paint on the floor, the unsteady perspective, Jean's foot, even the bowl of the toilet, tilts towards the artist.

Jean's isolated, contained and vulnerable position at the centre of the work is particularly striking. It is known that the Bratbys had a volatile and often violent relationship; he makes reference to this in the thinly veiled autobiography of his novels from the late 1950s and early 1960s (such as *Breakdown* or *Brake-Pedal Down*) in a tone that sits ambiguously and disturbingly between self-mythology and apology.[2] This violence seems to be written onto Bratby's thickly painted canvases and cannot be ignored in any discussion of his art.[3] Here, the fully clothed Jean, head bowed and gaze directed into the sink, is effectively trapped in the cramped space, as just another object in the room. Later in his career, Bratby would reflect: 'I sometimes painted my wife Jean Cooke as a particular person, not with affection. She was someone to paint', and this attitude certainly seems to apply here.[4] Fittingly, another contemporary critic found something similar in Bratby's paintings:

> I observe he is excited by only two kinds of objects: himself and expended objects. Himself is by no means expended, but he always seem to take it out of everything else. Those women are all expended, and what's left is all debris, and the apparatus for disposing of debris, sinks, lavatories ...[5]

Jean may well have the look of an expended object here, but she is also a crucial component of Bratby's home life and his sense of self. He seems to hover in the mirror above, as if to survey everything he has momentarily managed to contain on canvas.

The extent to which the artist's self is one of the primary concerns of this painting becomes clear when the way in which the painting is constructed around particular looks is analysed. *Jean at the Basin*, in its use of a mirror reflection, clearly quotes from Velazquez's *Las Meninas*. In Michel Foucault's influential analysis of *Las Meninas*, he highlights how the painting represents the difficulties of representation; how the array of looks between the artist, the court and the (not visible) King and Queen highlight how the image cannot 'present in a full light both the master who is representing and the sovereign who is being represented'.[6] However, in conflating the role of painter, painted subject and spectator in *Jean at the Basin* by depicting himself in the act of painting in the reflection in the mirror, Bratby attempts to gain an element of mastery over the difficulties of representation. Crucially, Bratby's raised right arm, with his brush in hand, comes in line with the edge of the mirror, as if to emphasize that he is literally forming the scene in front of him himself. Jean, meanwhile, takes no part in this exchange of looks; Bratby the artist looks at Bratby the subject and vice versa, over a scene that he both physically encloses (by standing in the doorway) and claims to form.

If the self is one focus of Bratby's art, then the location of the formation of that self – the home and the family – is another. In a review of one of Bratby's first shows at Helen Lessore's Beaux Arts Gallery in London in 1954, John Berger was quick to note this preoccupation, commenting that 'to enter the Beaux Arts Gallery is to enter Bratby's home'.[7] Bratby's family-focused art in many ways reflects wider concerns at the time. In 1955, Britain was still engaged in a process of post-war recovery in which the nuclear family had become central. The 1942 Beveridge Report, with its aims of eradicating the five 'Giant Evils' of want, disease, ignorance, squalor and idleness, and upon which the post-war welfare state was built, placed the family and full employment at the centre of its vision of a more equal, democratic society. Within the family unit itself, however, distinctions of gender remained: Stephanie Spencer has highlighted the duality of language in the Beveridge Report and subsequent legislation, where 'membership of the welfare state was couched in terms of reciprocal duty between husband and wife and between citizen and nation state'.[8] As a result, the husband and wife, and by extension the family, became the basis for the welfare state, with their roles divided between working/father husband and housewife/mother wife.[9] Arguably then, assumptions, trends and ideals about the gendered structure of the family unit that had been emerging in Britain since the nineteenth century became explicitly built into government policy and its vision for the future.

Inevitably, the gendered ideal of the family unit was not always observed by real families: some women worked, usually in order to supplement their family's income, while fathers were also increasingly likely to take on domestic chores and elements of child-rearing. However, the ideal remained and one discipline that focused intensely on the relationship between the ideal family and the realities of family life was psychoanalysis. The rise in popularity in Britain of child psychoanalysis, largely as a result of Melanie Klein's work before and during the Second World War, reached a peak after the war. Juliet Mitchell views government policy and psychoanalysis as essentially interlinked at this point, with post-war reconstruction built on the family-oriented welfare state, which was itself influenced and developed by psychoanalytical theories.[10] It is possible, then, to trace a psychoanalytic lineage in Britain from Klein's extension of Freud's theories on the Oedipus complex and family development, and the development of Klein's work by figures such as Susan Isaacs and, more popularly, John Bowlby and D.W. Winnicott (and, beyond the 1950s, to R.D. Laing). Alongside the ideals of government policy, this branch of psychoanalysis explains, polices and attempts to put right the deviations from the norms of family life.

Crucially, at the heart of the work of all of these figures is the relationship between the mother and child, the view of which was heavily dictated by Melanie Klein's ideas. Klein's mother and child relationship is based on the constant interaction of the child's feelings of love and hate towards the mother in the early stages of life. At this point, the mother forms the basis and centre of the child's world. The child experiences feelings of love towards the mother when it receives food, gratification and comfort; when these are denied, it experiences feelings of hate. These feelings trigger phantasies in the child that are imagined to have actually occurred, so feelings of hate are accompanied by aggressive, destructive phantasies towards the mother, which are then replaced by loving phantasies of reparation, out of fear and anxiety of having done damage to a loved one. This intermingling of love and hate in the child's relationship with the mother profoundly influences the development of an individual's emotional life from then on, through the aggressive and loving phantasies of the Oedipus complex, and on to relationships with other people and their own children in later life. The careful negotiation of these unconscious phantasies by the child is crucial to the successful formation and development of the self. For Klein, then, the vital, formative relationship between the mother and child is steeped in aggression as much as love, and a particular path through these two extremes needs to be worked out.[11]

In his 1957 essay, 'The Mother's Contribution to Society', D.W. Winnicott made the implications of Klein's work more explicit. He framed the 'long and exacting task' of parenthood as providing 'the only real basis for society, and the only factory for the democratic tendency in a country's social system'.[12] Furthermore, he places the mother at the centre of this task:

> [E]very man or woman who is sane, every man or woman who has the feeling of being a person in the world, and for whom the world means something, every happy person, is in infinite debt to a woman. At the time when as an infant (male or female) this person knew nothing about dependence, there was absolute dependence.[13]

With Winnicott's viewpoints, the connections between psychoanalysis and the post-war welfare state are perhaps a little clearer, as is the manner in which Klein's theories led to an overwhelming focus on the mother–child relationship as the foundation not just for the sociality of individuals, but also for the successful continuation of democratic society. This strand of psychoanalysis does not explicitly address the nuclear threat until the 1960s, though it seems pertinent to draw parallels between the psychoanalytic preoccupation with the family, the development of the welfare state, a move to post-war democratic consensus and the ongoing Cold War. As Barry Curtis has noted, Britain, due to its shifting international status at this point, found itself somewhere between the competition of the East–West superpowers, gradually embracing the family-focused consumer boom (so aggressively celebrated as fundamentally democratic by the USA) while also held back by a post-war austerity that gripped all the European nations directly affected by conflict.[14] In this light, where Cold War anxieties and conflicts were perhaps less openly expressed in Britain than elsewhere (more 'subdued and repressed' in Curtis' terms) but still present, the family serves as a site for Cold War ideals (democracy and consensus) as well as anxieties (breakdown of society, destruction of the self, annihilation through nuclear war).[15] In many ways, the domestic arena is a refuge from the battles of the Cold War, but it is also the locus of the internalization of the many anxieties from which Britons hoped to escape.

If the home was not only the site where the self was formed and the area where the ideological wars of the Cold War period were played out, it was also literally a shelter, the space that British families would use to protect themselves in the event of a nuclear attack. Melissa Smith has highlighted how, after beginning work on constructing and testing their own hydrogen bomb in 1954, a decision was made by the British government to focus on civil defence as a means of deterring attack, rather than planning for survival. Essentially, this meant going against officially commissioned advice from the 1955 Strath Report (which made clear the terrible destructive possibilities of an H-bomb attack in Britain and recommended that large public shelters be built on the outskirts of large towns), and instead issuing pamphlets and general information, 'giving people advice on how to improvise fallout protection in their own houses, using furniture and household items'.[16] At a time when Britain was still recovering from the Second World War, this approach was cheaper, easier to implement and allowed the government to build on many of the methods and messages employed at the height of the war. A 1957 poster, 'The Hydrogen Bomb' (Figure 6.1), demonstrates this advice in action: beneath a large image of a mushroom cloud from Britain's own atomic tests in that year and information on the immediate and long-term effects of an H-bomb attack, there is an advice section entitled 'What You Could Do'. Here, alongside relatively cheerful images of a nuclear family sitting in their sandbagged living room and a housewife whitewashing her windows to keep out the extreme heat of an imminent explosion, the public is told how to prepare for attack in their own homes:

6.1 Detail of 'The Hydrogen Bomb', Civil Defence Poster, 1957, The National Archives: INF 13/281.

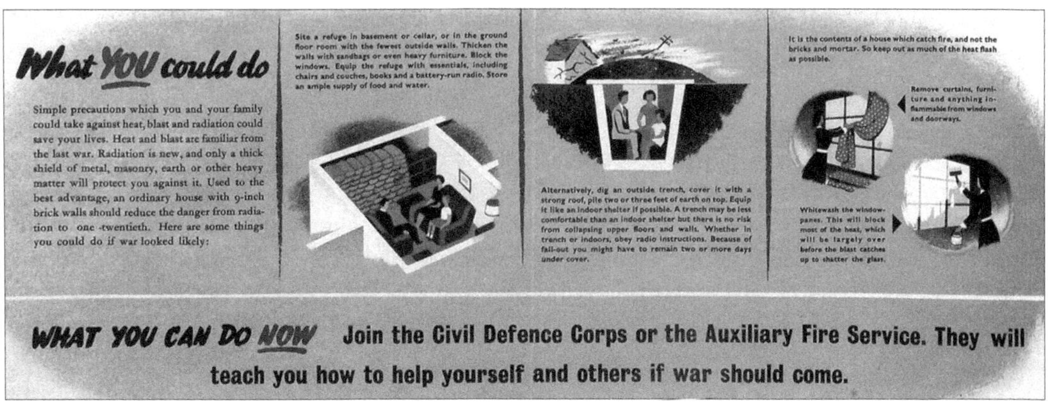

> Site a refuge in a basement or cellar, or in the ground floor room with the fewest outside walls. Thicken the walls with sandbags or even heavy furniture. Block the windows. Equip the refuge with essentials, including chairs and couches, books and a battery run radio. Store an ample supply of food and water.[17]

A pamphlet released in the same year goes into more detail, outlining the specific, horrific effects of a nuclear explosion – the intense heat, the powerful blast and the chilling effects of radiation poisoning. Despite this bleak picture, the home was still posited as the best place of defence: a makeshift fallout room like the one outlined above is suggested again, while advice on decontamination in the home is also given, using (improbably) washing machines, vacuum cleaners and pre-stored water. As far as the wider public was aware then, the home, if properly prepared, could be adapted for protection in the face of nuclear attack; as the pamphlet concludes, 'the household would have to be organised as the basic unit of survival'.[18] It was not only an arena for the complex formation of the self, but it was also a kind of front line – an ideological battleground and an adaptable shelter from possible destruction.

On the surface, the interior concerns of contemporary psychoanalysis and the vast, exterior threats of the Cold War and the H-bomb do not seem to overlap, but it would appear that the home – the space governed, in a complex manner, as much by interiority as external forces and influences – is the space in which these two elements can be seen to converge, particularly as it is represented in the fraught, emotive and often problematic paintings of John Bratby.[19] Returning to *Jean at the Basin*, these dual concerns can be located uncomfortably in one image. On the one hand, there is Bratby's sense of self, temporarily solidified in his reflection in the mirror. However, Jean, at the centre of the canvas, is also a crucial element here, the wife and expectant mother, the pivotal family figure. To turn to Melanie Klein again, her concept of symbol formation plays an important role in placing this image in the context of contemporary psychoanalysis. Symbol formation is the process in early childhood of leaving phantasies of attack and aggression towards the mother's body behind, projecting these phantasies on other objects, and thus coming to engage with the world beyond the parents and beginning to form a better understanding of the self.[20] In the domestic space, one of the primary spaces for the development of the infantile self, it is domestic objects that often assist and play a role in symbol formation.[21] Bratby's alignment of Jean with the domestic fixtures of the bathroom (particularly the sink, something that holds and contains, rather like the body of the visibly pregnant Jean) speaks to Klein's process of symbol formation, something that she considered to continue throughout an individual's life, beyond infancy. Jean's presence alongside the bathroom's domestic objects makes explicit the important status of woman in psychoanalytic interpretations of the development of selfhood – she is both mother and wife, present as Bratby takes in his own reflection, captured in the act of forming his own domestic space. No wonder, then, that Bratby appears defiantly above the scene and Jean bows her head, almost as if a particular battle has been won, at least temporarily. His position as artist/spectator in the doorway of the bathroom adds to this effect, as well as contributing to an atmosphere of containment that, along with the low, sloping ceiling of the bathroom, turns the domestic interior into a closed-off space. The parallels with the home shelter recommended in government advice pamphlets are striking – this windowless, sealed-off bathroom, with a sink for holding vital water, could protect Bratby and his pregnant wife if he stepped inside and closed the door. It is apt, then, that the same space, the home, seems to provide momentary protection here, both from the neuroses and phantasies of the self in the family as understood in psychoanalysis and from the threat of destruction from an H-bomb; but does it? Undertones of potential disintegration remain: disintegration of the self and of the domestic space, both captured unsteadily on canvas, not to mention the threat of domestic violence that is uneasily present in Bratby's art. A temporary balance may well have been found here, but by what means, and for how long? Protected, closed off from the outside, but still fraught with its own conflicts and battles: this is family life quietly imbued with the aims and anxieties of the Cold War.

In 1956, Bratby painted his only series of paintings that explicitly reference the threat of nuclear war in a number of canvases depicting individuals affected by radiation (see Figure 6.2). These images, largely taking himself or

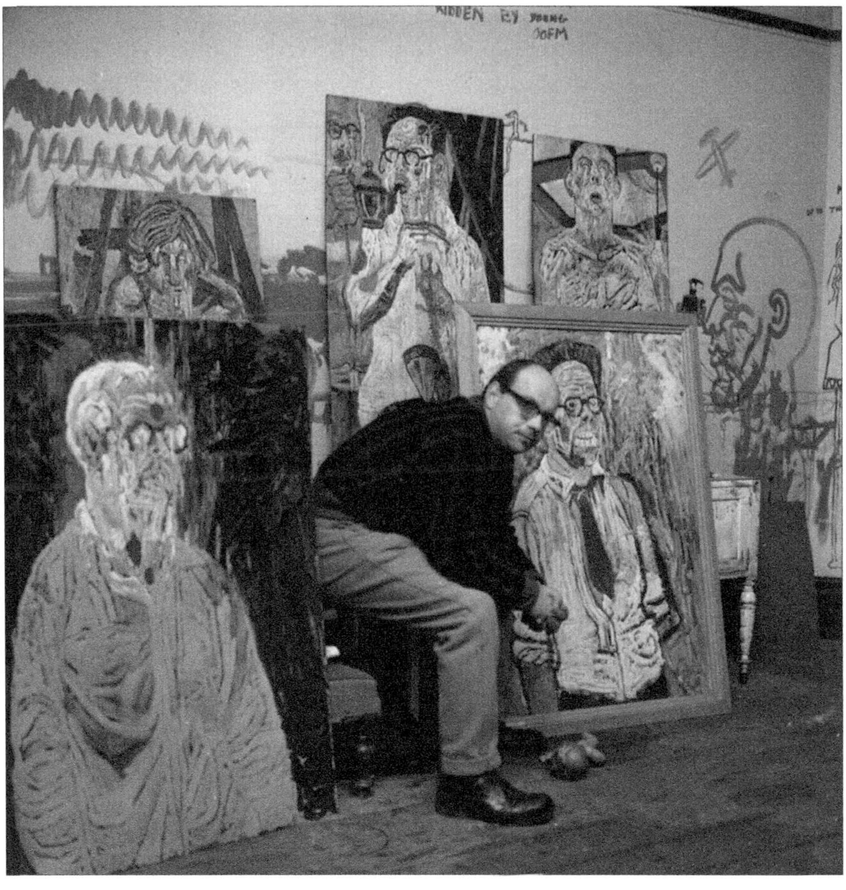

6.2 John Bratby with some of his paintings from an exhibition on the effect of the explosion of a hydrogen bomb on the human figure © Hulton-Deutsch Collection/ CORBIS.

members of his family as their subjects, were conceived as a protest against nuclear armament, and mark a very rare occurrence of Bratby actively engaging in political issues through his art.[22] *Hiroshima Horror: Self Portrait in White Shirt* (1956) – visible to the right of Bratby in the photograph – is part of this series and focuses on Bratby himself (another of his recurring subjects), while imagining the horrific effects of an atomic bomb blast on the body. Bratby's head and torso are visible here, the sleeves of his white shirt rolled to the elbow, as they often were while he painted, and a black tie knotted loosely around his neck. Instead of a close, thickly painted study of the artist's face, the head has been gripped into a blank, ghostly, horrific grimace. The lips of the mouth are parted to reveal the teeth clenched in agony, as if they are barely suppressing a Bacon-esque scream, giving the overall effect of making the head appear skull-like. Bratby allows the luminous white of the shirt to appear again in the creases and undulations of his own skin in his face, and this deathly pallor is offset by fleshy, bloody pinks that are also allowed to bleed under the shirt. These pinks and whites are repeated again in the background of the canvas, where the interior, a permanent feature of Bratby's other paintings, has been bleached out to nothing but this pale, cold backdrop. The overall tonal unity of the work alongside the stark, stiff pose of the figure gives the impression that we are witnessing an artist caught at work as an atomic bomb hits, somewhere between the sudden, blinding flash of light and heat, and the gradual disintegration through radiation poisoning into death.

In effect, *Hiroshima Horror: Self Portrait in White Shirt* makes visible the anxieties that sit uncomfortably beneath the surface of Bratby's paintings of his family in the 1950s. In an interview with Lawrence Thompson, Bratby acknowledged the way that painting in the home allowed him to address his own concerns and anxieties:

> I paint my face a lot, worrying about my problems. It shows in my face, and I'm interested in how everything shows, how changing thoughts show. Living in this house is very reassuring to a person like me. I can lean on it. I work very well in this house.[23]

In bringing annihilation upon himself and his home, the terrifying possibilities that are gestured to in works like *Jean at the Basin* come, momentarily, into concrete reality. These two paintings, initially, seem to be worlds apart from each other. In one sense they are, as while *Jean at the Basin* confines itself to one small, sheltered room, momentarily allaying anxieties with a look, a reflection, then *Hiroshima Horror: Self Portrait in White Shirt* bleaches out the protective domestic space with a flash of nuclear white to come face-to-face with the disintegrating self: everything Bratby has sought to avoid elsewhere. However, the two also feel intimately connected, partly due to this initial distance. *Hiroshima Horror: Self Portrait in White Shirt* merely makes visible all that *Jean at the Basin* tries to capture, hide, protect and control. In a way, it is unsurprising that Bratby chose to directly engage with nuclear threat in his art by imagining the effects of a nuclear blast on himself, his wife and others, as these are the subjects that obsess him, that he paints over and over again. Furthermore, with the home as the focus of psychoanalysis at this time, it is significant that Bratby erases the domestic interior, the family realm and the arena of the self in this series of portraits. In *Hiroshima Horror: Self-Portrait in White Shirt*, we see them disintegrate right before our eyes and, in the process, see the imagined and real threats to family life in post-war Britain given unsentimental, unflinching form.

In the Kitchen, Sheltered: The Home

Once the home is established as the area where psychoanalysis and the nuclear threat overlap, Bratby's other works of this period can be considered in a similar light. Not only do these paintings trace the events and people that shaped his domestic life but they also give constant insights into the motivations, anxieties and relationships that drove family life at this time. Bratby's tabletop still life paintings take on the objects of day-to-day family life in this dual way, revelling in the banal nature of their subject matter while also gesturing to outside threats to ordinary life. In *Still Life with Chip Frier* (1954; Figure 6.3), the contents of what looks to be an entire set of kitchen cupboards have been arranged on the kitchen table. The table juts across the canvas at a diagonal in order to display a tightly arranged sprawl of domestic foodstuffs, apparatus and clutter. There are food and condiments recognizable by their branded labels, such as Kellogg's cornflakes, Lea & Perrins Worcestershire sauce, Saxa table salt and Colman's mustard, alongside other, more anonymous bottles and jars. There are items for the preparation of food: a whisk, a can opener, cutlery, a colander, an orange juicer, a roasting dish, a tall metal kettle and the chip frier of the title, hung on the back of one of the chairs pulled into the table. There are even other domestic objects less obviously associated with the kitchen, such as a vase, paperclips or nail scissors, squeezed into the few gaps on the table's surface.

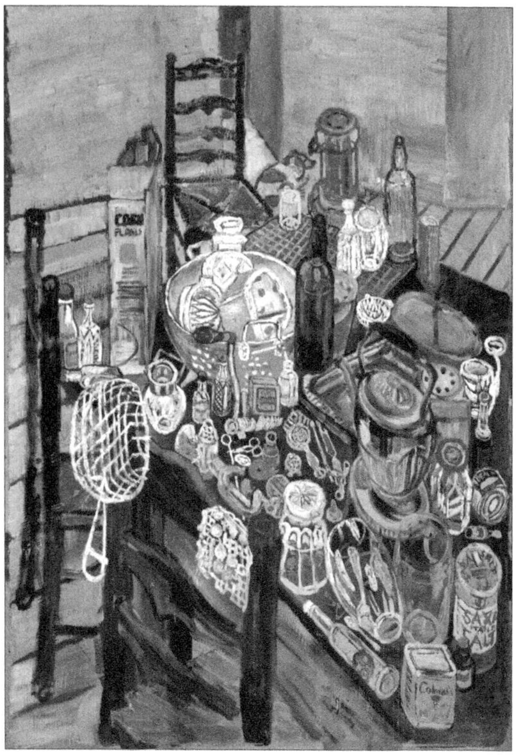

6.3 John Bratby, *Still Life with Chip Frier*, 1954, oil on board, 131.4 × 92.1 cm, Tate, London © The Estate of John Bratby. All rights reserved 2010/Bridgeman Art Library.

The sheer abundance of objects in this finely tuned chaos often caught the eyes of critics. In his early review of Bratby, David Sylvester commented on the helplessness of the image, stating that: 'the painter seems to be saying, "There's so much in the world, one can't hope to stop and organise it: there's only time to fix it, to trap the whole mess as it is".'[24] More recently, Martin Harrison has considered Bratby's tabletop still lives as 'the honest outpouring of a "hungry", ambitious artist ... a celebration of the immediate onset of a more materially affluent society', and it is worth noting that this particular painting emerged in the same year that rationing in Britain was finally coming to a complete end.[25] At the time, however, Robert Melville viewed these images less as celebratory and more as leaving 'an indelible impression of a family living entirely on cereals, salt and beef-extract'.[26] This observation raises the question of whether Bratby's overflowing tabletop might also be considered in connection with anxieties about nuclear threat and the family. While *Jean at the Basin* has an enclosed, shelter-like quality, here the at once frantic and careful recording of domestic items reads like a catalogue for survival, a pooling together of all the supplies in the house on the kitchen table, the communal centre of the home and under which, in previous wars at least, families had created makeshift bomb shelters. As it is here, the home, never mind the kitchen table, would not save you from the direct hit of a nuclear bomb, nor from the effects of radiation, but the sheer range of objects that Bratby brings together here anticipates lists of supplies that government pamphlets advised householders to hoard, as the threat of the H-bomb loomed into the 1960s.[27] The abundance of objects here also speaks to a number of experiments – conducted in the US and the UK in the 1950s and reported in newspapers – on how successfully families and groups of people could survive in fallout shelters. In 1959, for instance, the Powners, a US family from New Jersey, entered a specially created shelter at Princeton University, emerging after two weeks to inform the press that they had managed to endure the solitary confinement with 'the aid of tranquilizer pills for the children, a bottle of whiskey for themselves, and a well-stocked library that included a copy of the unexpurgated "Lady Chatterley's Lover"'. Mrs Powner recommended the shelters ('These shelters are practical, and I think every one should build one in his home'), and a doctor, who had been listening to the family over the fortnight with the aid of a hidden microphone, concluded that 'they have demonstrated people can be isolated in shelters for two weeks with no hazard to their health and well-being'.[28] These experiments continued into the 1960s – in 1965, a group of British women had a slightly less happy two-day experience in a shelter in York, complaining of lethargy and causing a newspaper reporter to note that 'after the first 24 hours even that basic feminine impulse to make frequent cups of tea deserted them'.[29] Clearly, there were concerns that the fallout shelter should allow the smooth continuation of familial, domestic life as far as possible, and Bratby's obsessive display of objects reflects this strange combination of survival, readiness and banal, everyday detail.

Aside from this house-as-fallout-shelter reading, *Still Life with Chip Frier* also references threats from inside the home, gathering together an inventory of domestic objects as a display of provisions, though essentially ambiguous as to who has provided them, and who they are being provided for. In

Melanie Klein's work, she argues that the child's dependence on the mother in infancy has to be carefully negotiated as the child matures; the breast and its product 'come to stand in his mind for love, pleasure and security' and has to be replaced with other food.[30] This transference of loving feelings to other objects in tandem with the successful development of independence (alongside an acceptance of previous dependence on the mother) is crucial for the formation of the individual self. If this does not happen correctly, it can lead to fear of dependence or its opposite, overdependence and greed.[31] The conflicting possibilities of dependence can be recognized as concerns in *Still Life with Chip Frier* and the figure of Jean haunts this work, despite her not appearing in it. A lack of dependence could be argued in the eager display of domestic objects, as if Bratby is showing what he has been able to provide for himself and his family, brand names and all. However, this ignores the way that the objects on the table, particularly the transparent, sleek vases, the empty glass jars and bottles, the delicate scissors and doily, or the lean, elegant kettle, have a feminine quality. They are either vessels to be filled or items for making or decorating and many of them look empty and a little tired, as if they have been over-used or used up completely. Though not pictured, the female figure of the household feels present here in the objects themselves. In Bratby's hands, the still life also becomes a reflection on dependence, on who provides in the home and how the self is both reinforced and disrupted by this abundance of goods in the domestic interior.

In the Womb and at War: Jean, the Wife and the Mother

A number of Bratby's other works also navigate the relationship between a masculine sense of self and the aspects that threaten it, whether the feminine figure in the home or the nuclear threat outside of it. While it is tempting to seek out reflections of these threats in Bratby's paintings, there appears to be something much more complicated going on: internal and external threats intermingle and overlap, helping to create images that are at once familiar and strange. If, in *Still Life with Chip Frier*, survival and dependence merge in one image, while also being drawn apart and undermined by anxiety, greed and fear, a painting such as *Jean and Still Life in front of a Window* (1954; Colour plate 11) explores these themes in a perhaps more explicit and interconnected way. Here, Jean sits at the head of the Bratbys' kitchen table, nude. Her expression is stiff and humourless – she gazes out at us with wide, unflinching eyes and looks up slightly. It is not hard to imagine her eyes meeting Bratby standing at the other end of the table, where we, as viewers, also now stand. The table partially obscures her, though the top half of her torso and the right hand side of her body – breast, stomach, arm and leg – are visible, while her skin is picked out in pinks, raw reds and sickly yellows. In front of her, the kitchen table is once again crammed with objects: cereal boxes, sugar, table salt, jam and mustard, as well as silverware, pint glasses, vases and bowls, all arranged on top of a chequered tablecloth. It seems no accident that the table almost appears to unfurl with a steep perspective away from Jean's half-concealed torso, so that her breast and just-visible stomach or womb are visually equated with foodstuffs and kitchen equipment. This is an image of maternal roles conflated on canvas in that sense: the maternal body on display with one rounded breast emphasized, a clear symbol of fecundity and the possibility of children, alongside a table bursting with food, an abundance of things tumbling towards you, colourful and shining. Bratby has painted everything that sustains him and his sense of self in the domestic sphere.

However, *Jean and Still Life in front of a Window* hardly feels like a celebration of family life and its potential, as it combines the ordinary with a number of unsettling elements that are difficult to ignore. Jean looks visibly uncomfortable, perched on a wooden chair, her cheeks and nose red and her steady gaze betraying weariness as well as a sense of fear and almost anger that feels disconcerting. This painting was awarded the Guggenheim Prize in 1956, and Bratby composed a public statement called 'How I Painted the Picture that Won the Guggenheim Prize' that purported to explain the process behind the work's production and Jean's important, demanding role:

> Jean, my wife, was that morning at the Royal College of Art, so I waited until she got in. I fixed a huge fire, got in some beer which I couldn't afford, to sustain her in her posing ordeal to come, and I resolved to take her to the pictures when she had finished posing … Jean came in, grimaced when I told her what she had to do, and prepared to pose … I had started that dinner time, both of us without any dinner, because my inspiration might go if I didn't exploit it immediately. Afternoon went through and the evening passed as I worked and worked. We snatched a hasty meal and worked in the late evening until past 12pm … A horrible ache in my stomach, and a dry, searing feeling in my stomach tormented me, as I worked in that still lighted room. Jean understood and suppressed her weariness for the sake of the work. We went for a quick walk at 2am, got back about an hour later and worked until 9 when the house was woken into life.[32]

Bratby's statement frames the process as physically demanding and straining, though largely for him. Furthermore, not only does Bratby sacrifice comfort for the sake of his art, but Jean, and the normal routine of daily domestic life, stops for him too. These elements of control in his account are echoed in Bratby's painted representation: Jean is crammed crudely and tightly alongside the kitchen objects, with the table seeming to jut into her body as if she cannot escape, while the blind of the window behind her has not been lowered, seemingly displaying her nakedness to neighbours and passers-by. While this image aligns the potential maternal figure with foodstuffs and utensils so that they all become objects related to Bratby's selfhood in the home, there are also undertones of violence: Jean's gaze betraying tiredness, with the table brutally pushed up against her body.

The strange duality of Jean as potential subject of attack as well as potential mother finds parallels elsewhere in 1950s British culture, particularly in John Osborne's 1956 play *Look Back in Anger*. The play is built around the relationship between Jimmy and Alison, a married couple living together in their Midlands flat with their lodger, Cliff. Critics have often framed Jimmy and Alison's relationship in terms of class difference, but gender difference is also a constant theme in the play, often expressed by Jimmy himself. His views on Alison, and women in general, veer from fear and scorn to complete dependency. So, early on in the play, after Alison has likened him to a child, Jimmy embarks on a rant against women where he accuses them of being violent and destructive. To Jimmy, Alison jumps on the bed 'as if she were stamping on someone's face' and draws the curtain in a 'casually destructive way … like someone launching a battleship'. At their dressing table, meanwhile, women sit 'dropping their weapons and banging down their bits of boxes and brushes and lipstick' so that 'when you see a woman in front of her bedroom mirror, you realise what a refined sort of a butcher she is'.[33] Immediately after this rant, however, Jimmy starts playing and wrestling with Cliff, knocking Alison over and scalding her with her iron in the process. She sends Jimmy out like a child while Cliff tends to her.[34] Following a brief sulk, Jimmy apologizes and the two make up, retreating into a childish game of squirrel and bear ('You're very beautiful. A beautiful, great-eyed squirrel', 'Well, you're a jolly super bear, too').[35]

If Jimmy and Alison's relationship veers between violence and childishness, threat and dependence, then there are also moments when these two aspects overlap. At the end of act one, Jimmy (unaware that Alison is pregnant at this point) reacts to the news that her friend Helena is coming to stay with them with a particularly violent outburst. He wishes out loud for her to have a baby and for it to die, to 'wake you out of your beauty sleep', then likens himself to a baby in Alison's womb, as a victim of her sexual passion:

> She has the passion of a python. She just devours me whole every time, as if I were some over-large rabbit.
>
> That's me. That bulge around her navel – if you're wondering what it is – it's me. Me, buried alive down there, and going mad, smothered in that peaceful looking coil. Not a sound, not a flicker from her – she doesn't even rumble a little … She'll go on sleeping and devouring until there's nothing left of me.[36]

Jimmy's equation of himself with a child buried in Alison's womb is the point at which the various aggressive and childish strands of their relationship come together in one disturbing image of the husband as 'smothered' infant in the mother's womb. This complex view of Jimmy and Alison's relationship speaks to the similarly multi-layered *Jean and Still Life in front of a Window*. Here, Jean becomes both a figure to be attacked from the threatening viewpoint above her, with the table pushed in to contain her, but also, perhaps, a figure to be feared. If, for Jimmy, Alison is potentially able to consume a man, Jean becomes, for Bratby, not just an object of consumption herself alongside the other domestic objects, but also a figure that consumes. In placing her at the head of the table, she looks insatiable, capable of devouring not just everything in front of her, but also, in the blank look of tiredness and menace in her eyes, Bratby himself. The glimpse of her breast and rounded belly hints at the very fears that Jimmy expresses in *Look Back in Anger*. Roles of mother, wife and lover, as well as father and child, are allowed to overlap here, with their underlying aggression, anxiety and dependence prominently displayed.[37]

At the centre of this fluid yet clearly problematic interpretation of familial roles are two figures – the mother and the masculine self. Klein's theories of the early love–hate relationship between mother and child, and the concepts of symbol formation and reparation that emerge from this, have shown the connections and interactions between these two figures. In a later essay, 'On the Theory of Anxiety and Guilt', she clarifies that anxiety (so vital to the individual's development) is rooted in the death instinct. Anxiety about ultimate annihilation, for Klein, lies at the heart of the mother–child relationship, as it is a vital drive towards reparation and symbol formation after unconscious aggressive drives towards the mother's body, and continues to be present in all individuals throughout life.[38] It becomes clearer, then, how such contradictory images of a female figure, particularly in relation to motherhood, can arise in *Look Back in Anger* and *Jean and Still Life in front of a Window*. The female is attacked, feared and loved at once, vital to the development of a sense of self through unconscious struggle and anxiety. Thus, in Bratby's painting, we are made to loom over Jean, to threaten her, just as she is made to return our gaze with her own ambiguous glance, somewhere between defiance, tiredness and fear, as if the battle is ongoing. Her fecundity and the possibility that her body can contain something are emphasized in her depiction: just as she can contain and love a child (or a husband), she is also capable of consuming and devouring.

This idea of the mother as a vital and loving figure, while also someone to fear, is repeatedly addressed by D.W. Winnicott. In his view, knowledge that we were all once dependent on a mother can lead to 'a vague fear of dependence', sometimes taking the form of 'a fear of WOMAN, or a fear of a woman' driven by a fear of domination.[39] Seeking out danger helps to combat this fear of dependence and reach maturity, and men engaging in risk-taking, specifically going to war, is one way of doing this. Women are also able to engage in risk-taking, as Winnicott equates men undertaking war with women undertaking pregnancy and childbirth. However, with the development of the atomic bomb, the opportunity for men to fight in a war appeared to have disappeared, with individuals essentially powerless in the face of nuclear war as the threat of destruction hung over everyone equally. So, 'with no more wars, men find themselves high and dry', while women continue in their risk-taking roles.[40] Later, in 1963, Winnicott would suggest that one of the effects of the nuclear bomb on the psyche was to create a permanent state of adolescence. By denying men the chance to seek maturity in war, 'the atom bomb affects the relationship between adult society and the adolescent tide which seems to be for ever coming in'.[41] Winnicott's picture of men as having their possibilities of independence compromised by the atomic bomb finds a voice in one of Jimmy Porter's most famous speeches in *Look Back in Anger*:

> There aren't any good, brave causes left. If the big bang does come, and we all get killed off, it won't be in aid of the old-fashioned grand design. It'll just be for the Brave New-nothing-very-much-thank-you. About as pointless and inglorious as stepping in front of a bus. No, there's nothing left for it, me boy, but to let yourself be butchered by the women.[42]

No wonder then that Osborne's character of Jimmy and Bratby's image of his wife in *Jean and Still Life in front of a Window* both evoke images of women that are contradictory and ultimately intensely problematic. In the post-war British family, the woman, and particularly the mother, had become a central figure in the formation of individuals within the family, while also being the subject of unconscious attack within that realm. As a result, and filtered through interpretations of Melanie Klein's theories on symbol formation and anxiety, an individual (particularly a male individual) had to negotiate a state of independence while also recognizing their initial and in many ways recurrent dependence on women. Alongside this, the family became the location of defence against the nuclear threat – a makeshift shelter, a democratic ideal – just as it was becoming an arena of attack and defence between individuals. This contradictory view of the home – as defence against new external threats, while also riddled with internal ones – brings about a situation that actually seems to undermine psychoanalytic definitions of how an individual can come to be formed inside and outside the family, while continuing to posit it as the only legitimate possibility. As a result, these male figures make explicit what had previously remained in the unconscious of the family. Jimmy Porter's rants and threats against his wife and women in general resonate as attacks and forms of defence against something much larger, while Bratby's images of Jean – brimming with a barely suppressed violence, while also retaining elements of affection and need – emerge from a similar uncertainty.

Is it possible, then, that the post-war nuclear threat in the 1950s intensely undermined the formation and definition of the masculine self and, by extension, the so-called nuclear family? Are Bratby's paintings affected both by the intense psychoanalytic focus on the family in the 1950s and the overwhelming anxiety that emerged with the atomic bomb? While the widely popularized psychoanalytic theories of Melanie Klein made the home into a site where the individual self is formed through a serious of unconscious battles, D.W. Winnicott explicitly made links between the mother, the self and the threat of nuclear war, implying that the exterior threats to the home might lead to the family unit's eventual breakdown. Thus, Bratby's images such as *Jean at the Basin* and *Still Life with Chip Frier* simultaneously evoke the closed-off family home and fallout shelter, while also being built primarily around one male artist's struggle for the definition of a sense of self. Furthermore, *Jean and Still Life in front of a Window* makes the contradictions of family life and selfhood under nuclear threat (gestured to in Bratby's obsession with self-portraiture) reside in the body of the mother – threatened and threatening, loved and feared, containing goodness while also possibly containing something terrifying, something on which to depend while also something from which to escape. Perhaps most horrifically amongst these paintings is Bratby's *Hiroshima Horror: Self Portrait with White Shirt* as part of a series where the nuclear threat is made real, the domestic arena obliterated and the battle for the self lost. In all of these paintings, the possibility of nuclear war permeates the self-contained domestic space, making shelters out of living spaces, turning relationships into life or death battles, and making the possibility of the loss of self akin to ultimate destruction.

Acknowledgement

I would like to thank the Paul Mellon Centre for Studies in British Art for assistance with the costs of reproducing the images for this essay.

Notes

1 The circumstances surrounding the Bratbys' marriage are recounted in Maurice Yacowar, *The Great Bratby: A Portrait of John Bratby RA* (London: Middlesex University Press, 2008), 21–24.

2 Yacowar also references the violent nature of the Bratbys' marriage in his study (*The Great Bratby*, 24–26). Bratby's novels, written from the late 1950s onwards and published throughout the 1960s, are largely autobiographical, despite his claims to the contrary (probably out of fear of law suits from people he had represented in a less than flattering light). For example, in *Breakdown*, the Bratby-esque main character, James Brady, is tempted to stray from his wife by an unstable female art teacher, Esmerelda: 'With a somewhat mad, or shall we say unbalanced, female threatening him, it is surely more likely that he would go on breaking up, going from degradation to depravity', see John Bratby, *Breakdown* (London: Hutchinson, 1960), 80. He seeks to apologize or explain his behaviour in these novels. The obituaries published after Jean's death also go some way to confirming this view of their marriage.

3 It has been shown that men's accounts of domestic violence are inherently connected to patriarchal assumptions and are often built on narratives of 'inner violence, social ownership, confession'. It is perhaps productive to think about Bratby's paintings more broadly as narratives of domestic violence in this way. See Jeff Hearn, *The Violences of Men: How Men Talk About and How Agencies Respond to Men's Violence to Women* (London: Sage, 1998), 144–145.

4 Quoted in Yacowar, *The Great Bratby*, 139.

5 David Piper, 'The New Realism', *Architectural Design* (April 1957): 135–137.

6 See Michel Foucault, 'Las Meninas', in *The Order of Things* (London: Routledge, 2001), 3–18.

7 John Berger, 'John Bratby', *New Statesman and Nation*, 25 September 1954, 358.

8 Stephanie Spencer, *Gender, Work and Education in Britain in the 1950s* (Basingstoke: Palgrave Macmillan, 2005), 30.

9 See Spencer's chapter on the gendered implications of the welfare state, in *Gender, Work and Education in Britain in the 1950s*, 22–47. Also, Elizabeth Wilson, *Only Halfway to Paradise: Women in Post-War Britain 1945–1968* (London: Tavistock, 1980); and Denise Riley, '"The Free Mothers": Pronatalism and Working Women in Industry at the End of the Last War in Britain', *History Workshop Journal*, vol. 11, no. 1 (1981): 59–118.

10 See Juliet Mitchell, *Psychoanalysis and Feminism* (London: Allen Lane, 1974), 231.

11 Melanie Klein and Joan Riviere, *Love, Hate and Reparation* (New York: W.W. Norton and Company Inc./The Norton Library, 1964), 57–67 in particular. Also, Melanie Klein, 'Early Stages of the Oedipus Conflict', in *Love, Guilt and Reparation and Other Works, 1921–45* (London: Hogarth Press and the Institute of Psychoanalysis, 1981), 186–198.

12 D.W. Winnicott, 'The Mother's Contribution to Society', in *Home Is Where We Start From: Essays by a Psychoanalyst*, comp. and ed. Claire Winnicott, Ray Shepherd and Madeleine Davis (Middlesex: Penguin, 1986), 124.

13 Winnicott, 'The Mother's Contribution to Society', 125.

14 Barry Curtis, 'War Games: Cold War Britain in Film and Fiction', in *Cold War Modern: Design 1945–1970*, ed. David Crowley and Jane Pavitt (London: V&A Publishing, 2008), 122. Though Britain found itself less immediately affected by the Cold War, the climate of tension and fear was still felt, as Robert Hewison has noted – see Robert Hewison, *In Anger: Culture in the Cold War 1945–60* (London: Weidenfeld and Nicolson, 1981), 21–22.

15 Curtis, 'War Games', 122. Early on in the catalogue, David Crowley and Jane Pavitt frame the domestic sphere as a political arena and symbol of US freedom in the Cold War (12). Beatriz Colomina has also considered the domestic sphere as both shaped by the Second World War and affected by the debates and battles of the Cold War – see Beatriz Colomina, *Domesticity at War* (Barcelona: Actar, 2006).

16 Melissa Smith, 'Architects of Armageddon: The Home Office Scientific Advisers' Branch and Civil Defence in Britain, 1945–68', *British Journal for the History of Science*, vol. 43, no. 2 (June 2010): 159. See also Melissa Smith, '"What to Do If It Happens": Planners, Pamphlets and Propaganda in the Age of the H-Bomb', *Endeavour*, vol. 33, no. 2 (June 2009): 60–64.

17 'The Hydrogen Bomb', Civil Defence Poster, 1957, INF 13/281, The National Archives, UK.

18 'The Hydrogen Bomb', Civil Defence Pamphlet (London: Her Majesty's Stationery Office, 1957).

19 The idea of the family as being negotiated and defined by interior and exterior forces concerned with the policing of sexuality is put forward by Michel Foucault and Jacques Donzelot – see Michel Foucault, *The History of Sexuality I: The Will to Knowledge* (London: Penguin, 1998), 24–26 in particular; and Jacques Donzelot, *The Policing of Families*, trans. Robert Hurley (London: The Johns Hopkins University Press, 1997), 3–6.

20 Melanie Klein, 'The Importance of Symbol Formation in the Development of the Ego', in *Love, Guilt and Reparation and Other Works*, 219–228.

21 This is illustrated by Klein's explanation of her analysis of Dick, an autistic child, in the same essay. See Klein, 'The Importance of Symbol Formation in the Development of the Ego', 217–218.

22 Yacowar, *The Great Bratby*, 110–111. Yacowar includes an image of Bratby alongside several of these canvases, and recounts how the artist also contributed to an anti-nuclear exhibition at Bromley Public Library in 1960, and signed Henry Moore's petition to restore China's seat at the UN in February 1961. Despite this, Yacowar also records Bratby's slightly ambivalent stance: 'When I have been asked to help, and oblige, it doesn't mean to say I agree. You see I'm a co-operative chap … I would not throw away my life for the cause although I do not disagree with their policy' (*The Great Bratby*, 111).

23 This quote comes from the unpublished interview with Bratby by Lawrence Thompson, 'Portrait of the Artist as a Displaced Person', found in a selection of the artist's papers shown to me at Julian Hartnoll Gallery, London.

24 David Sylvester, 'The Kitchen Sink', *Encounter* (December 1954): 63.

25 Martin Harrison, *Transition: The London Art Scene in the Fifties* (London: Merrell, 2002), 88.

26 Robert Melville, 'Exhibitions: Paintings and Sculpture', *Architectural Review*, vol. 116, no. 696 (December 1954): 400.

27 A list of items for equipping a fallout room for at least a week is given in a 1963 Civil Defence Handbook and included food, water, bedding, a table and chairs, plates, cutlery and utensils, a stove, a radio set, a clock, books and magazines, a first aid kit and a box of personal papers. See *Advising the Householder on Protection Against Nuclear Attack* (London: Her Majesty's Stationery Office, 1963), 11.

28 This article, '5 Live 14 Days in 8x9 Ft. Shelter', taken from an August 1959 edition of *The New York Herald Tribune*, was included in a Home Office file for research into the effects of living in fallout shelters for long periods of time ('Shelter and protected accommodation: Living in fall-out shelters experiment', HO 338/104, The National Archives, UK).

29 The results of this York experiment were published at the time – see 'Lethargy Troubled 3 Women Shut in Fall-Out Refuge', *The Times*, 23 March 1965. It was also included in the Home Office file for research into the effects of living in fallout shelters for long periods of time ('Shelter and protected accommodation: Living in fall-out shelters experiment').

30 Klein and Riviere, *Love, Hate and Reparation*, 90.

31 Ibid., 84–86.

32 John Bratby, 'How I Painted the Picture that Won the Guggenheim Award', *Art News and Review*, 27 October 1956, 9.

33 John Osborne, *Look Back in Anger* (London: Faber, 1975), 24.

34 Ibid., 26–34.

35 Ibid., 33–34.

36 Ibid., 37–38.

37 Bratby consistently depicts Jean as someone to be feared, controlled and overcome: *Jean in Bed* (1954) and *Jean on the Sofa with a Teddy* (1954) – illustrated in Yacowar, *The Great Bratby*, 24–25 – are two powerful, disturbing examples that I was unable to reproduce here.

38 Melanie Klein, 'On the Theory of Anxiety and Guilt', in *Envy and Gratitude and Other Works, 1946–1963* (London: Hogarth Press and the Institute of Psychoanalysis, 1987), 28–29.

39 Winnicott, 'The Mother's Contribution To Society', 125.

40 D.W. Winnicott, 'This Feminism', in *Home Is Where We Start From*, 193.

41 D.W. Winnicott, *Deprivation and Delinquency*, ed. Claire Winnicott, Ray Shepherd and Madeleine Davis (London: Tavistock, 1984), 150.

42 Osborne, *Look Back in Anger*, 84–85.

7

Covert Resistance: Prunella Clough's Cold War 'Urbscapes'

Catherine Spencer

Reviewing Prunella Clough's first retrospective at the Whitechapel Art Gallery in 1960, the critic David Sylvester deemed her most recent paintings to be 'private and abstract', epigrammatic telegrams hinting at 'what goes on behind an intensely private face in public places'.[1] This visual and conceptual reticence has become a critical commonplace in discussions of Clough's work, frequently correlated with the artist's personal and professional reserve.[2] It has been linked to Clough's relative neglect in histories of post-1945 art in Britain, and more dubiously invoked to support readings that frame the apparent modesty of her work as a stereotypically 'feminine' trait.[3] While Clough's works are undeniably hermetic, this quality demands careful handling to ensure that it does not write over, or write out, the wider socio-cultural implications of her paintings. This chapter proposes that the secretive effect generated by Clough's sophisticated fusion of realist ciphers with ambiguous abstraction during the late 1950s and 1960s can be approached as a highly nuanced response to the uncertainties of the Cold War context. As she shifted from realist studies of dockworkers and factory labourers to increasingly abstracted canvases during this period, Clough developed a supple visual vocabulary that could concatenate multiple references to industrial, scientific and military processes, resulting in deliberately elusive but charged images. Focusing on three central components of this vocabulary – the map, the aerial view and the interlinked forms of cell and atom – this chapter argues for a reappraisal of Clough's work as intimately connected to the psychogeography of Cold War Britain.

Despite her sustained engagement with abstraction, as late as 1982 Clough insisted: 'I see my subject matter mainly as landscape.'[4] Elsewhere, in an apt neologism alighted on by the art historian Margaret Garlake, Clough defined the object of her painterly attention as the 'urbscape'.[5] This coinage beautifully captures Clough's interest in the expanding urbanized fabric of post-war Britain, with its grey macadamized loops of motorways, airstrips, power plants and factories. The paintings Clough executed during the late 1950s and throughout the 1960s manifest this concern, notably *Electrical Installation* (1959; Figure 7.5), *Bypass I* (1960; Colour plate 12), *Electrical Landscape* (1960; Colour plate 13), *Brown Wall* (1964; Figure 7.4) and *Untitled* (1965; Colour plate 14). Although a lifelong Londoner, Clough travelled extensively throughout Britain's industrial and urban zones. On these trips, often accompanied by the sculptor Ghisha Koenig, Clough recorded her impressions using a notebook and camera.[6] Clough also maintained a house in Southwold on the east coast for many years, making repeated trips there and to Lowestoft. While Clough has rightly been recognized for her unique portrayal of the industrial and increasingly post-industrial environment, the urbscape that fascinated her was also marked by the military, scientific and technological concerns of the Cold War.

Although much of this urbscape's infrastructure and its wealth of pre-abstracted forms – canals, shale pits and mining areas – had a long industrial history, other parts were a legacy of the more recent industrialization spurred on by the Second World War, which were assimilated into Britain's Cold War framework as it developed its own hydrogen bomb during the 1950s (see Figure 7.1).[7] The USA located its early warning system against attack from the USSR in the East Anglian landscape that Clough crossed on her Southwold visits, while the nuclear reactor at Sizewell, constructed between 1961 and 1967, sat 15 miles down the coast from the town.[8] The geographer W.G. Hoskins balefully charted such industrialized, militarized scenery in his 1955 book, *The Making of the English Landscape*: 'airfields have flayed it bare wherever there are level, well-drained stretches of land … Over them drones, day after day, the obscene shape of the atom-bomber, laying a trail

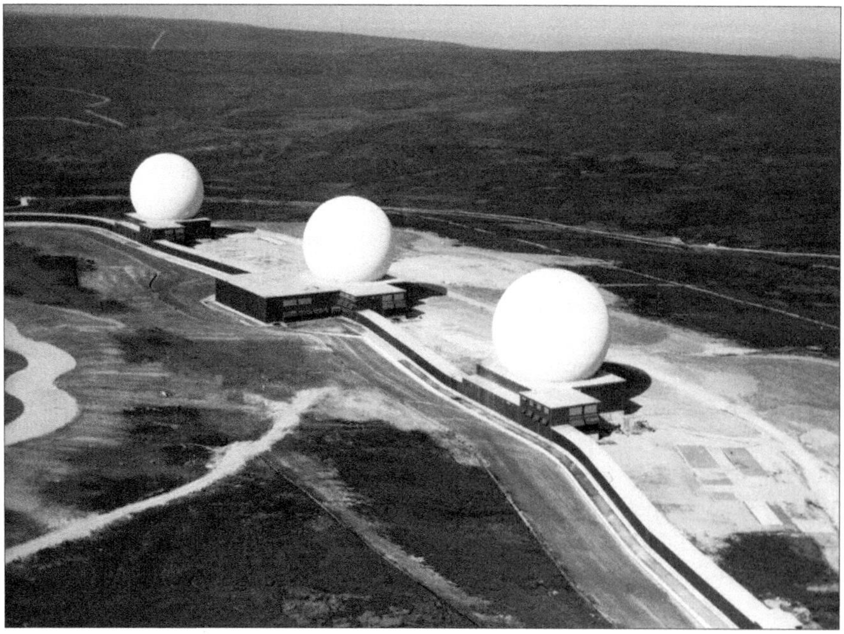

7.1 Unknown photographer, aerial view of RAF Fylingdales Early Warning Radar Station, 16 September 1963, black and white photograph © Imperial War Museums (HU 69120).

like a filthy slug upon Constable's and Gainsborough's sky.'[9] For Hoskins, this was 'England of the Nissen hut, the "pre-fab", and the electric fence, of the high barbed wire around some unmentionable devilment; England of the arterial by-pass ... of the bombing range wherever once there was silence'.[10] Paintings like *Bypass I* and *Electrical Landscape* depict a similar terrain in which the supposedly binary poles of city and country have collapsed, supplanted by a network of urban, industrial and military conurbations sectioned up into discrete but intersecting areas.

The indeterminacy permeating Clough's urbscapes corresponds with Britain's ambivalent political position in the Cold War. Overshadowed and rendered effectively obsolete by the overarching stand-off between the USA and USSR superpowers, Barry Curtis observes how 'in Britain, the gladiatorial confrontation of West and East was present in the popular imagination as remote events at frontiers and flashpoints, coinciding with a withdrawal from Empire and a sustained remoteness from Europe ... The signs and scenarios of Cold War conflict were relatively subdued and repressed.'[11] While the reticence of Clough's canvases can be correlated with this atmosphere of repression, their engagement with industrial and scientific imagery also registers the uncertain potentialities for augmentation and destruction posed by technological developments.[12] However, in contrast to Hoskins' mournful manipulation of a nostalgic nationalist pastoral as a foil to nuclear modernity, signalled by the term 'England' and his apocalyptic visions of 'high explosive falling upon the prehistoric monuments of Dartmoor', Clough's paintings celebrate the urbscape.[13] Although communicating the uneasiness instilled by the post-war landscape's electric fences, barbed wire, radiation and pollution, they also invest it with moments of redemptive poetry and beauty.

The disorientating effect of Clough's canvases, engendered by her use and subversion of cartographic, aerial and microscopic viewpoints, becomes a form of covert resistance, countering the crude dividing lines of nationalist, imperialist and technocratic geographies. Instead, Clough's paintings prioritize the overlooked, the doubtful and the hybrid, positioning these states as central to experience. Through close readings of *Bypass I* and *Electrical Landscape* from Clough's 1960 Whitechapel retrospective – a show that marked Clough's move away from realism – this chapter begins by considering the allusions to cartography in her work. It demonstrates how the map allowed Clough both to convey the instability of the Cold War landscape and to negotiate politicized debates around realism and abstraction. The following section links the map with the aerial view evoked by several of her canvases, situating this in relation to surveillance technologies and the threat of attack from the air. Throughout, the chapter roots Clough's sensitivity to bordered, contained landscapes in her formative experience of conflict during the Second World War. Finally, it addresses the ramifications of the map and the aerial view for later works from the mid-1960s, finding in Clough's engagement with physical and metaphorical boundaries a questioning of technological developments.

Battle Lines

Throughout her career Clough kept densely detailed visual notes: she also wrote poetry.[14] One draft of a poem, entitled 'Prospect', opens:

We whose days are traced by maps – rivers and railways
Linking in the long afternoons, symbols for industry
And oil wells[15]

Clough's paintings have frequently drawn comparison with the planar layout of the map or diagram. As early as 1953, before Clough's work adopted the flatbed arrangement of her abstract canvases, the critic John Berger playfully imagined that 'her sketches must be blue-prints', attributing the final paintings with 'the tension and the integrity of purpose of working models'.[16] The visual affinity between Clough's paintings and maps can be linked to her work during the latter part of the Second World War as a cartographer for the Military Railways division of ETOUSA (European Theater of Operations, United States Army).[17] Although the pressures of war necessitated this work, 'Prospect' indicates how the map infiltrated Clough's thinking, shaping her understanding of how the relationship between individual and environment might be mediated. The contours and symbols that constitute the map provided an organizing structure that gestured both towards experientially verifiable topography and to a more metaphorical, semiotic mode of communication, blurring representation and abstraction. As Cold War tensions replaced those of the Second World War, the map's military associations, ranging from its role in defining boundaries and borders to establishing targets for destruction, entailed that it formed a powerful means of channelling nuclear doubt.[18]

The influence of cartography underpins *Bypass I*'s composition from a series of rounded, creamy-grey tessellated forms ringed with contour-like fine lines (see Colour plate 12). Other, stronger lines cut like roads, rivers or railways – Clough's 'symbols for industry' – across the canvas. A cluster of shapes conglomerates in a seam at the canvas's centre, as if the coordinates of a map have been shaken up and left to settle with kaleidoscopic effect. The painting's title sets these elements in play to infer the tarmac lines of a motorway bypass skirting a city, or striking across land on an overcast, cloudy day. Indeed, while the bypass, 'treeless and stinking of diesel oil, murderous with lorries', had been a feature of the landscape since the early 1920s, the first motorway bypass had opened in 1958, so that *Bypass I* speaks to a very recent moment of Britain's urbanization.[19] Yet even as the painting confers the bird's-eye perspective of the map onto this territory, and apparently proffers markers through which it could be deciphered, it refuses to grant any surety about the landscape traversed. The coordinate-like shapes act as a kind of double bluff, raising the possibility of interpretation, but unable to act as clear symbols without a readable legend or key. Cartographic theorists Denis Wood and John Fels argue that the map 'is nothing more than a vehicle for the creation and conveying of authority about, and ultimately over, territory'.[20] Yet although the configuration of *Bypass I* references map-work, it makes no authoritative claim to place, for all its evocative, atmospheric qualities, particularly in the use of greens, blues and greys to suggest geographic and weather formations. Clough exposes the map, to use Wood and Fels' terminology, as 'a field of concepts', a slippery, unstable entity reflecting an urbscape that cannot easily be defined or navigated.[21]

Clough's exploration of urban, industrial subjects, and her movement towards abstraction, placed her work at the heart of artistic debates shaped by Cold War politics. This is demonstrated by the rapid shifts in the interpretation of her work during the 1950s. Berger, who was one of Clough's earliest champions and a lifelong friend, identified the paintings in Clough's 1953 Leicester Galleries exhibition, including *Lorry with a Cabledrum* (1953) and *Milk Lorry* (1950), as the productions of 'fundamentally a still-life painter', while interpreting their quotidian materiality as a humanist celebration of the worker.[22] Berger insisted that, even when not depicting the human body, Clough's 'interest in machines is fully human because she never deifies them and is intelligent about their function. She finds in them, even when deserted, evidence of human ingenuity.'[23] Clough's decision to exhibit with the left-leaning Artists International Association (AIA) early in her career might seem to confirm Berger's implicit alignment of her work with social realism and Socialist politics.[24] It is possible to detect echoes between realist works like *Lorry with a Cabledrum, Cooling Tower II* (1958) and *Bypass I*, whose circles, squares and crosses conjure the nuts and bolts of some complex machine as much as geographical features. By Clough's 1960 retrospective, however, Sylvester could assert that, while the subjects of her most recent works like *Bypass I* might masquerade as 'those of a social realist', her paintings were in actuality highly abstract.[25] He maintained that:

> I can't, indeed, think of many paintings which strike me as more abstract. They don't evoke the things named in their titles, they don't evoke other things either, and they don't leave me with any feeling that there's something which they ought to be evoking. What they do evoke is a certain way of seeing.[26]

This 'way of seeing', *Bypass I* infers, is contingent and elliptical.

Like Berger, Sylvester was a proponent of realist painting, but as James Hyman observes, he championed an alternative mode of 'Modernist realism' informed by the continental discourses of existentialism and phenomenology.[27] Sylvester also engaged rigorously with the impact of American Abstract Expressionism, and in his review of Clough's Whitechapel retrospective linked works such as *Bypass I* and *Electrical Landscape* with artists including Mark Rothko and Robert Motherwell, both of whom had featured in the 1956 exhibition *Modern Art in the United States* at the Tate Gallery.[28] Selected from the New York Museum of Modern Art's collection, the show included a large section of Abstract Expressionist works, which the catalogue contextualized with the assertion that 'since the beginning of the 1940s, abstraction more than any other trend has released the creative forces of American art'.[29] A second Tate show in 1959, entitled *The New American Painting*, focused specifically on Abstract Expressionism. The location of Clough's retrospective at the Whitechapel was also significant in this respect: Bryan Robertson, the director and a long-term friend of Clough's, staged a series of exhibitions showcasing American painting including a Jackson Pollock exhibit in 1958 and a Rothko display in 1961.[30] In his review of the 1956 Tate exhibition, the painter and critic Patrick Heron, also a friend of Clough's, celebrated Abstract Expressionist paintings for their 'flatness, or rather, their spatial shallowness ... Their constant denial of illusionistic depth'.[31] Clough's *Bypass I*, and other works like *Electrical Landscape*, comparably emphasized the flatness of the picture plane with their cartographic, diagrammatic lines. Yet despite their use of abstraction, that very map-like quality ensures that they continue to gesture, irresistibly, to an experienced terrain. It feels premature to call these works 'private and abstract': instead, they fall between the battle lines drawn by Berger and Sylvester.

In 1952 the AIA organized a show entitled *The Mirror and the Square*, which sought to counter simplistic formal and political distinctions between abstraction and realism: although Clough did not participate directly in the show, she was a member of the selection committee.[32] The catalogue text expanded on the two symbols chosen by the AIA:

> the Mirror, held up to nature, presents the spectator with a literal image of the visible world. ... The square serves the draughtsman as a precision instrument ... it serves here as a symbol for those artists whose pictures rely upon the presentation of 'abstract' forms, and the delicacy and precision of their relationship. Between these poles ... lies the labyrinth of contemporary styles in painting and sculpture through which the enquiring spectator must find his way ... Beneath the skin of every great mirror image lies a formal 'abstract' and every abstract picture is governed by natural laws, whether the forms delineated are organic in origin, or as inorganic as the crystal.[33]

This rejection of definitive classification denotes a refusal of reductive political equations between realism and Socialism or Communism, or abstraction and capitalism.[34] Specifically, it denies any attempt to associate either realism or abstraction with the coveted position of artistic freedom – and of the geographic alignment of the former with the USSR and the latter with America. The rhetoric of *The Mirror and the Square* catalogue moreover implies that the distinction between realism and abstraction is a false one. While Clough's early works might have been considered 'mirrors' and her subsequent ones 'squares', paintings such as *Bypass I* comparably decline to be drawn on the fraught question of formal and political alliance, instead embracing hybridity.

Clough's formal evasiveness powerfully relates to the physical and psychological topography of the Cold War. Michael Middleton, who wrote the catalogue essay for Clough's 1960 retrospective, acutely described how Clough's paintings are attuned to 'the signs and winking lights of the bypass, the dark debris of the industrial waste-land, the structures of mysterious purpose that provoke no twist of the head from the uncurious'.[35] Middleton's account has strong parallels with the Cold War landscape charted by the cultural geographers Wayne Cocroft and Roger Thomas, composed from 'military bases, surrounded by wire mesh fences and constantly lit up at night by floodlights', and characterized by divisions and bureaucratic secrecy.[36] *Bypass I*, as its title emphasizes, traces an area literally passed by on the way to another place, a grey elsewhere that incorporates the industrial park and the military zone. Clough's diary entry for 5 June 1960 reads 'finish bypass, cont. rework slag etc.', the painting's execution mirroring its relationship with a declining, fragmented industrial scene glimpsed from the motorway.[37] This dislocation in turn reverberates with Cold War anxiety. Although 'a specialist's confrontation, not a peoples' conflict', Peter Hennessy notes that it nonetheless 'aroused fear on a wide scale'.[38] The British government worked on its own hydrogen bomb throughout the 1950s, as announced in the February 1955 Defence White Paper, culminating in the Grapple trials in the Pacific from May 1957.[39] The equivocal location of works such as *Bypass I* becomes analogous with the geography of Cold War Britain, schizophrenically formed from the disjunction between bureaucratic, metropolitan centres envisaged as the targets of potential devastation, and distant sites of power generation in neglected areas, together with even remoter island testing grounds in former colonies.

Clough's combination of abstraction and realism through the map device can be conceptualized as what Kobena Mercer has termed 'discrepant abstraction', which he defines as 'hybrid and partial, elusive and repetitive, obstinate and strange: it includes everything that does not neatly fit into the institutional narrative of abstract art as a monolithic quest for "purity".'[40] Clough's filtered cartographies can be compared to those of other painters working in Britain during the 1960s and early 1970s, such as Frank Bowling and Rita Donagh. Clough's references to maps were, however, never as explicit, or used specifically to engage with contested issues of nationalism, as they were in the work of these artists.[41] Yet Clough similarly employed cartographic processes only to question them, challenging territorial certainties. The battle lines of Clough's paintings in the 1950s and 1960s respond to two conjoined areas of concern: the aesthetic machinations between abstraction and realism for political ends, and the technological, industrial and military geopolitical concerns that these artistic tools might embody.

The View from Space

Clough's use of cartography's supporting scaffold as she moved into abstraction is closely connected with the aerial viewpoint, intrinsic to the map's propositional logic and a vantage point that continually flickers between microcosmic and macrocosmic. *Electrical Landscape*, which was also completed in 1960 just in time to be shown at the Whitechapel retrospective, sees Clough working through many of the formal and thematic issues fuelling *Bypass I*. The linear brushwork and precisely applied oil paint again invite comparisons with a map freshly unfurled for inspection, while vertiginously pitching the viewer into the adoption of a bird's-eye perspective (see Colour plate 13). A buff-coloured slab of paint dominates the centre-left of the canvas, bounded by sections of deep, inky blue like the sea lapping a harbour or port on the right. The central section is divided and then re-divided into a mosaic by a series of internal lines, while from it radiate slender paths of creamy-grey. As in *Bypass I*, the title simultaneously implies and denies a specific location. Certain shapes, such as the powder blue crescent, defy immediate identification. Others seem to offer interpretative coordinates, notably the nest of ovals at the centre, and the delicately criss-crossed network of filament-like lines above. These initially prompt a rush of connotations including wires, fuses, transistors, pylons and power cables, before they start to shuttle dizzyingly between different possible scales. The electrical landscape here might be that of a circuit board held up for inspection under a magnifying glass, as much as an electrical plant surveyed from above.

Electrical Landscape's precise, almost mathematically calibrated, bisecting lines crackle with static charge, like the build-up of atmospheric pressure before a storm.[42] This tension also derives from the aerial view which, like the map, is implicated in the military strategies of reconnaissance spy planes, circling bombers, homing missiles and satellites, all targeted on and relaying information back to bureaucratic centres. Another of Clough's poems, entitled 'The Landscape Near an Aerodrome', assumes the perspective of passengers in an 'air-liner with shut-off engines' as it 'glides over the suburbs', imagining how they:

... let their eyes trained by watching
Penetrate through dusk the outskirts of this town
Here, where industry shows a fraying edge.
Here they may see what is being done.[43]

Clough's concentration on the urbscape – the zones of industry at the town's outer limit, where things are 'being done' – involves a celebratory refocusing of attention on overlooked and aesthetically disregarded manufacturing activity, even as that 'fraying edge' captures both production zones bleeding into residential areas and economic decay. At the same time, there is an ominous quality to the circling, strategic scrutiny imagined by the poem, which comparably informs the aerial view in *Electrical Landscape*.

Clough's war work would have provided many an object lesson in the overlap between the map, the aerial view and military technologies of observation and destruction, particularly given the relationship between reconnaissance photography and map-making.[44] In her diaries from the mid-1940s, appointments for map-work sit alongside neatly jotted notes recording the impact and severity of London bombings, while *Electrical Landscape*'s

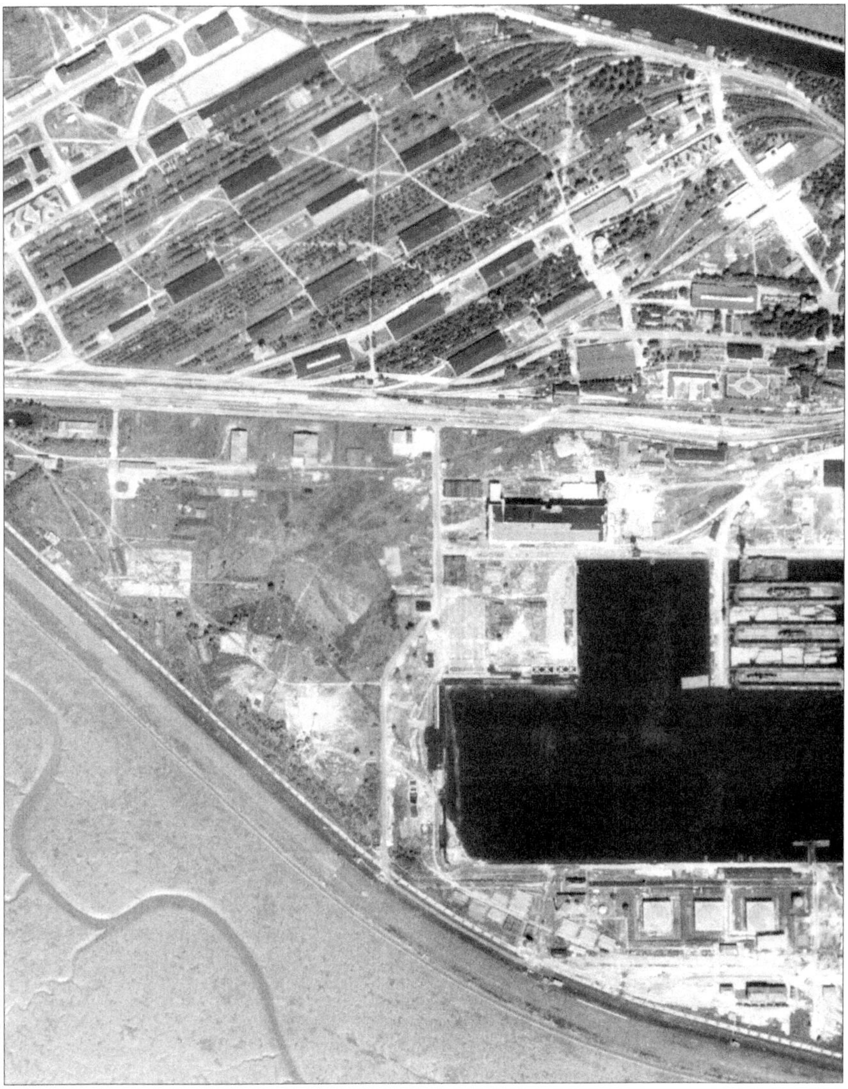

7.2 No. 541 Squadron RAF, part of a vertical photographic reconnaissance aerial taken over Wilhelmshaven, Germany, showing the navel ammunition depot at Mariensel, prior to attacks by the USAAF (27 January 1943) and RAF Bomber Command (11/12 February 1943), black and white photograph © Imperial War Museums (C 3406).

layout has strong compositional affinities with reconnaissance photography of German dockyards taken by the RAF during the Second World War (see Figure 7.2).[45] Paul Saint-Amour traces the emergence of aerial photography to its military use in the First World War, noting that 'the techniques and reflexive discourse of aerial photography got exported out of their originary military contexts into their commercial ones'.[46] This intersection between mapping, reconnaissance and the aerial view continued during the Cold War, as the aerial view became linked to spying and information gathering. The covert, privileged glimpse of knowledge relayed by the spy plane would assume a pivotal role in the 1962 Cuban missile crisis, sparked partly by aerial images from an American U-2 reconnaissance flight over Cuba that revealed missile sites under construction. Paralleling the enhanced role played by the

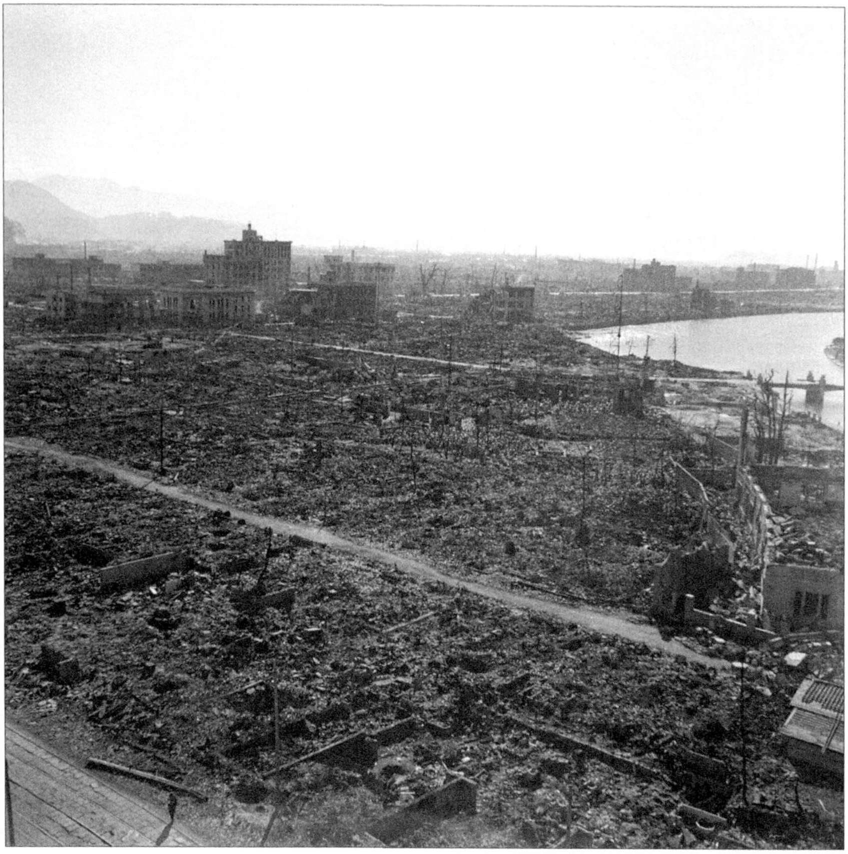

7.3 Unknown photographer, view looking across the centre of damage at Hiroshima, c. 1945, black and white photograph © Crown Copyright, Imperial War Museums (Q (HS) 832).

secret service in Cold War intrigues, and the related boom in spy fictions, the aerial view offered by *Electrical Landscape* is permeated with a sense of clandestine observation.

Although Scott Kirsch persuasively demonstrates that the images of vertical mushroom clouds erupting heavenwards from nuclear test sites dominated Cold War visual culture, circulating through mass periodicals including *Life* in America and *Picture Post* in Britain, it was also informed by elevated views of the decimation at Hiroshima and Nagasaki (see Figure 7.3).[47] A special issue of *Picture Post*, published after the atomic bomb drop, paired photographs of the vertical columns of smoke rising over Hiroshima with an image of Tokyo from above after an earthquake in 1923, in an attempt to visualize the effects of the bomb on the city fabric.[48] The subsequent photographs taken of Hiroshima and Nagasaki indicate how reconnaissance and destruction merge in the aerial perspective, offering documentary witness while placing the viewer in the position of the bomber. The connection between information gathering and the threat of subsequent devastation infuses *Electrical Landscape*'s insistent flatness: the painting affords the viewer the luxury of pouring over, picking out and pinpointing details, but in so doing bestows an awareness of the landscape's vulnerability. This is emphasized by the stark contrast between the intricacy of the painting's left-hand section and the void-like blankness of the deep blue to the right, as if directly contrasting evidence of industrious activity with its absence. Clough herself used photography throughout her career to capture the patterns and geometries of urban and industrial landscapes. Sacha Craddock connects Clough's familiarity with photography to the cropped, flat qualities of many of her paintings, identifying a repeated compositional device whereby 'the image or picture is held still as if caught frozen by the constantly ranging eye of the surveillance camera'.[49] Clough's work repeatedly demonstrates the impact of photography's changes in scale and shifts in perspective from the frontal to the aerial.

Together with military surveillance, Clough's interest in aeriality also relates to the equally militaristic vistas of Earth generated by the space race between the USA and the USSR.[50] On 4 October 1957 the USSR launched their Sputnik satellite, beating the USA into space and affording them, in the words of historian Lorna Arnold, 'a profound psychological shock'.[51] Clough's diaries follow the space race: an entry on 13 September 1959 reads 'moon hit USSR', marking the first moon landing by the Soviet Luna 2 spacecraft.[52] Clough recorded the 'space flight' on 12 April 1961, when Russian cosmonaut Yuri Gagarin became the first man in space.[53] Throughout 1969 Clough tracked America's ultimate ascendancy in the space race with Neil Armstrong and Edwin Aldrin's moon landing.[54] This enthusiasm for space and astronomy is evinced by Clough's poem 'To the 200-inch Telescope', the title of which refers to the Hale telescope in America's Palomar mountains, which opened in 1949 and remained the world's most powerful telescope for the next 45 years.[55] *Electrical Landscape*'s shifts between the near and far mirror the action of a high-powered telescope, or a camera zooming in on the Earth, collapsing distance and revealing detail.

Berger praised this element of Clough's style, noting that 'however large the spatial and structural implications of her painting, she preserves in a remarkable way the intimacy of every object. To a geologist an aspect of a mountain seen through a telescope can be as intricately known as a stone in the hand.'[56] Yet Clough employs the aerial perspective to veer between moments of almost geological clarity and passages of paintwork deliberately resistant to exegesis.[57] Despite the details revealed in *Electrical Landscape* and *Bypass I,* and in other works that invoke the aerial view such as *Mining Area* (1954–1956) and *Landscape Through Glass VI* (1958–1959), those less intelligible elements continue to trouble the viewer. Equally, the ability of the more legible sections to generate multiple interpretations refuses to grant legitimacy to a single one. Saint-Amour identifies precisely such a conflict between extreme readability and confusion at the heart of aerial photographs during the First World War, observing that, although they 'possessed the conceit of laying reality bare by their deadly accuracy', this 'was mitigated by an accompanying insistence on the de-familiarizing power of the vertical view'.[58] *Electrical Landscape* operates in an analogously de-familiarizing manner, slipping constantly between microcosm and macrocosm, transmuting the military associations of the aerial perspective into a meditation on uncertainty.

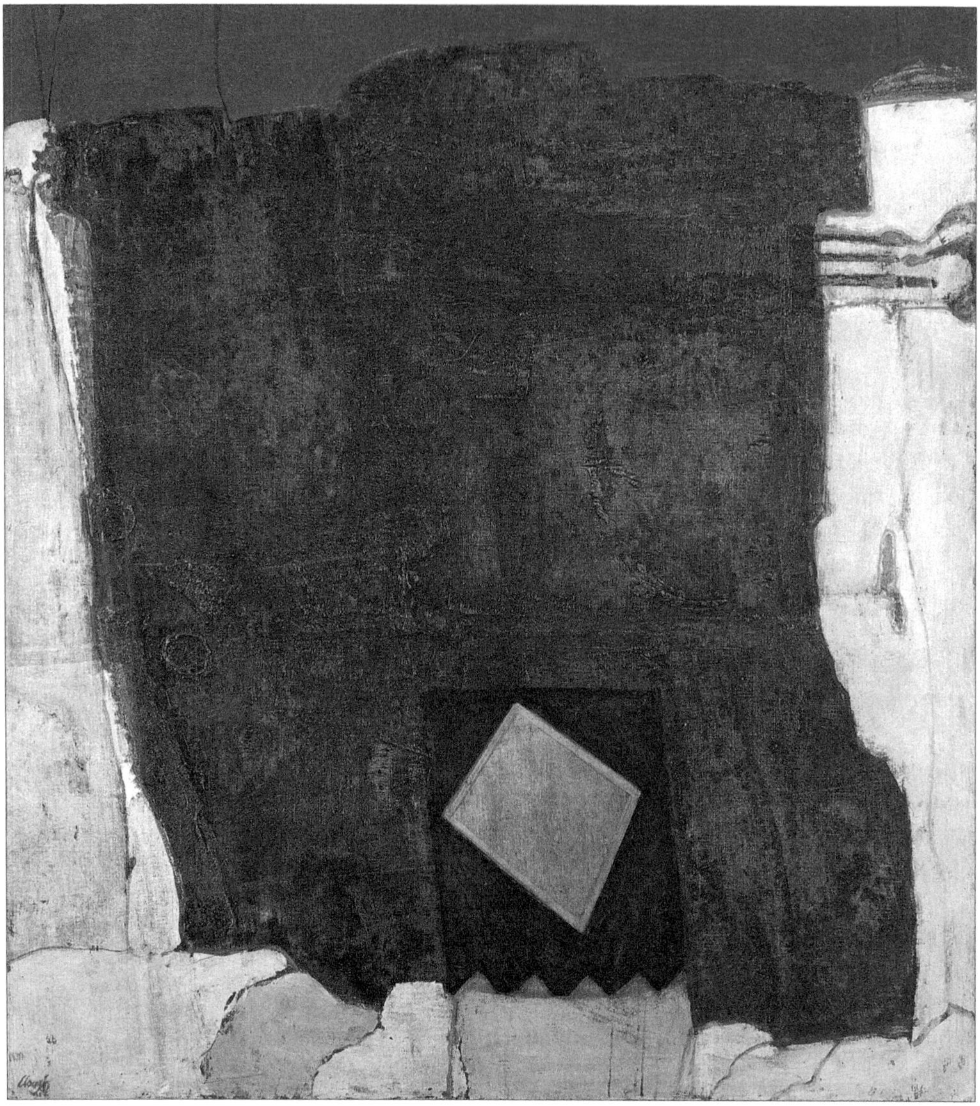

7.4 Prunella Clough, *Brown Wall*, 1964, oil on canvas, 131.7 × 121.6 cm, Pallant House Gallery, Chichester (Wilson Gift through The Art Fund)

Invisible Threat

The sense of occlusion in many of Clough's paintings also stems from her predilection for margins and edges, attested to by the most recent works displayed at the Whitechapel Art Gallery in 1960, and in the pieces that she would go on to produce as the 1960s progressed, like *Garden Hedge* (1965) with its spiky divisions. Perhaps the most overt example occurs in *Brown Wall* (1964). *Brown Wall*'s schematic composition reflects the influences of the map and the aerial view, yet while it could conceivably be said to unfold a series of carefully demarcated fortifications seen from above, the stronger sense is of a vertical barrier, blotting out the horizon (see Figure 7.4). At the bottom of the painting an orange diamond embedded in a brown square, underlined by a saw-tooth chevron in warning red and arranged so that its corner points upwards, exacerbates the impression of enclosure. In the early oil painting *Boats in Winter and War Defences* (1941–1942) and the closely related watercolour *Shore-line and War Defences* (c. 1941–1942), Clough studied the etiolated, geometric forms of defence structures lining the coast in anticipation of attack during the Second World War.[59] The eerie quality of these landscapes, apparently uninhabited but marked with human traces, carries over into *Brown Wall*, which resonates subtly but powerfully with the barriers erected during the Cold War, from the physical marker of the Berlin Wall to the psychological distinction between East and West. For Garlake, *Brown Wall* is 'ripe for demolition, the site of a thousand posters ... scraped and luxuriously over painted; a long social history lurks within the density of the paint'.[60] This crumbling, battered texture offers hope that the claustrophobic confinement imposed by the wall might not last. Clough was deeply engaged by the edge as a formal device, one that could separate but simultaneously join, like a Barnett Newman 'zip'. As Frances Spalding writes: 'at a more metaphorical level, she also loved paintings that were "on edge", on the edge of not being balanced or on the edge of becoming something else.'[61] The edges of *Brown Wall* might not necessarily be non-negotiable, but wavering on the verge of disintegration.

While Clough's exploration of edges and boundaries indicates the longstanding influence that map-making had on her practice, it can also be interpreted as a reflection on the limitations attendant on perception and knowledge. A slightly earlier image from 1959, which can be seen as a precursor to *Electrical Landscape*, implicitly imbues scientific and technological forms with uncertainty. *Electrical Installation* is composed from a constellation of gracefully delineated oval shapes connected by a frenetic but fluid network of fine lines (see Figure 7.5). The forms of the resulting image correlate with the intricate wirings of an electrical system, as well as plant stems and seeds. This conjunction of technological and organic references links *Electrical Installation* to the wider interest in biomorphic imagery during the post-war period, which Brooke Kamin

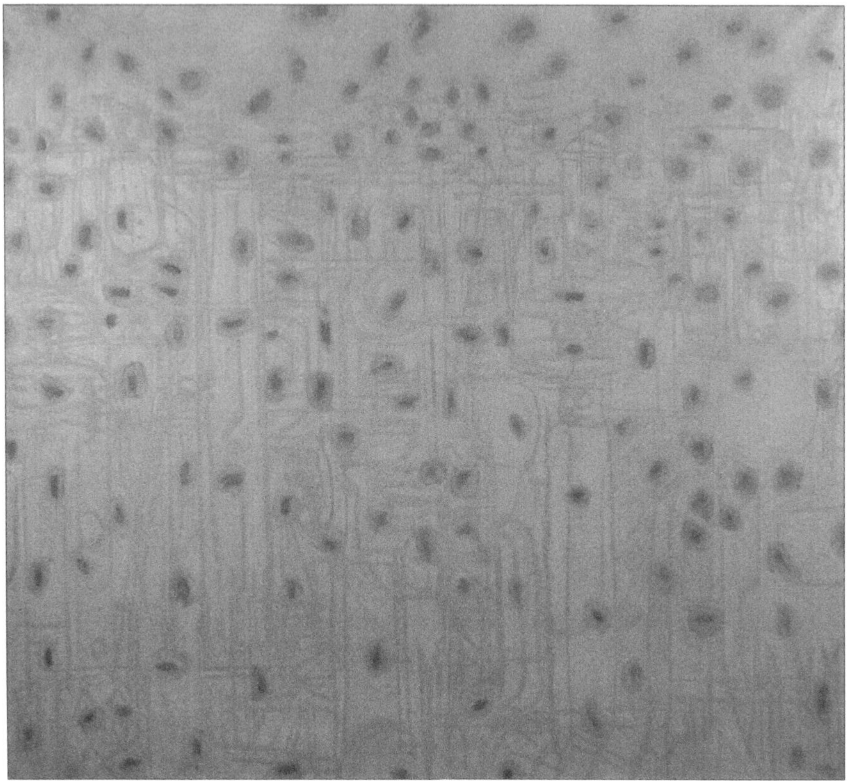

7.5 Prunella Clough, *Electrical Installation*, 1959, oil on canvas, 115 × 127 cm, private collection

Rapaport and Kevin Stayton identify as 'vital forms', shaped by scientific developments in microscopy.[62] The soft blue ovals distributed across the creamy-coloured field of *Electrical Installation* are haloed by lighter aureoles of the same hue, redolent of radioactive glow. These small, particle-like forms resemble the molecular patterns revealed by microscopy, composed in turn from arrangements of atoms. *Electrical Installation* alludes both to electrical wiring and to the movement of particles that creates electrical charge – its energy might well be generated by the atomic reactions of nuclear power.[63] In an episode that demonstrates the pragmatic edge to Clough's interest in this new environment, she related how she had been 'to Bradwell (Essex) where I once came down in an aeroplane and which they propose to "ruin" by building an atomic power station. (It seems very suitable and would look O.K. if they'd only let one get near).'[64] The painting's configuration embraces rapid alternations in scale between microcosmic and macrocosmic, establishing a chain of associations that progresses from the atomic level, through electrical systems and the gridded city lights powered by those systems, to the arrangement of stars and galaxies.

As in *Electrical Landscape*, the different possible scales encompassed by *Electrical Installation* cast doubt on what is shown – or, perhaps more precisely, on what the viewer thinks they see. While *Electrical Installation*'s conflation of the biomorphic, technological and atomic is at once dazzling and dynamic, it also echoes the destabilizing effects of ostensible scientific advance critiqued by the philosopher Hannah Arendt. Arendt prefaced her 1958 book *The Human Condition* with a reflection on the launch of Sputnik the year before. For Arendt, Sputnik was paradigmatic of the 'newest experiences and ... most recent fears' that, together with older technologies including the telescope and the microscope, fostered alienation from the world.[65] In *The Human Condition,* telescopic and microscopic views paradoxically execute a radical curtailment of perspective, recursively conflating the arrangement of the solar system with molecular structures. In this shift, Arendt finds, 'we deal only with the patterns of our own mind, the mind which designed the instruments and put nature under its conditions in the experiment'.[66] A similarly qualified approach can be detected in Clough's assimilation of imagery derived from the microscope and the telescope into her work during the 1950s and 1960s.

Like the forms in *Electrical Installation,* the hexagonal shape in Clough's *Untitled* (1965) follows in a lineage of images integrating organic and scientific references (see Colour plate 14). It mimics the central nucleus and surrounding membrane of a cell, or stylized visualizations of atoms with their cloud of electrons. In this respect, it echoes the earlier popularization of scientific designs during the 1951 Festival of Britain. This imagery was created by a group of scientists led by Dr Helen Megaw, Assistant Director of Research at Cambridge's Cavendish Laboratory, in collaboration with designers and print-makers. Megaw had the inspiration that the patterns revealed by X-ray diffraction photographs of crystals, which she was using to determine the atomic structures of molecules, could be transferred onto wallpaper, textiles and ceramic dinner services. *Untitled* resembles images such as Dr Dorothy Hodgkin's drawing of *Insulin 8.27,* which was made into wallpaper by William J. Odell and John Line & Son, with its repeated hexagonal structure derived from insulin molecules.[67] Both *Untitled* and *Electrical Installation* similarly replicate the aerial perspective afforded to the eye by the barrel of the microscope. Megaw used the analogy of the map to elucidate the diagrams used for the Festival designs, explaining how the crystallographer 'will often use different colours for the different kinds of atoms; this is the same sort of convention as that used by the map-maker when he paints his seas blue and the British Empire red'.[68] Megaw's connection between mapping and scientific diagramming, albeit inadvertently, anticipates the Festival's underlying neo-imperialism. Jo Littler describes how, in the space-age futurism of the Festival, particularly palpable in the *Dome of Discovery,* 'imperial mapping is extended in and through scientific discovery: British imperial expansion on the earth naturally progresses into the exploration of space' in an attempt to recompense for, or rather disguise, the disintegration of Empire.[69] In distinct contrast to the empirical confidence of the designs produced by the Festival Pattern Group, *Untitled,* as

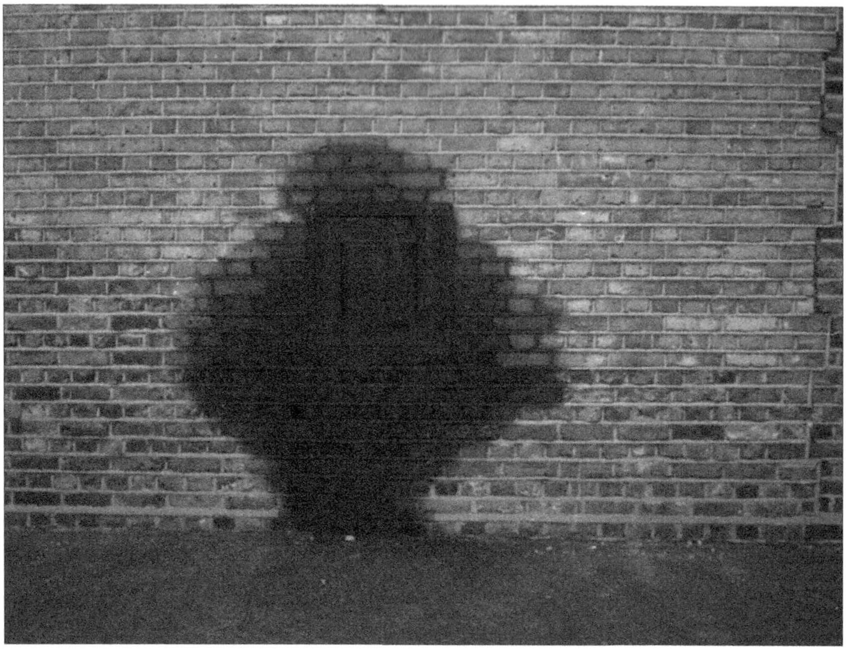

7.6 Prunella Clough, untitled photograph, date unknown, Tate Gallery Archives

Clough's title emphasizes, declines to classify its subject, just as Clough used the map to undermine conceptions of territorial ownership.

Although small, the pulsing quality of the brushwork at the centre of the hexagonal shape in *Untitled* and the sickly green hue that saturates the whole composition combine to emit a quietly menacing quality. The work's coloration could be described using a reference that Clough made in a different context to a particular shade of 'pollution green'.[70] The skeletal shadow-outline of a companion figure to the left of the canvas, a diamond that echoes the stronger hexagonal formation on its little plinth, underscores the prevailing inference of contamination and erasure, as if a partner shape has been destroyed. The dominant figure in *Untitled* and its ghostly counterpart correspond with an undated photograph taken by Clough on one of her many urban peregrinations. This photograph shows a stain blossoming over a wall, an oily residue that spreads across the brickwork like a petrol mark on a garage driveway, or the cloud of an exhaust pipe (see Figure 7.6). The sense of seeping pollution in this photograph and *Untitled* recurs throughout Clough's depictions of industrial landscape, exemplified by *Chemical Works III* (1960) with its virulent band of acidic orange earth to the left of the canvas (see Figure 7.7). The contrast between *Untitled*, *Chemical Works III* and the 'vital forms' imagery promoted by the Festival denotes a shift from earlier post-war official narratives of optimistic scientific advance, to the growing fears during the late 1950s and 1960s concerning the impact of chemicals and radiation on the environment.[71]

In her 1962 book *Silent Spring*, the American scientist Rachel Carson warned that:

> Along with the possibility of the extinction of mankind by nuclear war, the central problem of our age has … become the contamination of man's total environment with such substances of incredible potential for harm – substances that accumulate in the tissues of plants and animals and even penetrate the germ cells to shatter or alter the very material of hereditary upon which the shape of the future depends.[72]

Carson's book unfolded a devastating vision of the impact that human activities were having on the environment, and as a result on bodily cells and tissues. Many of Clough's images balance on a knife-edge between a desire to recognize the beauty of the urbscape and a sense of elegy for the damage and loss accompanying its industrial and technological transformations. Nonetheless, the layered, feathery brushstrokes of emerald green and ruby red paint in *Untitled* also create a vivifying quiver, investing the central form with movement, life and lapidary beauty. As with *Brown Wall, Bypass I, Electrical Landscape* and *Electrical Installation,* the work's ambiguity precludes hyperbolic reading, preventing the image – and the individual who looks at it – from becoming fixed to one or other viewpoint. Composed increasingly from motifs that seem half-remembered or half-recognized, and impressions that have become detached from their moorings and reconfigured, Clough's canvases from the late 1950s onwards, through to the 1990s, enact the modulation of experience through memory. While Clough's paintings often retain a root in the observable world, they encourage subjective associations, refusing to dictate what is seen and sharing their strategies of covert resistance with the viewer.

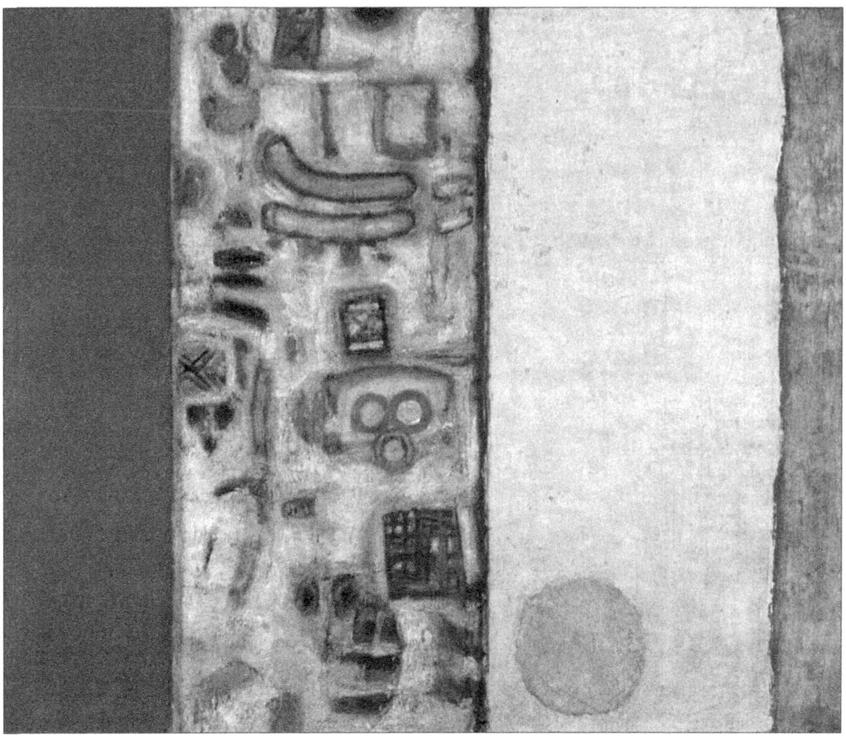

7.7 Prunella Clough, *Chemical Works III*, 1960, oil on canvas, 62.5 × 75 cm, Government Art Collection, UK © Estate of Prunella Clough 2013. All rights reserved DACS.

As she moved from realism and figuration to an increasingly abstracted mode, Clough employed the already semi-abstract visual strategies of cartographic motifs and the aerial perspective, together with references to scientific and technological imagery, as a means of responding to the infrastructure of her surroundings. In doing so, Clough quietly but firmly questioned the apparent sureties of these ways of seeing, in order to register the instabilities – physical, political and metaphorical – of the Cold War terrain. Her blend of referential clarity and determined opacity both apes and undercuts new technologies of making visible, constituting an indirect counterpoint to the future Labour Prime Minister Harold Wilson's celebratory 1963 'white heat' rhetoric.[73] Her influences, however, to use Sylvester's words, are always 'thoroughly digested', at once elusive and generous in their frame of reference and potential meaning.[74] *Bypass I*, *Electrical Landscape* and *Untitled* express the pervasive ways that Cold War doubts, stemming from repressed and sometimes intangible fears, inflected artistic responses to the built environment produced under the conflict's tensions. Like the white noise of televisual static, the boundaries and anxieties of the Cold War form the background to many of Clough's transitional canvases from the period of her Whitechapel retrospective, so that they constitute compelling distillations of the confrontation's effects. Ultimately, in Clough's paintings from the late 1950s and early to mid-1960s, the canvas forms a site at which the contradictions and ambiguities of this environment could be explored and, if not dispelled, then perhaps savoured and briefly cherished.[75]

Acknowledgements

I would like to thank Tate Archive, particularly Allison Foster, Jennifer Mason and Adrian Glew, for kindly allowing me to access Prunella Clough's uncatalogued papers. I would also like to thank Lisa Tickner, under whose guidance the ideas for this essay originally took form, and Jo Applin for her encouragement.

Notes

1 David Sylvester, 'Exhibitions', *New Statesman*, vol. LX, no. 1542, 1 October 1960, 471–472, at 471 and 472.

2 Clough's dealer, Annely Juda, described Clough as 'very private' and her work as 'very … personal, very quiet … you couldn't really push it into any category. But it was full of integrity and personality.' Annely Juda, interview by Monica Petzal, London, 12 February 2003, British Library National Life Stories Collection: Artists' Lives, C466/151, F14525.

3 Bryan Robertson felt that Clough 'had to survive a subtle degree of "marking down", of not quite being allowed the full treatment at official levels', and 'paid a certain price for being a woman of her particular generation in terms of her

career', even as he somewhat unhelpfully proposed that her work instanced 'a power that I can only call feminine'. Bryan Robertson, 'Happiness Is the Light', *Modern Painters*, vol. 9, no. 2 (Summer 1996): 16–20, at 20 and 18. Three years before Clough's death in 1999, Norbert Lynton drew attention to her critical neglect in his article 'Artist in Need of a Tate Show', *Royal Academy Magazine*, vol. 50 (Spring 1996): 19. This was subsequently remedied: see Ben Tufnell, ed., *Prunella Clough*, exh. cat. (London: Tate Britain, 2007). For the first valuable book-length biographical study of Clough, see Frances Spalding, *Prunella Clough: Regions Unmapped* (London: Lund Humphries, 2012).

4 Prunella Clough, interview by Bryan Robertson, London, 1982, in Robertson, *Prunella Clough: New Paintings 1979–82*, exh. cat. (London: Warwick Arts Trust, 1982), unpaginated.

5 Clough's diary entry for 24 November 1959 reads: 'brown urbscape finish'. In her diaries, Clough usually refers to 'landscapes', or uses the abbreviation 'lsc', but after this 'urbscape' also appears. Prunella Clough, 1959 Diary, Prunella Clough Papers, Tate Gallery Archive, 200511 (hereafter TGA 200511). Garlake persuasively pinpoints Clough's use of the word 'urbscape' to her 'development of a less literal approach to her subject [which] coincided with a shift to the non-industrial urban scene', Margaret Garlake, 'Fishermen and Velvet Kebabs: Prunella Clough's Subjects', in *Prunella Clough*, ed. Tufnell, 95–109, at 104.

6 Clough later recalled that, ironically, her and Koenig's gender allowed them a privileged vantage point on these occasions: 'we found that two strange women "sketching" were soon ignored and left to wander around.' Prunella Clough, quoted by Judith Collins, 'A Biography', in Sacha Craddock and Isabelle King, *Prunella Clough*, exh. cat. (London: Camden Arts Centre, 1996), 5–9, at 8. While Clough made extensive written notes and took photographs, she rarely sketched outside: 'since I do not draw directly in landscape, it is the memory or recollection of a scene, which is also a whole event, that concerns me.' Clough, interview by Robertson, unpaginated.

7 Clough stressed: 'the landscape which preoccupies me happens to be in its nature fairly geometric, like … the crossed bars of a gate or the circular shape of an oil drum seen head on.' Clough, interview by Robertson, unpaginated.

8 The East Anglian network of early warning missile defence systems, while cordoned from public view, nonetheless became an important point in the contemporary imaginary. The playwright J.B. Priestley invoked the site in a condemnatory 1957 article for the *New Statesman*, which pleaded for Britain to renounce nuclear weapons and catalysed the establishment of the Campaign for Nuclear Disarmament: 'the world was polarised when, without a single protest from all the noisy guardians of our national pride, parts of East Anglia ceased to be under our control and became an American air base.' J.B. Priestley, 'Britain and the Nuclear Bombs', *New Statesman*, vol. LIV, no. 1389, 2 November 1957, 554–556, at 555.

9 W.G. Hoskins, *The Making of the English Landscape* (London: Hodder & Stoughton, 1955; Harmondsworth: Penguin, 1970), 299. I am indebted to Barry Curtis's insightful essay on Cold War visual culture for this reference. Barry Curtis, 'War Games: Cold War Britain in Film and Fiction', in *Cold War Modern: Design 1945–1970*, ed. David Crowley and Jane Pavitt, exh. cat. (London: Victoria and Albert Museum, 2008), 122–127, at 122.

10 Hoskins, *The Making of the English Landscape*, 299.

11 Curtis, 'War Games', 122.

12 Crowley and Pavitt argue that the Cold War mentality entailed that 'utopia and dystopia were not only concurrent: they could even be present in the same object'. Crowley and Pavitt, 'Introduction', in *Cold War Modern*, 10–23, at 14. Paul Boyer comparably employs the word 'schizophrenic' to 'comprehend early postwar America in a single adjective'. Paul Boyer, 'The United States, 1941–1963: A Historical Overview', in *Vital Forms: American Art and Design in the Atomic Age, 1940–1960*, ed. Brooke Kamin Rapaport and Kevin L. Stayton, exh. cat. (New York: Brooklyn Museum of Art, 2001), 38–75, at 39.

13 Hoskins, *The Making of the English Landscape*, 299.

14 Clough's father was a civil servant but also a published poet: the carefully typed manuscripts in Clough's archive reflect a shared passion.

15 Prunella Clough, 'Prospect', undated manuscript, TGA 200511.

16 John Berger, 'Machine-Life Painting', *The New Statesman and Nation*, vol. XLV, no. 1154, 18 April 1953, 454–455, at 455.

17 Collins observes that 'during the war … [Clough] drew charts and graphs of imports and maps of military railway lines. The maps were drawn with ruling pens on linen for making blueprints; most of them were never used.' Collins, 'A Biography', 6. Clough's diary for 15 May 1943 records: 'appointed temp draughtswoman'. Subsequent entries make multiple references to 'mapwork' (13 July 1943), 'maps' (17, 29 and 30 January 1944; 14 March and 15 April 1944) and 'finish large chart' (28 February 1944). Clough, 1943 and 1944 Diaries, TGA 200511.

18 Jeff Nuttall identified doubt as a central element of 'bomb culture', asserting that, after the atomic bomb was dropped, 'moral values, thought absolute, were now seen to be comparative, for all social entities around which morality had revolved were now called into doubt and nothing of morality remained'. Jeff Nuttall, *Bomb Culture* (London: MacGibbon & Kee, 1968), 21.

19 Hoskins, *The Making of the English Landscape*, 299.

20 Denis Wood and John Fels, *The Natures of Maps: Cartographic Constructions of the Natural World* (Chicago: University of Chicago Press, 2008), 7.

21 Ibid.

22 Berger, 'Machine-Life Painting', 454.

23 Ibid., 455.

24 Clough exhibited several times with the AIA during the 1940s and 1950s, including the shows *Young Painters Working in Britain* (1944), *Pictures to Be Enjoyed* (1953) and *Contemporary Trends* (1955). For a history of the AIA, see Lynda Morris and Robert Radford, *AIA: The Story of the Artists International Association 1933–53*, exh. cat. (Oxford: Museum of Modern Art Oxford, 1983).

25 Sylvester, 'Exhibitions', 471.

26 Ibid.

27 James Hyman, *The Battle for Realism: Figurative Art in Britain During the Cold War, 1945–1960* (New Haven and London: Yale University Press, 2001), 16.

28 Sylvester reflected in his review of the 1960 Whitechapel retrospective that 'one suspects the influence of Americans like Rothko, Kline, and Motherwell', but acknowledged that Clough's images were painted 'years after she must have

first come into contact with these artists' work. She is a painter who remains open to influences but she uses them in her own time, so that … they are always thoroughly digested.' Sylvester, 'Exhibitions', 472.

29 Homer [Holger] Cahill, 'American Painting and Sculpture in the Twentieth Century', in *Modern Art in the United States: A Selection from the Collections of the Museum of Modern Art, New York*, exh. cat. (London: Tate Gallery, 1956), 11–28, at 20.

30 According to Margaret Garlake, Robertson's Whitechapel exhibitions 'provided the fullest source of information on the New York School to be found anywhere this side of the Atlantic'. Garlake, *New Art, New World: British Art in Postwar Society* (New Haven and London: Yale University Press, 1998), 80.

31 Patrick Heron, 'The Americans at the Tate Gallery', *Arts* (New York) (March 1956), in *Patrick Heron: Painter as Critic*, ed. Mel Gooding (London: Tate Publishing, 1998), 100–104, at 102. Clough later recalled the excitement of the 1960s and 'the changes in the air … because people had been to America'. Prunella Clough, interview by Bridget Jackson, London, date of recording unknown, British Library National Life Stories Collection: Artists' Lives, C466/44/01, F13491.

32 In her letter accepting the AIA's invitation to act on the selection committee, Clough warned that she could not submit to *The Mirror and the Square* because she was 'keeping paintings' for her 1953 show at the Leicester Galleries. Prunella Clough, Letter to Mrs Meninsky, 25 September 1952, Artists International Association Papers, TGA 7043/4/30/51.

33 Artists International Association, *The Mirror and the Square*, exh. cat. (London: New Burlington Galleries, 1952), unpaginated.

34 The political and institutional manipulations of abstraction and realism have been well rehearsed: although I will not repeat them here, they are important for thinking about Clough's work. See Eva Cockcroft, 'Abstract Expressionism, Weapon of the Cold War', *Artforum*, vol. 12, no. 10 (June 1974): 39–41; and Serge Guilbaut, *How New York Stole the Idea of Modern Art: Abstract Expressionism, Freedom, and the Cold War* (Chicago: University of Chicago Press, 1983).

35 Michael Middleton, *Prunella Clough: A Retrospective Exhibition*, exh. cat. (London: Whitechapel Art Gallery, 1960), 7.

36 Wayne Cocroft and Roger Thomas, *Cold War: Building for Nuclear Confrontation 1946–1989*, ed. P.S. Barnwell (Swindon: English Heritage, 2004), 3. Eric Hobsbawm identifies the Cold War mentality as inherently binary, describing how the conflict 'polarized the world controlled by the superpowers into two sharply divided "camps"'. Hobsbawm, *Age of Extremes: The Short Twentieth Century, 1914–1991* (London: Michael Joseph, 1994), 238.

37 Prunella Clough, 1960 Diary, TGA 200511.

38 Peter Hennessy, *The Secret State: Whitehall and the Cold War* (London: Allen Lane, 2002), 1.

39 The USSR detonated its first atomic bomb in 1949, spurring the USA's development of increasingly powerful thermonuclear weapons such as the hydrogen bomb. Although the USA first tested a hydrogen bomb in 1952, it was with the test of the Bravo bomb in March 1954, and the resulting severe fallout and media frenzy, that nuclear weapons testing became a public concern. See Lorna Arnold and Katherine Pyne, *Britain and the H-Bomb* (Basingstoke: Palgrave, 2001).

40 Kobena Mercer, 'Introduction', in *Discrepant Abstraction*, ed. Mercer (London and Cambridge, MA: The MIT Press, 2006), 6–29, at 7.

41 In Bowling's *Night Journey* (1968–1969), for example, the silhouettes of Latin America and Africa float in a field of highly saturated colour streams – red, green, golden yellow and blue. For Mercer, *Night Journey*'s erasure of the map's outer edges 'perform[s] an act of post-colonial rewriting that is directed towards a critique of Eurocentricism'. Kobena Mercer, 'Black Atlantic Abstraction: Aubrey Williams and Frank Bowling', in *Discrepant Abstraction*, 182–205, at 200.

42 Clough's diaries overflow with observations on weather and climate, such as 'wet cold' and 'cold grey' (29 and 30 March 1960). That these appear in diaries mainly devoted to social appointments and noting works in progress marks an especial sensitivity to atmosphere. References in Clough's 1959 diary corresponding to the development of a 'large electric' painting (19 August 1959) coincide with observations like 'stuffy', 'cosmology' (8 August) and 'violent storms' (21 August). Prunella Clough, 1959 and 1960 Diaries, TGA 200511.

43 Prunella Clough, 'The Landscape Near an Aerodrome', undated manuscript, TGA 200511.

44 Aerial photographs taken by military spy planes were pieced together and used to create strategic maps pinpointing enemy targets. See Hanna Rose Shell, 'Mending the Net: Camouflage, Serial Photography, and the Suture of Self-Effacement and Reconnaissance', in *Hide and Seek: Camouflage, Photography, and the Media of Reconnaissance* (New York: Zone Books, 2012), 77–125.

45 Map-making engagements sit alongside elliptically expressive notes such as 'v. heavy bombing' (27 April 1944) and 'fly bombs all day' (1 July 1944), demonstrating how Clough's involvement in military mapping paralleled a very real experience of the devastation it could be used to wreak. Prunella Clough, 1944 Diary, TGA 200511. As well as reconnaissance photographs, *Electrical Landscape*'s layout also has links with other artists' deployment of aeriality and cropping in their works with disorienting effect, such as Marcel Duchamp and Man Ray's collaborative photograph, *Dust Breeding* (c. 1920).

46 Paul Saint-Amour, 'Modernist Reconnaissance', *Modernism/Modernity*, vol. 10, no. 2 (April 2003): 349–380, at 353.

47 See Scott Kirsch, 'Watching the Bombs Go Off: Photography, Nuclear Landscapes and Spectator Democracy', *Antipode*, vol. 29, no. 3 (1997): 227–255.

48 'Man Enters the Atom Age', *Picture Post*, vol. 28, no. 8 (25 August 1945): 11. Clough's diary entry for 6 August that year records: 'tonight announced that the first atomic bomb = 20,000 tons H. E., has been dropped, this morning, on Japan.' Clough subsequently reflected: 'one just knows that from now on, or rather from way back, things are different.' Prunella Clough, Diary entries for 6 and 11 August 1945, quoted in Spalding, *Prunella Clough: Regions Unmapped*, 54. On 15 October 1945, Clough noted that she 'listened to part of [John] Hersey's Hiroshima dispatch (New Yorker)' on the radio, which was later published in *The New Yorker*. Prunella Clough, 1944–1947 Diary, Prunella Clough Papers, TGA 201215/1/1. See also John Hersey, 'Hiroshima', *The New Yorker*, 31 August 1946, 15–68.

49 Sacha Craddock, 'Prunella Clough: Paintings', in *Prunella Clough*, exh. cat. (London: Camden Arts Centre, 1996), 15–22, at 20.

50 Stephen Petersen recounts how photographs generated by the space race between the USA and the USSR constituted 'a new class of image …

both representational and abstract, uniquely exalted and … increasingly commonplace in the picture press and cinema'. Stephen Petersen, *Space Age Aesthetics: Lucio Fontana, Yves Klein, and the Postwar European Avant-Garde* (Philadelphia: Pennsylvania University Press and The Getty Research Institute, 2009), 7.

51 Arnold and Pyne, *Britain and the H-Bomb*, 199.

52 Prunella Clough, 1959 Diary, TGA 200511.

53 Prunella Clough, 1961 Diary, TGA 200511.

54 The run of entries begins on 26 May 1969, peaking with the excitedly capitalized 'MOONSHOT ALL NIGHT. TV' for the televised broadcast of the landing on 20 July. Prunella Clough, 1969 Diary, TGA 200511.

55 Prunella Clough, 'To the 200-inch Telescope', undated manuscript, TGA 200511.

56 Berger, 'Machine-Life Painting', 455.

57 Clough was acutely conscious of the difficulties involved in translating the visual into the verbal: when asked about her teaching practice (Clough taught at Chelsea School of Art and Wimbledon School of Art for much of her working life), she reflected that 'teaching how to make a piece of art is a nonsense, you can only come to it obliquely … it's challenging … to find any words that will do. I think the verbal challenge is always interesting and never solved.' Clough, interview by Jackson.

58 Saint-Amour, 'Modernist Reconnaissance', 356.

59 An element of seaside Surrealism also infuses this image, comparable with Paul Nash's use of Surrealist juxtaposition to convey the melancholy strangeness of war in landscape paintings such as *Totes Meer (Dead Sea)* (1940–1941).

60 Margaret Garlake, 'Prunella Clough, Cambridge' [Review of *Prunella Clough* at Kettle's Yard], *The Burlington Magazine*, vol. 141, no. 1159 (October 1999): 639–640, at 639.

61 Spalding, *Prunella Clough: Regions Unmapped*, 161.

62 Rapaport and Stayton, 'Introduction', in *Vital Forms*, 22–35, at 22.

63 Bridget Riley's 1963 painting *Fission* provides a particularly illuminating point for comparison with *Electrical Installation*: a field of spots seems to emerge from, or disappear into, the centre of the canvas, which, combined with the title *Fission*, suggests the energy of nuclear experiment.

64 Prunella Clough, Letter to David Carr, undated, TGA 200511/1/1/23/54.

65 Hannah Arendt, *The Human Condition* (Chicago: University of Chicago Press, 1958), 5.

66 Ibid., 286.

67 The key role played by female scientists in the Festival Pattern Group represents a powerful example of how scientific imagery might be utilized by women in an act of covert resistance against gendered stereotypes, while promoting the power of the so-called 'decorative' in a way that anticipates later feminist reclamations of craft.

68 Helen Megaw, 'Patterns in Crystallography', unpublished essay commissioned by the Design Research Unit, November 1946, in Lesley Jackson, *From Atoms to*

Patterns: Crystal Structure Designs from the 1951 Festival of Britain – The Story of the Festival Pattern Group, exh. cat. (Shepton Beauchamp and Yeovil: Richard Dennis in association with the Wellcome Collection, 2008), 94–95, at 95.

69 Jo Littler, '"Festering Britain": The 1951 Festival of Britain, Decolonisation and the Representation of the Commonwealth', in *Visual Culture and Decolonisation in Britain*, ed. Simon Faulkner and Anandi Ramamurthy (Aldershot: Ashgate, 2006), 21–42, at 26.

70 Prunella Clough, Notes headed 'Wembley/Park Royal, 1962', undated notebook, TGA 200511.

71 A sense of this continues in Clough's later canvases such as *False Flower* (1993), in which an eerily iridescent bloom glows like oil on water against a black background.

72 Rachel Carson, *Silent Spring* (1962; London: Penguin, 2000), 25. *Silent Spring* was first published in the UK in 1963.

73 In his speech to the Labour Conference in October 1963, entitled 'Labour and the Scientific Revolution', Wilson famously asserted: 'the Britain that is going to be forged in the white heat of this revolution will be no place for … outdated methods.' Harold Wilson, *The Times*, 2 October 1963, quoted in Dominic Sandbrook, *White Heat: A History of Britain in the Swinging Sixties* (London: Little, Brown, 2006), 4.

74 See note 28.

75 The word 'cherished' is Sylvester's – he writes of the inference in Clough's work that 'there was something there the sight of which has been quietly, wryly, cherished'. Sylvester, 'Exhibitions', 471.

8

The Aesthetic of Scientific Authority in a Nuclear Age: Jacob Bronowski and Feliks Topolski

Kate Aspinall

Introduction

At 8:15am on 6 August 1945 the American B-29 Superfortress bomber *Enola Gay* deployed the first atomic weapon, a 15-kiloton uranium bomb, on the Japanese city of Hiroshima. The bomb was codenamed 'Little Boy'.[1] The plutonium bomb dropped three days later on Nagasaki had the codename 'Fat Man'.[2] Three months after the bombings this curious, though classified, anthropomorphism of the weapons was redoubled publicly by English writer and critic Cyril Connolly, who wrote in *Horizon*, the literary arts periodical that he edited: 'The race is now on between the atomic bomb and its desire to explode.'[3] In Connolly's cynical eloquence, the Bomb adopts an imaginative agency – *its* will to explode is what matters most, not *ours* to explode it. The Bomb takes on the guise of an authority, a status not limited to bombs themselves in the nuclear age, but also directed toward their creators: scientists took on an increasingly prominent role in public discourse. Their status as authorities, however, was more tenuous, with the awe shown for them during the Second World War turning in part into Cold War distrust in the 1950s, as imminent nuclear danger made debates about moral responsibility pitched, evident in Bertrand Russell's heated 1956 words: 'It has become evident that, if scientific skill is allowed free scope, the human race will be exterminated, if not in the next war, then in the next but one or the next but two – at any rate at no very distant date.'[4]

From within this context the celebrity historian of science Jacob Bronowski advocated for a fuller public engagement with nuclear authority in lectures, broadcasts, publications and television serials. This engagement entailed a more nuanced understanding of scientists, and particularly a clearer understanding of the nature of their authority and of their judgement. Bronowski first encountered the ramifications of nuclear weaponry as a delegate on the British Chiefs of Staff Mission to Japan, where he collected data from Hiroshima and Nagasaki just three months after the strikes, though his experience with nuclear science pre-dated the war, encountering it indirectly while studying mathematics at Cambridge at the same time that chemist and physicist Ernest Rutherford was leading groundbreaking research in the Cavendish Laboratory.[5] Bronowski's experiences in Japan, however, represent a turning point, something he described in *Science and Human Values* (1961) as a humbling revelation of his imagination's inability to visualize the ramifications of this new science.[6] Implicit in his writings is an ethical mandate to understand in full the components of nuclear science, not merely to gloss over them with facts. In order to grasp the violent reduction of a functional city into an inert statement, an individual needed not only to understand mathematical and scientific language, but also to be able to visualize this understanding. Throughout his life, and particularly over the three post-war decades, Bronowski refined this impulse towards using images as tools for understanding, honing in on visual models that illustrated the essence of an abstract concept, in this case the reality of nuclear atrocity.[7]

Bronowski's quest reached its apogee with the influential television series *The Ascent of Man*, which began development in 1969 and aired in 1973,[8] a project that his widow, sculptor Rita Bronowski, described in 1985 as the culmination of a 'lifetime interest in cultural and anthropological evolution'.[9] Written and narrated by Bronowski, it was intended to chronicle an integrated view of the artistic, technological and ideological development of mankind in 13 thematic episodes. In the one devoted to the ethical aspects of nuclear science, Bronowski opens the discussion with the portraits drawn and painted by London-based émigré artist Feliks Topolski (see Colour plate 15 and Figure 8.3).[10] Topolski's vibrant linear aesthetic displayed a dynamic synthesis of subjective observations and experiences. For Bronowski, this not only demonstrated the slapdash, trial-and-error similarity of creative and scientific processes (which, he thought, often subconsciously informed each other) but also illustrated, in parable fashion, how the public ought to view authority: fallible and subjective but also in some sense comprehensible; just as Topolski seems to in his art.

In this way, Topolski's paintings and drawings, whatever their original inspiration, take on a philosophical meaning when understood as an exhibit within Bronowski's engagement with scientific authority in the nuclear age. As a visual embodiment of Bronowski's ideal epistemological approach, they become instructive rather than merely aesthetic.

Atomic and Hydrogen Bombs

Understanding nuclear science requires engagement with the nuclear products, such as the bomb, as well as with the scientists who create them. Little Boy's deployment did not mark the beginning of the story of atomic hopes and fears, and it did not mark the end, but it signalled a jump from an imagined possibility to an unimaginable reality. Before addressing the dynamics of nuclear science and authority and their relation to Bronowski and Topolski, it is useful to briefly summarize what thermonuclear weaponry entailed and how it became publicly visible.

Atomic bombs involved several elements that might have come directly out of science fiction – an invented element (plutonium, produced from uranium), an explosive chain reaction (first proposed only in 1913) and measurements with sensational new terminology (such as, for casualties, 'megadeath').[11] One pound of the raw material of nuclear explosives (roughly a one inch cube of uranium), if fully converted to energy, corresponds to 1,350 tons of coal or 5,000 barrels of crude oil.[12] In atomic weapons the atom is split (nuclear fission), thus rapidly releasing energy in an explosive. In this reductive process the magnitude of a blast is inherently limited. Hydrogen weapons, in contrast, release energy by joining atoms (nuclear fusion) and so the magnitude is theoretically infinite.[13] The potential for annihilation of life on Earth through a large chain reaction was theoretically possible with hydrogen bombs, a possibility addressed in relation to military policy by an American strategist at the RAND Corporation, Herman Kahn, who coined the idea of a 'Doomsday Machine' or 'Homicide Pact' – whose detonation would extinguish life on Earth – in *On Thermonuclear War* (1960).[14] This weapon was a defensive one. Through being frightening, inexorable, persuasive, cheap (relative to non-nuclear arsenals), non-accident prone and controllable, it would direct its threat to the enemy's imagination.[15]

The Cold War was more than a single struggle over domestic and foreign imaginations, and it was not entirely cold.[16] In addition to 'periphery' conflicts precipitated or fuelled by the atomic superpowers (the US and the USSR), such as the Korean, Vietnam and Afghani Wars (to name a few), the nuclear arms race presented a serious threat to world health. Beginning in the 1940s and continuing through to the early 1960s, the United States, followed by the Soviet Union, Britain, France, India and China, conducted atmospheric nuclear weapons tests.[17] A rain of radioactive particles (radionuclides)[18] would fall around a blast site, presenting the immediate threat of radiation poisoning within 48 hours of the blast, and trace radiation would persist far longer. Radionuclides, which rapidly permeated the atmosphere, were distributed by air currents and gradually fell to earth far from blast points, where they infiltrated an ecosystem's soil and food chains. The prospective gradual global distribution of carcinogenic particles first attracted the concern of scientists, but soon became a public issue. In 1955 the United Nations initially appointed a committee of scientists from 15 nations to investigate the effect of radiation on humans.[19] The United States and the Medical Research Council of Great Britain independently began studying the effect of trace doses of radiation. Though the unglamorous but real threat of fallout did not eclipse more pitched public concerns over potential warfare (especially with theoretical weapons, such as Doomsday Devices), the concern was great enough to help spark the environmental movement in the United States by the early 1960s, and in Britain, the Campaign for Nuclear Disarmament in January 1958.[20]

Official information releases about nuclear weapons were limited at first.[21] According to the Mass Observation evidence in Britain during the 1940s, public opinion was initially split between relief at the end of the war and apprehension that nuclear energy was initiating mankind into a stage of development prematurely.[22] Despite loud and furious objections from scientists and intellectuals, public protest only began to gain momentum after the botched US test of a hydrogen bomb in March 1954, codenamed Castle Bravo. Although a technological success, the Los Alamos scientists responsible for its construction underestimated the radius of the blast for Bikini Atoll. Surrounding residential atolls of Rongelap, Rongerik and Utirik were affected by fallout and evacuated, many residents suffering radiation sickness, and radioactive rain was measured as far away as Tokyo on 17 May. Most serious to American secrecy, however, was a Japanese fishing boat, the *Fukuryu Maru* (*Lucky Dragon*), which although outside the official danger zone (72 miles from the test site) was exposed to a shower of radioactive ash. When the ship docked on 14 March, the 23 crewmembers were seriously ill with radiation sickness and hospitalized. On 24 September 1954 one crewmember, Aikichi Kuboyama, died of treatment-related hepatitis and was mourned in Japan as the first hydrogen bomb fatality. The effects on the atolls, on the *Maru*'s crew, as well as on the Japanese fishing industry were not militarily suppressible and sparked Japanese political protests, as well as American media outrage.[23] In April 1955 a government report on Bravo's fallout was publicly described by nuclear scientist Ralph Lapp in the *Bulletin of the Atomic Scientists* and then later in articles in the *New Republic*, which seized on projecting local effects of thermonuclear war.[24] The November 1954 issue of the Chicago-based *Bulletin of the Atomic Scientists* reported that on 17 April the radiation level on the *Maru*'s deck was 10-100mr, leading the article's author to suggest that the fishermen would have been exposed to hundreds of roentgens of radiation, a significant number as 400 roentgens was considered a fatal dose.[25] What the *Fukuryu Maru* established was the face of a hydrogen threat, which had previously played out largely in mathematical shorthand. These individual-scale, visualizable associations were necessary to illustrate the human costs for the public at large. Engagement with the public was thus not only a matter of relaying classified reports but also of demonstrating their significance.

Scientific Authority in the Nuclear Age

Nuclear and thermonuclear weapons were for the great majority of the public an abstract concept, and for years after the war what little that was known about their development, deployment and consequences was mediated by the scientific community.[26] The community itself was not homogenous, ranging from large networks of laboratories to myriad specialists who interpreted available information for public consumption. In this context, identifying authority – much less assigning accountability – becomes problematic.

The development of first nuclear (atomic) bombs and then thermonuclear (hydrogen) bombs precipitated significant changes in the scientific community and its methods. Particle physics experts who had been theoretical scientists, working predominantly in small-scale, academic laboratories, were incorporated into 'Big Science' – a general term for interdependent networks of large numbers of scientists who bring different bodies of expertise to a common research project.[27] Big Science arose to meet the structural needs of the United States' Manhattan Project, as well as British equivalents such as the Maud Committee, the Atomic Energy Research Establishment at Harwell and later the Aldermaston thermonuclear projects.[28] Within this framework, 'pure' theoretical and applied scientists, as well as engineers and technologists, accept each other's results on trust – and in doing so they rely on vicarious knowledge. Testimony becomes evidence, just as in non-scientific communities what passes for common knowledge is acquired by the testimony of others. The ideal in science that reliability is assured through independent repetitions partially breaks down within the testimony-as-evidence system, which has become a subject within philosophy of science.[29] A.G. Hester lamented in the September 1955 issue of *Atomic Scientists Journal* how this fracturing of testimony hampered the accurate assessment of the potentials (both negative and positive) of nuclear weapons:

> It has to be realised, however, that no simple statement of the danger is possible at this time. Certain scientific facts are known – though the work involved even in collecting and summarising these is formidable; other facts are still secret. Moreover, to draw conclusions from the facts requires judgement and extrapolation – and in the drawing of conclusions even scientists are found to differ, especially if they work in isolation.[30]

Facts and judgement are here distinguished as two separate elements of nuclear science. All of these factors complicate not only judgement in a scientific context, but also the authority of anyone or any institution to make decisions about atomic matters.

A division emerged between scientific expertise itself and the authority over it, and it is fruitful here to look at the implications of the term 'authority'. It represents power and final (or at least expert) judgement. As such it is intimately bound up with ideas of action, decision and pronouncement. A chronological spectrum emerges: in the build-up to a given decision, authority turns testimony into conclusive testimony (or expertise). In the aftermath of the decision, authority involves first responsibility for the past and then obligations to the future. Within Big Science, laboratory decision makers who mandated levels of secrecy hold authority, but they often hold little expertise in the specific scientific discoveries that they control and deploy. The scientists, though far from helpless in relation to points of procedural order, are dislocated from authority over their intellectual property. By agreeing to share knowledge only within a project, they are bound into a web of expertise in which their own consciences are irrelevant. Whoever owned the rights to the project owned not the intellectual property of the researchers but also how their individual expertise combined into conclusive knowledge. Bronowski decried the effects of this compartmentalization in his 1955 address to the meeting of the Atomic Scientists Association, titled 'Responsibilities of Scientist and Public': 'We have contrived weapons and policies with our public conscience, which each of us individually would never have undertaken with his private conscience. Men are only murderers in large groups.'[31]

Judgement was removed from the individual scientist, and yet responsibility for scientific products was still largely accorded to the scientist. Calling the Cold War a 'war of nerves' that exacerbated a feeling of helplessness in *The Common Sense of Science* (1951), Bronowski admonished the scientific community for allowing itself to become a generic symbol of clinical omniscience:

> Science, the fact that science is there, mysterious, powerful; the fact that most people are impressed by it but ignorant and helpless – all this seems to me to have contributed to the division in our minds. And scientists cannot escape the responsibility for this. They have enjoyed acting the mysterious stranger, the powerful voice without emotion, the expert and the god.[32]

This was avoidable, contended Bronowski, through embracing judgement in the form of 'popular science'. This entailed the scientist as expert educator and translator – embodying authority in the public eye.

In the late 1940s BBC Home Service programmes, such as the Radio Doctor, provided public information about wartime health concerns arising from rationing and limited medical care.[33] 'Science', as encompassing dispenser of vital expertise, increasingly became a reassuring part of the British domestic experience during a time of national crisis. As technology developed, especially the time-saving home appliances, so too did science's image. Faith in scientific knowledge or 'scientism' – particularly that it could solve any problem, given sufficient time – became a system of belief.[34] With the development of space programmes, 'popular science' seemed to provide far-flung exploits of ingenuity. Popular conceptions of the generic character of the scientist changed from the dowdy don to that of a daring, intellectual explorer. The scientist came to embody the man of genius as a public authority.

By the time of the Castle Bravo crisis, however, scientists had also begun to be vilified. Bronowski opened his 1955 address to the Atomic Scientists Association: 'People hate scientists. There is no use in beating about the bush here. The scientist is in danger of becoming the scapegoat for the helplessness which the public feels.'[35] As Bronowski warned in *The Common Sense of Science*, by acting God, scientists had:

> … failed to make themselves comfortable in the talk of people in the street; no one taught them the knack, of course, but they were not keen to learn. And now they find the distance which they enjoyed has turned to distrust, and the awe has turned to fear; and people who are by no means fools really believe that we should be better off without science.[36]

Within the scientific community a rift developed between Big Science experts and popular science authorities. As secretive military research projects recruited specialists, the two sciences diverged – the popular authority with a moral mandate to communicate, and the Big Science expert with a legal obligation to maintain silence. The pre-Second World War ideal of the egalitarian and international brotherhood of scientists unified by their common goal of truth and discovery was divided, as those responsible for cutting-edge, technological and military wonders were driven behind closed doors. Rather than security loosening again after the Second World War, the Cold War's arms race secrecy locked the rift in place – on fear of persecution or imprisonment – for the subsequent two decades.

Bronowski and Scientific Authority

Bronowski became more than his common epithet 'historian of science' suggests in the three post-war decades: he became a celebrity popular scientist, the embodiment of expertise and authority. As physicist Charles Shapiro described him in 1985, he was an 'exemplar of the broadly educated scientist-citizen'.[37] More prosaically, blogger Geoff Coupe reflected in 2010: 'When I was growing up, Jacob Bronowski was a presence on the telly. He was the scientist, the boffin, who could be relied upon to explain science to the rest of us.'[38] His personal authority grew from his versatility and popularity, and his call for the public was inspirational: they had to understand their own accountability for nuclear science and thus what was required in order to rise to this challenge, and what its consequences could be.

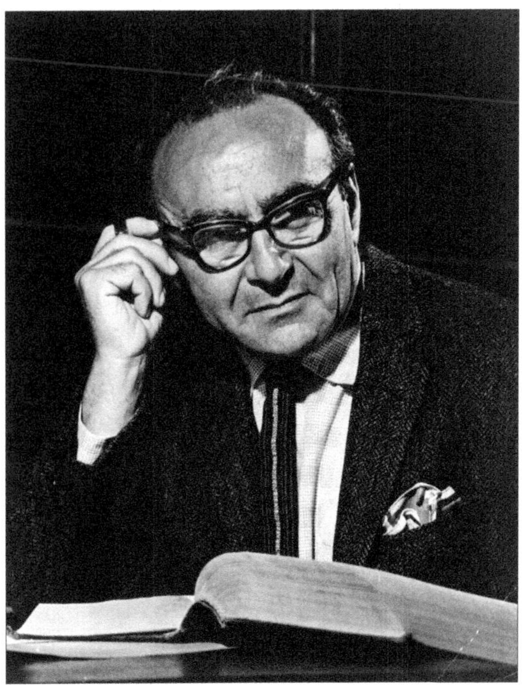

8.1 Unknown photographer, *Jacob Bronowski*, 1969, photograph, 23.9 × 18.9 cm, National Portrait Gallery, London © National Portrait Gallery, London.

Originally a mathematician working in the field of algebraic geometry (now called 'topology'), Bronowski headed the Operational Research Units for the British Ministry of Home Security and the Joint Target Group in Washington during the war. His scientific interests, however, began to develop towards philosophy of science – specifically studying cultural and anthropological evolution, or as he termed it, 'human specificity'. Working alongside biologists at the Salk Institute in San Diego, California from 1963, he explored what makes man unique, how he changes his environment and ideologically evolves.[39] Alongside these interests, he remained committed throughout his life to poetic language, co-editing the art magazine *Experiment* during his student days at Jesus College as well as several subsequent creative projects, including *Spain 1939: Four Poems* (1939) and *The Face of Violence* (1954), a radio play. The great majority of his numerous publications, despite their varied topics (including mathematics, poetry, theatre and literary criticism), pertain to the nature of knowledge, and often specifically scientific judgement.[40] Despite this impressive intellectual range, however, Bronowski's ideas were rarely agenda-setting. His theories en masse can appear quilted together from pieces of other philosophers, scientists and historians of science – aspects of Bertrand Russell's common sense and humanism, faith in the scientific community from Karl Popper, and fragments of Thomas Kuhn's paradigm-busting anomalies.[41] Bronowski's influence lay in his gift for rhetoric and his passion for his subject. His persona became part of his advocacy, just as the biochemist Paul Saltman, in ruminating on why *The Ascent of Man* was so effective, wrote: 'We saw the personification of the holistic human being, holistic in the sense that art and science were personified as one, physics and metaphysics were related and symbiotic.'[42] Bronowski held himself up as a man capable of bilingualism – of engaging in both arts and science through respecting their fundamental expression of human creativity.

Bronowski's involvement with art and science is frequently aligned with 'the two-cultures debate', begun in 1959 between the physicist and novelist Charles Percy Snow and the literary critic and academic Frank Raymond Leavis over the fracturing of culture into two polarizing bodies, the arts and the sciences. Snow is said to have acknowledged Bronowski as an inspirational early advocate of demystifying the scientific process,[43] but to read Bronowski solely against this agenda obscures his society-wide goal of spurring the public into an empowered and responsible relationship with science.[44] There were not two cultures for Bronowski in any meaningful sense, just the one. In a democratic society, delegation created professional divisions rather than there existing different emergent systems of values. Much of his work is prefaced on the fiercely held belief that artists and scientists share not only values but also a creative process.[45] To exert great energy on discussing the communication breakdown between the spheres would only add, Bronowski thought, an emotional weight to a superficial fissure. It was only because they were propelled by the complexity of physical discoveries that scientists adopted a method of communicating theoretical quantities with each other, and this made these discussions, encoded in their specific vocabularies, almost entirely incomprehensible to a scientifically illiterate population, for whom they were, despite this, immediately relevant. Bronowski's sensitivity to the banality of scientific jargon is evident in the Report of the Mission to Japan. Paragraphs of dry calculations suddenly give way to rhetorical flourish:

> Such figures could be multiplied and become meaningless: the reader who finds them so may prefer a summary analogy. This is that the scale of destruction expected [from an atomic bomb] would be that which would befall a model town built to the scale of Gulliver's Lilliput, 1 inch to the foot, if there were exploded above it a bomb more than twice as large as the largest British 'blockbuster', which with its case weighed about six tons.[46]

This is fundamentally a translation exercise, and a necessarily clumsy one. The evocative reference to Swift's *Gulliver's Travels* is an offering to help and encourage readers to develop scientific literacy.

The tension between public and private consciences was not only a matter for scientists, but also an important behavioural influence on the wider public. Just as Bronowski admonished scientists to understand their position and (dis)empowerment within Big Science, he equally advocated for the public's acceptance of their shared responsibility for the misuse of nuclear power. The social responsibility of the educated and interested citizen recurs in Bronowski's writings as a talismanic practice against 'all forms of acquiescence, of indifference, and all imposition of secrecy and denial'.[47] Judgement, if understood as revisable critical participation in the world, is the cornerstone of the ethical citizen. Ideally the layman would listen to scientific discussions (secrecy is abolished in Bronowski's ideal world) and then express his or her private judgement through voting for politicians and decisions that would in turn support freedom of a liberated community of scientists. The latter, once re-endowed with authority, would also exercise his own private judgement in how he directed his research. Bronowski's spreading of accountability across the whole of society, regardless of whether the citizen wanted it or was engaged in scientific discoveries, becomes explicit in *The Ascent of Man*. Recounting how he heard someone remark in physicist Leó Szilárd's presence that 'it is the tragedy of scientists that their discoveries were used for destruction', Szilárd replied, so Bronowski reports, that 'it was not the tragedy of scientists, but the tragedy of mankind'. Bronowski continues:

> It is said that science will dehumanise people and turn them into numbers. That's false, tragically false. Look for yourself. This is the crematorium in the concentration camp at Auschwitz. This is where people were turned into numbers. Into this pond were flushed the ashes of some four million people. And that was not done by gas … it was done by arrogance, it was done by dogma, it was done by ignorance. When people believe that they have absolute knowledge, with no test in reality, this is how they behave. This is what men do when they aspire to the knowledge of gods. Science is a very human form of knowledge, we are always at the brink of the known, we always *feel* forward for what is to be hoped. Every *judgement* in science stands at the brink of knowledge and is *personal*. Science is what we can know although we are *fallible*. In the end the words were said by Oliver Cromwell: 'I beseech you in the bowels of Christ to think it possible you may be mistaken.'[48] [my italics]

The responsibility for the abuse of scientific invention was not a scientific problem, but a human one. The nature of scientific judgement – as explorative, uncertain, humble and tested – ought instead to be a model for all judgement.

Private, tested judgements were necessary to heal the rift that Bronowski attributed to the public and private conscience. How a mob, a crowd or a collective changed the behaviour of any one participant had been a rising academic concern since urban centres blossomed during nineteenth-century industrialization.[49] In Bronowski's early radio play *The Face of Violence*, this is the explicit theme. A project begun in 1947, it was broadcast on the BBC Third Programme on 28 March 1950 and won the Italia Prize in 1951. Its 1954 published form is half a compilation of anthropological examples of how violence and societies developed symbiotically, and half a drama about a man seeking revenge on another man after the murder of a child in an unidentified prison or concentration camp. The protagonist fixates on his need to find one person to embody violence – to hold accountable for the crime of a collective – until he ultimately becomes no better than the man he seeks. The play is a fable about collective accountability. Just as Nazi propaganda alighted upon an image of Jewish physiognomy to contain economic hardship, the play's important sub-theme is acknowledging the human need to find visual forms for abstract sufferings.

The *Face of Violence* calls upon its listeners to make political and personal judgements. To make scientific judgements, however, involves a literacy in scientific terminology. Images provided one way to open up a dry concept. Bronowski harnesses (rather than vilifying) this need to visualize, using it as an aid for conceptualizing scientific language. In his oeuvre, visuals take two forms: poetic imagery and visual equivalents. In *Science and Human Values*, Bronowski voices the desire for 'the ashy, clinical sea of rubble' of Nagasaki harbour to be preserved as a monument to the leaps of imagination necessary for informed judgements. This becomes a rhetorical trick, which inspires the reader to imagine himself in a conference room amid the rubble. The intention of conveying the importance of a full imaginative understanding is enforced through visual association. Bronowski's experimentation with visual equivalents included referencing artists and artworks. In an address to the Institute of Contemporary Arts (ICA), London in August 1951, he invoked artistic movements to illustrate scientific revolutions, mentioning Henri Matisse, Paul Cézanne, Pablo Picasso, John Piper, Graham Sutherland and Henry Moore, who display a synchronicity with scientific discoveries.[50] Rather than looking at the mechanical structure of things, science had become preoccupied with 'underlying structures which bear no relation to their obvious surface mechanical function'.[51] Germanium's crystal shape was composed of molecular hexagonal prisms; so too were Moore's 'shelter drawings' – such as *Tube Shelter Perspective* (1941) – created from 'squiggles'.[52] The ICA talk aligns science and art in order to tease out the singularity of human judgement, but it does not provide an illustration of how human judgement functions. Implicit in Moore is a style rather than a methodology. Producing drawings for autonomous consumption was an unusual activity for him and one that waned after the Shelter

Drawing series. In 'The Grain in the Stone' episode of *The Ascent of Man*, Moore reappears alongside footage of him carving. He is represented artistically by the monumental and disturbingly organic solidity of *Knife Edge Two Piece* (1962–1965). Though Moore can be utilized to argue for a synchronicity with scientific invention or as an example of human tactile senses and achievements, *Knife Edge Two Piece* does not create a parable of judgement. It does not denote the subjectivity and fallibility necessary to remain humble and curious.

Bronowski and Feliks Topolski

The subjectivity at the heart of both scientific and artistic projects could be proved through recourse to almost any artist, but understanding the nature of judgement, as emergent from a synthesis of subjective experiences, requires a particular aesthetic. Topolski's drawings and paintings play up his own subjectivity in synthesizing related experiences and repetitive lines, layering upon themselves to create a febrile image. Both Topolski's self-characterization as a pseudo-documentary artist and his pictorial style reiterate a visual equivalent for judgement that is at the 'brink of the known', feeling forward, personal and fallible. Topolski's artwork was deployed as just such a parable in Bronowski's unscripted narration for *The Ascent of Man*.

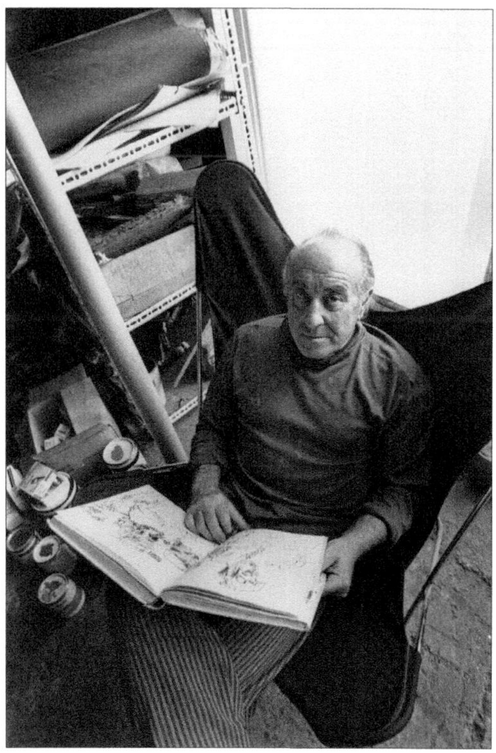

8.2 Bob Collins, *Feliks Topolski*, 1971, photograph, National Portrait Gallery, London © Estate of Bob Collins/National Portrait Gallery, London.

The 'Knowledge or Certainty' episode of *The Ascent of Man* opens with two palms approaching the camera; they spread to reveal a female face behind them and then, in a sequence of close-up fragment shots, her eyes flicker and her fingers run along the face of an elderly man. A narrator that we are led to believe is the woman begins to describe what she, being blind, can know of the man whose face her fingertips explore. When Bronowski begins to talk, it is to clarify: 'One aim of the physical sciences is to give an exact picture of the material world. One achievement of physics in the twentieth century has been to prove that that aim is unobtainable.' The man, presented standing in a gallery, is identified as Stefan Borgrajewicz, and his face is identified as an apparently graspable concrete object. The camera then pans around the enclosing gallery, focusing on a procession of portrait drawings and paintings of Borgrajewicz by Topolski. All three men are identified as Polish.

> We are aware that these pictures do not so much fix the face as explore it; that the artist is tracing the detail almost as if by touch; and that each line that is added strengthens the picture but never makes it final. We accept that as the method of the artist. But what physics has now done is to show that that is the only method of knowledge.[53]

The episode continues, encompassing the Holocaust and questions of responsibility for the bombing of Hiroshima and Nagasaki. What it makes explicit is not only the illustration of a point of history or of theory with a relevant picture or image, but how the drawing process of Topolski in particular becomes an aesthetic analogy for understanding how judgement is undertaken.

The painting that Bronowski lingers near as he speaks is a larger-than-life triple-headshot of Borgrajewicz painted in tans and beiges like an ink drawing, a style characteristic of Topolski. Layers of loose lines, sometimes in different colours but rarely colouristic in that little attention is given to the play of local colour, knot into a recognizable face. It is the lines, thick and glutinous, congregating at junctures that follow the creases in Borgrajewicz's skin, that the un-named blind woman admits to mistaking initially for scars, as if lines were traces of eye movement, just as the scars are traces of experience. Shadows cast by the tip of the nose become a purplish-tinged blue-grey as if the skin were eaten away, revealing a cavity in the face. The eyes are asymmetrical, as they are in all the portraits shown by Topolski. The mouth too is a jagged, clenched line and merging point for the sharp wrinkles converging in it. Borgrajewicz's right eye peers deadened and yet evaluating – confrontational. Topolski synthesizes his feeling for, knowledge of and

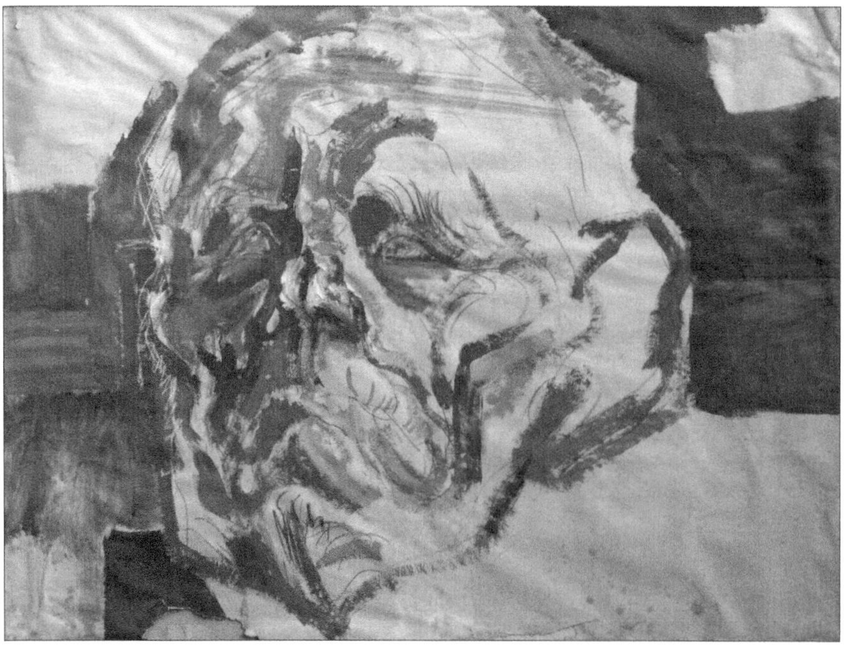

8.3 Feliks Topolski, *Auschwitz Inmate, Survivor (Stefan Borgrajewicz)*, undated, graphite, conté crayon and acrylic paint on paper, Estate of Feliks Topolski © By kind permission of the Estate of Feliks Topolski.

observations of Borgrajewicz in the final painting. Perhaps he is also reading his own past into his observations: the deadened eye an accusation from friends and family, who fell victim to the Holocaust in Poland – Borgrajewicz was a concentration camp survivor, a piece of information that Topolski would probably have found out or been told and is revealed to the viewer at the end of the episode. The fragile, shifting webs of lines speak to the fragility of bodies, time, chance and history. Embedded within *The Ascent of Man*, the painting loses its individual identity and becomes instead an exhibit in an external narrative, a monumental statement of Bronowski's ethics: its fragility becomes the fragile beauty of judgement itself.

The Ascent of Man is one of a range of influential 'authored documentaries' produced in the late 1960s and 1970s, which include Kenneth Clark's *Civilisation* (1969), *Alistair Cooke's America* (1972–1973) and David Attenborough's *Life on Earth* (1979).[54] It was first conceived as a 'history of science' reply to *Civilisation*, which the BBC's Head of Features Group, Aubrey Singer, suggested that Bronowski write and narrate.[55] Though Bronowski objected to Clark's insular approach to culture through fine art, he only agreed to take on the project after the producers agreed to use neither science nor history in the title.[56] Bronowski's stated goal, as recorded in a post-broadcast interview with a *Bioscience* reporter, was to examine the 'special faculties which make human beings different from other animal species' through an exposition of the ideas that shape human values at crucial historical moments.[57] The result received wide acclaim. Much of the efficacy of its episodes arises from the craft of the editing crew in pairing images of Bronowski as he delivers the continuous narrative among buildings or monuments with graphics and associative images. Abstract concepts are made tangible and evocative through visual aids.[58] The 'Knowledge or Certainty' episode is an ethical aside in the progression of the series. Although other artists are referenced in other episodes (Paolo Uccello, William Blake, Georges Seurat and Henry Moore, among others), they serve merely as examples of human ingenuity. Topolski's work within the discussion of knowledge, knowability and the ethics of gaining and testing knowledge takes on an emotional and moral weight missing in the other works.

Topolski was an obvious choice for inclusion in the *The Ascent of Man*. On one level he was experienced, prolific and versatile, producing portraits for the television series *Face to Face* from 1959 to 1962, among other activities.[59] His aptness for inclusion also arose from the intermingling of his personal history and his approach to art. Born in Warsaw in 1907, he immigrated to London in 1935 (roughly 15 years later than Bronowski, who was one year younger). In Britain he quickly established himself in a range of disparate disciplines: he achieved renown as a portraitist (both painting and drawing), contributed graphics to publications ranging from *Punch* to *Lilliput* to *Vogue*, designed theatre sets, and wrote prolifically on artistic and cultural subjects. During the Second World War, he worked as a War Artist and after its conclusion joined the broadening sphere of public art, fuelled by a mural he painted at the 1951 Festival of Britain. In 1953, he founded a broadsheet-inspired combination of text and illustration, *Topolski's Chronicle*, which circulated widely until shortly before his death in 1989. Through his various forms of art, he integrated his eclectic, history-orientated education with a desire to participate in contemporary events. He dabbled in many disparate art movements in Britain – borrowing phenomenological ideas from Modernist Realists, publishing drawings in Social Realist periodicals – though he publicly supported no school.[60] Instead, he abided by a self-formulated credo, which is resonant with the Magnum Photo Agency's insistence on documentation and reportage. He advocated neutrality for artists: the necessity of observing all that happens and creating a synthesis of impartial experience and artistic interpretation. To this purpose, Topolski sketched avidly while travelling around both London and the world, later working his sketches into complex murals, prints and paintings.

The idea that drawing processes were a synthesis of subjective experiences – probing chaotic vision for a coherent pattern – was fundamental to Topolski's working method and self-promotion. Writing of his experiences for *Face to Face* in his autobiography, *Fourteen Letters* (1988), Topolski reflected on Swiss psychiatrist Carl Gustav Jung's refusal to sit for a portrait before his interview: 'There was some reasonable explanations for this, but I was certain that this shrewdest shaman, who gladly faces photographers confident in his nimbleness at turning to them his best cheek, would have felt defenceless if exposed to an artist's circling and perhaps penetrating gaze.'[61] It was not just the gaze, however, but the act of physical witness and participation that provided insight. In a short polemic for *The Burlington Magazine for Connoisseurs* in July of 1943, titled 'Contemporary Comment and Caricature', Topolski described contemporary art's inability to rise to the challenges of war art as the result of fundamental approaches to art: 'The war artist of to-day does not seem to picture the very essence of war with its grandeur, horrors and tragedy but aims more at recording the incident.'[62] 'Synthesis' was the solution to 'incident'. It is the action of the artist at his or her most basic: 'Photography catches a fragmentary glimpse of truth; the artist selects, composes and gives a synthesis of truth, therefore a higher and truer picture than the camera.'[63] It is this 'synthesis' or grouping of elements gathered from experiencing the world that form a final work: one composed of incidents, but able to represent the essence of the subject. In order to do so, artists need to sketch while 'passionately observ[ing] the world not only with open eyes but also open hearts and minds'.[64] The sketching of details was combined with the experience of witnessing – or of being present in the moment with the entire being of the witness. No solitary fragment sufficed to record the essence of an experience; it needed the mind to select, synthesize and create an order to display true meaning.[65] Thus, rather than seeing composed works as tainted by artistry, they were able to 'synthesize' and process the full power of the moment.

Topolski's synthesis was antagonistic to illusions of absolute knowledge. It created a truer picture than a fragment, but in Bronowski's words, it did not so much 'fix the face as explore it ... each line that is added strengthens the picture but never makes it final.'[66] Topolski's style embodied subjective knowledge. As Bronowski suggested, in the portrait of Borgrajewicz, each line remains a note for the viewer, informing us about the habitual movements of the sitter, the rhythm created between the slant of an eye and the nose, the volumes created by their poise, or the confinement of their clothes. The result is stretches of calm smoothness as well as dark, crowded intersections whose balance creates expressive recognizability. Topolski's subjects appear in an arrested negotiation between a solid identity and their particularities. Each line could be seen as a single testimony, and through their conjunction an image emerges, but the image is messy, mobile, fragile. Topolski has privileged visual knowledge of Borgrajewicz, which emerged from numerous studies and sittings. And so, like a scientist experimenting with multiple trials, he is eligible for the privileges of authority: a particular ability to make informed (in this case visual) judgements. This expertise is limited, however: it is personal and only representative of the physical object it was intended to capture. It is a glimpse or an insight into a form created through composite explorations – as the fallible 'truths' of science were believed to emerge from repetitions of research. Just as the reader of *Science and Human Values* finds himself suddenly at a conference table amid the rubble of Nagasaki, in following Bronowski's gesticulation to gaze at the troubling fragility of Topolski's painting, the viewer finds himself feeling the fragility of testimony, of knowledge and of authority.

Topolski's aesthetics were presented by Bronowski as aesthetics of authority. By incorporating his subjectivity, his perceptions' fragility and yet also signs of visual expertise, Topolski exemplified what ethical, personal judgement could be. Rather than authority presented as a clinical, emotionless voice of the inhuman expert, Topolski's paintings portrayed an artistic process that was human, fallible and yet insightful. Pictures such as the portrait of Borgrajewicz represent a form of honesty, not only in its final state but also in the evidence of its construction – no line erased, just added; no polish, just a confrontation with the visual testimony of an observer – presenting not only what authority but also what uncertain knowledge could look like. From this position of humility, personal judgements as a form of critical agency could be made, whether ethical or otherwise. The fear of nuclear and thermonuclear weapons presented a threat to civilization, perhaps as much as the actual weapons did. If members of the general public and scientists reneged on their personal ethical judgements, very real threats from nuclear fallout, if not from warfare, could ensue. Adversely, disarmament threatened the momentum of discoveries, which bolstered the socio-political systems. As Bronowski described in 'Responsibilities of Scientist and Public':

> If an immense revulsion of public feeling does lead to the destruction of the scientific tradition, the world may again enter a dark age as it did after the Goths destroyed Rome. It is not impossible that the whole mechanical and intellectual society which we know could be abolished by a great wave of fanaticism.[67]

The need to understand the nature of mind, the nature of knowledge gaining and the nature of the resultant authority through visual aids thus became an imperative.

Notes

1 Frederic Solomon and Robert Q. Marston, *The Medical Implications of Nuclear War* (Washington, DC: National Academy Press, 1986), xiii.

2 Originally two types of bomb had been produced: a uranium bomb codenamed 'Fat Man' and a plutonium bomb, 'Thin Man'. The choice of names on an immediate level reflected a desire to confuse codebreakers by appearing to refer to Winston Churchill and Franklin Delano Roosevelt. 'Fat Man', a squat plutonium bomb, was successfully tested and a duplicate was dropped on Nagasaki. 'Thin Man' proved a dud and was replaced by 'Little Boy', a uranium-235 gun-type bomb, which was dropped on Hiroshima. Everett M. Rogers and Nancy R. Bartlit, *Silent Voices of World War II: When Sons of the Land of Enchantment Met Sons of the Land of the Rising Sun* (Santa Fe, NM: Sunstone Press, 2005), 230.

3 Cyril Connolly, 'Comment', *Horizon*, vol. 12, no. 72 (December 1945): 365.

4 Bertrand Russell, 'Science and Human Life', in *What is Science? Twelve Eminent Scientists and Philosophers Explain their Various Fields to the Layman*, ed. James R. Newman (London: Victor Gollancz, 1956), 8.

5 Margaret Gowing and Lorna Arnold, *The Atomic Bomb* (London: Butterworth, 1979), 14; Stephen T. Thornton and Andrew Rex, *Modern Physics for Scientists and Engineers*, fourth edition (Boston: Brooks/Cole, Carnegie Learning, 2006), 170.

6 'I did not know we had left the open country … Then suddenly I was aware that we were already at the centre of damage in Nagasaki. The shadows behind me were the skeletons of the Mitsubishi factory buildings, pushed backwards and sideways as if by a giant hand. What I had thought to be broken rocks was a concrete power house with its roof punched in. … The moment of recognition when I realised that I was already in Nagasaki is present to me as I write as vividly as when I lived it … On an evening like that evening, some time in 1945, each of us in his own way learned that his imagination had been dwarfed.' Jacob Bronowski, *Science and Human Values* (London: Hutchinson & Co., 1961), 13–14.

7 David Attenborough, who was Controller of BBC Two when *The Ascent of Man* was first commissioned, later reflected upon this, discussing his surprise when Bronowski began a discussion of dextrorotatory and levorotatory crystals not with a similarly shaped object, but with intricate tile patterns from the Alhambra. 'Specially Recorded Interview: Sir David Attenborough Remembers *The Ascent of Man*', *The Ascent of Man*, directed by Adrian Malone, Dick Gilling, Mick Jackson, David John Kennard (written and narrated by Jacob Bronowski) (1973; London: BBC Worldwide Ltd, 2005), 7:55.

8 Bronowski attributed the first outline to July 1969 and the last shot to December 1972. Jacob Bronowski, *The Ascent of Man* (London: British Broadcasting Corporation, 1980), 13.

9 Rita Bronowski, 'Bruno: A Personal View', *Leonardo*, vol. 18, no. 4 (1985): 224.

10 In the 1974–1980 editions of the book form of the series, a portrait drawing of Bronowski by Topolski comprises the back cover of the book jacket.

11 Megadeath is a measurement unit for casualties, referring to the death of a million people. It was used to estimate the possible effects of nuclear warfare. The *Oxford English Dictionary* sites its first use in the *Birmingham (Alabama) News*, 21 June 1953, E3/1, in reference to Herman Kahn and his circle of strategists dealing not in the numbers of atomic weapons but in kilotons and megadeath. 'megadeath, n.', OED Online, March 2012, Oxford University Press: <http://www.oed.com/viewdictionaryentry/Entry/252335> (accessed 12 May 2012).

12 Gowing and Arnold, *The Atomic Bomb*, 3.

13 Ibid.

14 Herman Kahn, *On Thermonuclear War* (New Brunswick, NJ: Transaction Publishers, 2007), 145; originally published by Princeton University Press, 1960. It would not be until 1981 that Paul Crutzen and John Birks would project the likelihood of a 'nuclear winter' as a result of a large-scale nuclear exchange of fire. Solomon and Marston, *The Medical Implications of Nuclear War*, xiii.

15 This is Kahn's list (in order) of desirable characteristics for a deterrent (cheap is relative to the cost of upkeep and production of non-nuclear arsenals and military equipment). Kahn, *On Thermonuclear War*, 146.

16 By Cold War I am referring to the period of hostilities (short of armed conflict and consisting of threats, violent propaganda and subversive political activities) that began after the summits between Roosevelt (USA), Churchill (UK) and Stalin (USSR) between 1943 and 1945, and continued until the thaw in relations in the mid-1980s.

17 Not to be confused with 'exoatmospheric' testing – which launches a bomb above the atmosphere – atmospheric testing is any explosion on the ground (rather than under it) whose force draws particles from the ground into a mushroom cloud in the atmosphere.

18 A nuclide (with either an artificial or a natural origin) that exhibits radioactivity.

19 Ironically it was in January 1955 that the British Minister of Fuel and Power, Geoffrey Lloyd, announced a ten-year plan to build twelve nuclear power stations, well above the number in development elsewhere in the world. And a month later a UK Defence White Paper informed the public of an earlier, secret decision to pursue development of a British hydrogen bomb just one year after the Bravo incident. David Kynaston, *Family Britain, 1951–1957* (London: Bloomsbury, 2010), 443; and Lorna Arnold and Katherine Pyne, *Britain and the H-Bomb* (New York: Palgrave, 2001), 113.

20 Though after the Cuban missile crisis in 1962, tensions eased as President Kennedy and Premier Khrushchev began to cooperate towards eventual disarmament, and a limited test ban was signed on 5 August 1963 between the USA, the USSR and Britain. Catherine Jolivette, 'Science, Art and Landscape in the Nuclear Age', *Art History*, vol. 35, no. 2 (April 2012): 264.

21 Robert A. Divine, *Blowing on the Wind: The Nuclear Test Ban Debate, 1954–1960* (New York: Oxford University Press, 1978), 9–13; Barton C. Hacker, *Elements of Controversy: The Atomic Energy Commission and Radiation Safety in Nuclear Weapons Testing, 1947–1974* (Berkeley, CA and London: University of California Press, c. 1994), 139–148; Lorna Arnold, *A Very Special Relationship: British Atomic Weapon Trials in Australia* (London: HMSO, 1987), 77–80.

22 For a detailed account of initial public responses, see David Kynaston, *Austerity Britain, 1945–1951* (London: Bloomsbury, 2007), 82–86.

23 Arnold and Pyne, *Britain and the H-Bomb*, 110; and Lord Peter Ritchie-Calder, 'The Sociological Consequences of Nuclear Energy', in *Economic and Social Consequences of Nuclear Energy*, ed. Roger Mellor Makins Sherfield (London: Oxford University Press, 1972), 79.

24 Arnold and Pyne, *Britain and the H-Bomb*, 111. Just as the Report of the British Mission to Japan had projected nuclear damage, based on evidence from Hiroshima and Nagasaki. *Report of the British Mission to Japan on the Effects of the Atomic Bombs at Hiroshima and Nagasaki* (London: HMSO, 1946).

25 James R. Arnold, 'Effects of the Recent Bomb Tests on Human Beings', *Bulletin of the Atomic Scientists*, vol. X, no. 9 (November 1954): 347–348, at 347.

26 As decried by science journalist Lord Ritchie-Calder in 1972: 'From the outset those concerned realised that scientific veracity had been impugned by circumstance. The scientist had been responsible for the achievement which had induced fears and he was being asked to explain away those fears. He was being called upon not only as the expert witness but as the spokesman, not only to give the facts, within the acknowledged limitation of his specialised knowledge, but to extrapolate these facts beyond his knowledge and to pass judgements and express opinions. One of the difficulties is that to the public "scientist" is a generic term; the ordinary person does not distinguish between a physicist, a chemist, or a biologist.' Ritchie-Calder, 'The Sociological Consequences of Nuclear Energy', 78.

27 Gowing and Arnold, *The Atomic Bomb*, 4; and Helen Longino, 'The Social Dimensions of Scientific Knowledge', in *The Stanford Encyclopedia of Philosophy* (Spring 2011 edition), ed. Edward N. Zalta, <http://plato.stanford.edu/archives/spr2011/entries/scientific-knowledge-social/>.

28 The Manhattan Project was taken over by the civilian organization, the Atomic Energy Commission (AEC), in 1946. It comprised 37 installations in 19 states as well as sites in Canada, had 37,800 contractor employees and represented an investment of over $2.2 billion (inflation adjusted, it is roughly equivalent to $27.3 billion in 2012). Merril Eisenbud and Thomas F. Gesell, *Environmental Radioactivity: From Natural, Industrial, and Military Sources*, fourth edition (San Diego, CA and London: Academic Press, 1997), 7; and Gowing and Arnold, *The Atomic Bomb*, 14–16.

29 Such as: John Hardwig, 'Epistemic Dependence', *Journal of Philosophy*, vol. 82, no. 7 (1985): 335–349, and idem, 'Evidence, Testimony, and the Problem of Individualism', *Social Epistemology*, vol. 2, no. 4 (1988): 309–321; Frederick Schmitt, 'On the Road to Social Epistemic Interdependence', *Social Epistemology*, vol. 2 (1988): 297–307; Michael Welbourne, 'The Community of Knowledge', *Philosophical Quarterly*, vol. 31, no. 125 (1981): 302–314. See also Longino, 'The Social Dimensions of Scientific Knowledge'.

30 A.G. Hester, Editorial, *Atomic Scientists Journal*, vol. 5, no. 1 (September 1955): 1.

31 Jacob Bronowski, 'Responsibilities of Scientist and Public', *Atomic Scientists Journal*, vol. 5, no. 1 (September 1955): 5.

32 Jacob Bronowski, *The Common Sense of Science* (Harmondsworth: Penguin Books, 1951), 145–146.

33 Jane Gregory and Steve Miller, *Science in Public: Communication, Culture and Credibility* (Cambridge, MA: Basic Books, 1998), 34.

34 Bryan Appleyard, *The Pleasures of Peace: Art and Imagination in Post-War Britain* (London and Boston: Faber and Faber, 1989), 34.

35 Bronowski, 'Responsibilities of Scientist and Public', 4.

36 Bronowski, *The Common Sense of Science*, 145–146.

37 Charles S. Shapiro, 'Jacob Bronowski: A Retrospective', *Leonardo*, vol. 18, no. 4 (1985): 219.

38 Geoff Coupe, 'My Father, the Bomb and Me', Geoff Coupe's Blog, 10 December 2010, <http://gcoupe.wordpress.com/2010/12/10/my-father-the-bomb-and-me/>.

39 R. Bronowski, 'Bruno', 224.

40 Bronowski's publications crest over a hundred, not including lectures, broadcasts and guest appearances on discussion panels. For a list, see 'Bio-Bibliography of Jacob Bronowski', *Leonardo*, vol. 18, no. 4 (1985): 282–287.

41 See specifically: Russell, 'Science and Human Life', 6–23; Karl Popper, *The Open Society and its Enemies* (Princeton, NJ: Princeton University Press, 1950); Thomas Kuhn, *The Structure of Scientific Revolutions*, second edition with postscript (Chicago: University of Chicago Press, 1970).

42 Paul Saltman, 'The Softest Hard Sell: Bronowski's Approach to Communicating Science', *Leonardo*, vol. 18, no. 4 (1985): 244.

43 Referenced in Bruce Mazlish, 'The Three and a Half Cultures', *Leonardo*, vol. 18, no. 4 (1985): 233.

44 For examples of works that address him within this context, see Mazlish, 'The Three and a Half Cultures'; and Becky E. Conekin, *'The Autobiography of a Nation': The 1951 Festival of Britain* (Manchester and New York: Manchester University Press, 2003), 58, 65–67.

45 Bronowski was not alone in this view: in 1956 a Harvard biologist, George Wald, contributed to a report on Harvard's visual arts, commenting: '… the same interplay of head and hand that goes into the production of a work of art. Just as the scientist has found his place within the university … just as the laboratory has become academically respectable, so the artist and the studio, given time and opportunity, should find their place.' George Wald's contribution, 'The Artist in the University', in the Report of the Committee on the Visual Arts at Harvard University, Cambridge, 1956, in Howard Singerman, *Art Subjects: Making Artists in the American University* (Berkeley, CA and London: University of California Press, c. 1999), 47–48, 72.

46 Though the report was co-authored by the delegates, the wording suggests the influence of Bronowski's then nascent agenda. *Report of the British Mission to Japan on the Effects of the Atomic Bombs at Hiroshima and Nagasaki*, 5.

47 Bronowski, 'Responsibilities of Scientist and Public', 11.

48 Szilárd had co-discovered the crucial Òneutronic reactor that enabled the first chain reaction and yet who also later lobbied (before 6 August 1946) for atomic weapons not to be used against populated targets. 'Knowledge or Certainty', *The Ascent of Man*, directed by Mick Jackson (written and narrated by Jacob Bronowski), 14 July 1973; London: BBC Worldwide Ltd, 2005, 43:28–43:53.

49 Such as Auguste Comte's *A General View of Positivism* (1848), Georg Simmel's *The Philosophy of Money* (1900), and the works of Karl Marx.

50 Jacob Bronowski, 'The Shape of Science in the Arts', Institute of Contemporary Arts, London, lecture transcript (2 August 1951), Tate Gallery Archive 955/1/7/21, 5–8. Interestingly he skips over artists such as Richard Hamilton, who had organized the exhibition 'Growth and Form' at the ICA, also in 1951, which took direct inspiration from crystallography among other scientific processes.

51 Bronowski, 'The Shape of Science in the Arts', 6.

52 Bronowski, 'The Shape of Science in the Arts', 8.

53 'Knowledge or Certainty', 3:12–4:00; and Bronowski, *The Ascent of Man*, 353.

54 The genre of 'authored documentary' was popularized by Clark's *Civilisation* and emphasizes the subjective view of the narrator, whose thesis emerges over several episodes.

55 Bronowski was a natural choice. Series producer Adrian Malone characterized him as having in England 'almost as many public appearances as the Beatles'. Quoted in Christine Russell, 'The Man Behind "The Ascent of Man": Jacob Bronowski', *Bioscience: American Institute of Biological Sciences*, vol. 25, no. 1(January 1975): 9.

56 Ibid., 12.

57 Quoted in ibid., 9.

58 Despite the power of this orchestration, since a television series is always a compromise between numerous parties, the sequences are not considered as a part of Bronowski's authorship.

59 *Face to Face* broadcast interviews by Major John Freeman with prominent public figures. Pencil portraits, often of close-up headshots made the day before in his studio, provided the first point of contact between viewer and interviewee – resonant of a sketch one might make while talking to someone and thus making the total experience both more intimate and penetrative. *Face to Face*, 1959–1962, London: BBC Worldwide Ltd, 2009.

60 'School' is used in the loose sense since critics and gallerists created the two in question. The distinction is adopted from James Hyman's characterization in *The Battle for Realism: Figurative Art in Britain During the Cold War 1945–1960* (New Haven and London: Yale University Press, 2001), 1–9.

61 Feliks Topolski, *Fourteen Letters* (London: Faber, 1988), unpaginated.

62 Feliks Topolski, 'Contemporary Comment and Caricature', *The Burlington Magazine for Connoisseurs*, vol. 83, no. 484 (1943): 166.

63 Ibid.

64 Ibid.

65 Ibid; Feliks Topolski, untitled draft regarding 'picturing' and the camera (c. 1950–1960), The Estate of Feliks Topolski Archive, uncatalogued, 1.

66 Bronowski, *The Ascent of Man*, 353; 'Knowledge or Certainty', 3:12–4:00.

67 Bronowski, 'Responsibilities of Scientist and Public', 4.

9

Painting the End: British Artists and the Nuclear Apocalypse, 1945–1970

Simon Martin

In 1952, *Time* magazine posed the rhetorical question: 'How should a modern artist react to the atomic age?'[1] The article was in the context of a review of the abstract artworks by the Italian Spatialists on show in the Venice Biennale, but it reflected a philosophical concern for all visual artists in the post-war years, for the question of a 'post-nuclear' art exposed the fundamental dichotomy between abstraction and figuration. Although the general atmosphere of unease that characterized the Cold War era was manifested in a non-specific manner in the work of many post-war British artists, relatively few chose to tackle the subject of the atom bomb directly, perhaps due to the difficulty of finding an artistic language with which to visualize an aesthetic of annihilation. The American Abstract Expressionist Jackson Pollock had reflected this conundrum when he told an interviewer in 1950: 'The modern painter cannot express this age, the airplane, the atom bomb, the radio, in the old forms of the Renaissance or any other past culture.'[2]

For those artists that did attempt to tackle the subject of the atom bomb, it was broadly a question of whether to focus on the abstract science behind the technology of the bomb, the spectacle of the mushroom cloud or on the actual (or potential) victims of a nuclear attack. The capacity for abstraction or figuration to reflect the existential issue of the atom bomb divided artists. In 1956, when John Minton gave his last 'Sketch Club' speech at the Royal College of Art, he bitterly satirized the young painters who were then painting abstract action-based art in relation to the nuclear threat:

> ... So you order your sack of cement and put on the Nescafe and begin to paint your board on the floor – not on the easel! That's original, no one else does that. Then you jump on it, off centre, that's to show you're sensitive. Then you paint a few dozen, pay someone to write a preface to the catalogue, number them, give them names ... You could call it anything, you could call it, *Eden Come Home* or ... *Vistavision Film Company*, you could build it for 2 million only – the price of an atom bomb or two films ... You sit in the espresso bar with your portfolio and wonder what you're doing

> ... Keep away, keep away, an atom bomb falls, [but] you're all alright, you're doing your abstract and problems are crowding around like the furniture, till 'put out the light' when you can't bear it no longer. Is it to do with being blinded?[3]

This tension between the impact on a human level and the abstract nature of destruction was embodied in a literal sense in the anthropomorphic names given to the first atomic bombs to be used in warfare: 'Little Boy', which was dropped on the Japanese city of Hiroshima on 6 August 1945 by an American B-29 Superfortress called the *Enola Gay*; and 'Fat Man', dropped on Nagasaki three days later. At least 148,000 people were killed in the bombings, and many more died over time from the effects of radiation. The first photographs of the bombed Japanese cities that were released shortly after the end of the Second World War were to become symbolic of the destructive power of the atom bomb in flattening the cities, but, significantly, images containing human bodies were not released until several years later.[4]

Following the dropping of these bombs, the immediate response of many people in Britain was inescapably ambivalent, for while the bombings of Hiroshima and Nagasaki arguably brought the Second World War to a swift conclusion in the Allies' favour, the scale of the impact of a single bomb was utterly chilling, even when compared to the devastation wreaked by mass bombings on European cities. John Minton, staying in Cornwall in August 1945, reflected on the wide conceptual gulf between the gentle British landscape and the bombing of Hiroshima when he wrote: 'here one feels safe. To read of the atomic bomb is like a strange fiction from outside, here close to the sea, what could be more improbable?'[5] Yet, one of the first visual artists to respond to the atom bomb was the abstract artist John Tunnard (1900–1971), who was living nearby in the equally remote Cornish fishing village of Cadgwith. In the summer of 1945 Tunnard recorded the creation of an oil painting called *Presage July 1945 – Atomic Bomb Flux* (code O.114) in his ledger. During the 1930s and early 1940s, he had been fascinated by the question of how to create abstract visual representations of sound, new scientific technology and engineering, and during the Second World War he had painted a number of works that sought to provide a non-literal representation of RADAR technology, reflecting his experiences as a coastguard in Cornwall close to several RADAR bases that were part of the Chain Home Low network along the coast. The fact that a nuclear explosion produces an electromagnetic pulse (EMP) was known since the atomic bomb was dropped on Hiroshima, not least because nuclear weapons emit a large amount of electromagnetic radiation as visible, infrared and ultraviolet light in a brilliant flash of light. Around this time, Tunnard had painted a number of works representing the regular oscillating frequencies associated with RADAR, and the preceding work in his ledger was a painting entitled *Magnetic Field* (1945, Ferens Art Gallery, Hull), an arrangement of forms in space that suggest the principle of scattered electromagnetic waves. Although it is likely that *Magnetic Field* was painted before the dropping of the atom bomb, it is interesting to note that in this work all sense of horizon has disappeared and the forms appear to radiate out from a central white light.

The first photographs released of a nuclear explosion were those of the blast from the Trinity test conducted by the United States Army on 16 July 1945 in New Mexico, and they had focused on the fireball itself, but later images primarily focused on the mushroom cloud that followed. In 1954, Tunnard designed the cover for a book entitled *Explaining the Atom* by Selig Hecht and Eugene Rabinowitsch, first published in 1947, but with the addition of chapters on the proliferation of bombs, the failure of bomb control and the emergence of atomic power added by Rabinowitsch after Hecht's death in 1947. Tunnard's design incorporated diagrammatic elements with the suggestion of a mushroom cloud over the globe, thus fusing the common image of destruction with the atom's positive uses. It reflected the artist's interest in the science of the atom, which he had explored in his mural inspired by crystallography for the Regatta Restaurant at the Festival of Britain in 1951 (see Chapter 4, Figure 4.3). Although his mural suggested a positive view of the atom, Tunnard's ambivalence was reflected in a brooding watercolour and gouache painting called *Cataclysm* (November 1946, private collection), suggestive of a bleak post-nuclear landscape with intimations of the Neo-Romantic aesthetic of decay, and the Surrealism of Salvador Dalí, who was himself fascinated by the atomic bomb, declaring that 'the atom was [his] favourite food for thought'[6] and producing paintings such as *Atomic Idyll and Melancholic Uranium* (1945, Museo Nacional Reina Sofía, Madrid).

For some of Tunnard's contemporaries the bringing together of the distinct artistic languages of abstraction and Surrealism provided a strategy for reflecting the dark impact of the bomb on the individual. Leslie Hurry (1909–1978), a committed pacifist who was horrified by the bomb's potential for massive annihilation, produced a group of paintings in 1945 with titles such as *The Bomb's All Yours*, *The Follies* and *Atom Bomb* (Figure 9.1). The latter work features a group of three violently contorted women, the elongated limbs of one of these framing a circular explosion. Her outstretched arms are a clear homage to the anguished figure on the right of Pablo Picasso's *Guernica* (1937, Museo Nacional Reina Sofía Madrid), as are the compositional devices of a victim's arm and outstretched foot at the bottom corners. Taking his lead from Picasso's Cubist abstractions, Hurry presents his figures with overlapping and multiple mouths, eyes and noses, both heightening the anguish and suggesting deformity as a result of the bomb.

It is perhaps inevitable that *Picasso*'s anti-war masterpiece, painted in response to the bombing of the eponymous Basque town during the Spanish Civil War, should be a point of reference for artists seeking a way to protest at the bomb. Just as Picasso had not witnessed the bombing of civilians first-hand, but instead had responded to press images, none of the artistic responses to the atom bomb were based on first-hand observation, but rather a personal sense of outrage and fear through the filter of media reports. Hurry combined his reference to *Guernica* with an inversion of the imagery

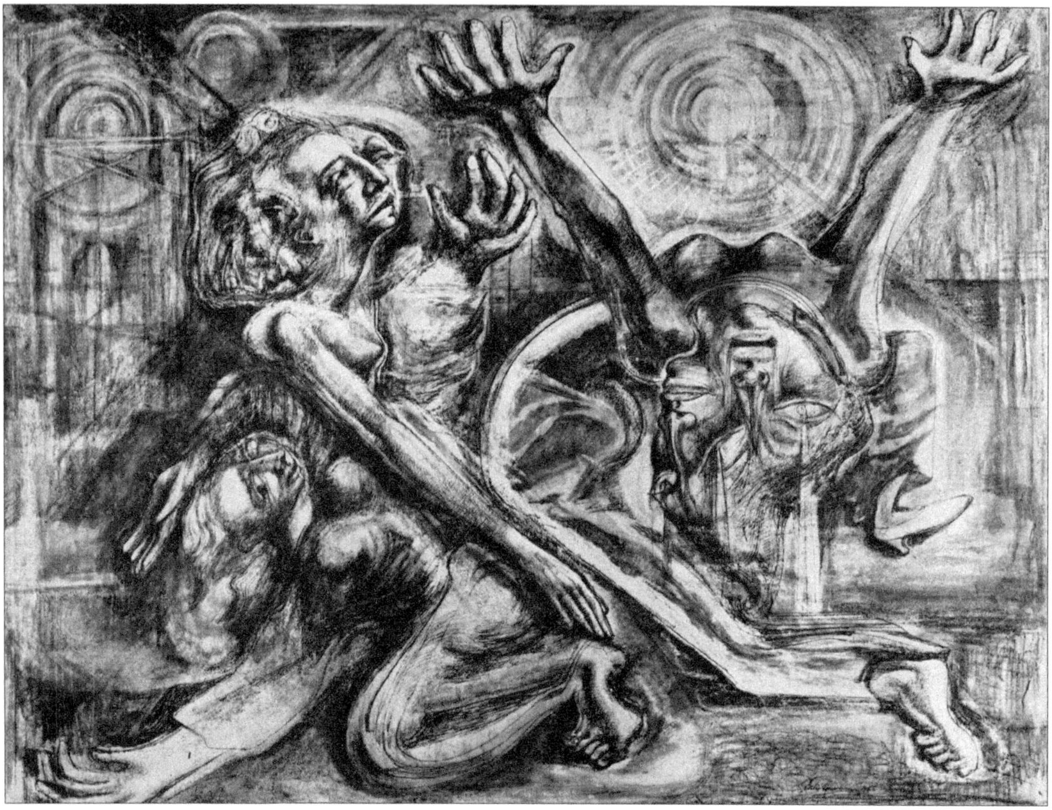

9.1 Leslie Hurry (1909–1978), *Atom Bomb*, 1945, watercolour and chalk on paper, current whereabouts unknown © The Estate of Leslie Hurry.

in William Blake's iconic engraving *Albion Rose*, in which Albion dances 'the dance of Eternal Death' with arms held out before a solar flare. The related painting *The Follies* (1945, private collection) presents a domestic bedroom setting in which a skeletal soldier, with his uniform flayed open to reveal his bony ribs, is surrounded by a whirl of ghostly cherub-like forms. The image of the fleshless skeleton of a soldier was a potent image also employed by Edward Burra (1905–1976) in his macabre watercolour *Skull in a Landscape* (Colour plate 16; 1946, Imperial War Museum, London). In this work a stark skeleton with bony fingers is presented before a decimated barren landscape and a blood red sky. Burra adorned the skull with a soldier's helmet and a beak-like nose, adding to the chilling quality of this apparently post-nuclear image. This painting gave a nod to Otto Dix's 1924 etching *Schädel* from *Der Krieg*, which depicts a worm-ridden skull to represent the death brought about by the First World War, but it also anticipated the iconic poster *Stop Nuclear Suicide* designed by F.H.K. Henrion (1914–1919) for the Campaign for Nuclear Disarmament (1963, Imperial War Museum), which features a photograph of a human skull superimposed on a mushroom cloud. Burra himself was relatively apolitical in his work, instead tapping into deeper anxieties of his age. However, in a letter to his friend William Chappell, he made the darkly humorous comment:

> I've given up dearie & never go out and can scarcely speak I dread going to London & put it off and cant see anyone very neurotic it seems to me but now everything is over except the ATOMS HE HE! I really feel like *death* and I cant go on.[7]

Edward Burra perhaps reflected the wider initial ambivalence of artists in the West, for it was the West, specifically America, which had dropped the weapon. The mixture of jubilation, horror, guilt and fear reflected this awkward position. In the 1959 film *Hiroshima Mon Amour*, by the French director Alain Resnais with a screenplay by Marguerite Duras, the female lead is asked by her Japanese lover what Hiroshima had meant to her in Europe. She replies: 'The end of the war completely. I mean, amazement that they dared, amazement that they brought it off. The start of a new and unknown fear too. Then indifference, and a fear of that indifference.' That new and unknown fear became increasingly acute, particularly after 1949 when it became clear that the Soviet Union possessed the same nuclear capabilities as the USA. In the subsequent years, the impersonal visual spectacle of the mushroom cloud came to define the Bomb in popular culture, as photograph after photograph was released by the Soviet Union, the USA, Britain, France and China as each nation sought to develop their own defence mechanism and flex their muscles in the ideological propaganda war.

Immediately after France had become the world's fourth nuclear power, when it successfully exploded its first atomic bomb, codenamed 'Gerboise Bleue' on 13 February 1960 near the Algerian town of Reggane, John Tunnard painted *Starting Point* (4 March 1960, private collection).[8] It seems that Tunnard's painting may have been inspired by the 100-metre high scaffolding tower from which the bomb was detonated in the sands of the Sahara Desert: the starting point for France's nuclear ambitions. Later that year he painted *Impasse* (August 1960, private collection), which depicts a barren and desolate landscape seen through a framing device of machine-line forms, with an area of white light on the horizon.

Tunnard's response to the apocalyptic potential of nuclear weapons was most strongly expressed in his gouache *Aftermath* (1962, private collection), which was painted on 3 June 1962, the same day that the nuclear bomb dubbed 'Bluegill' was tested on Johnstone Atoll in the Pacific Ocean. Bluegill was part of a series of nuclear tests conducted by the United States in 1962 called 'Operation Dominic', which were the last atmospheric tests before the Partial Test Ban Treaty was signed the following year in Moscow. Many photographs of the iconic mushroom clouds that resulted from the tests were published as part of the propaganda war with the Soviet Union, and in the weeks running up to the beginning of June 1962, 15 high-altitude nuclear explosions (HANE) were tested to determine the effects of the blast and radiation in the exo-atmospheric environment. In Tunnard's image, the brush has been dragged through dark watercolour or gouache to create the suggestion of a mushroom cloud over a dark horizon, seen from a distance. It is significant that the first

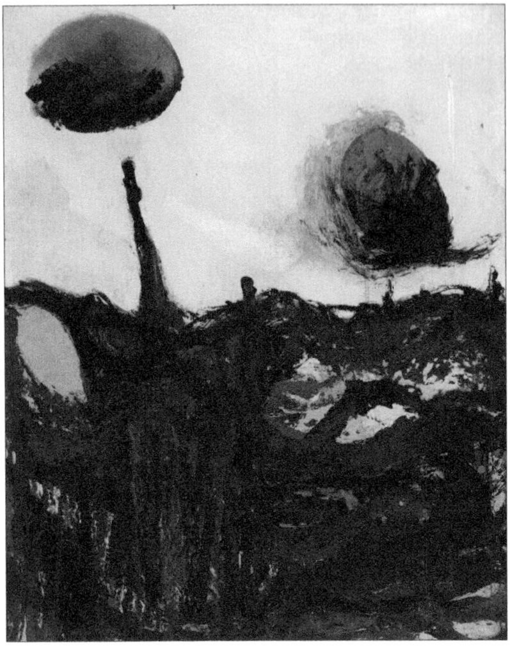

9.2 William Crozier, *Essex Landscape*, 1958–1959, oil on board, 77 × 62.5 cm, private collection © The Estate of William Crozier.

owner of the picture was the mathematician and science writer Jacob Bronowski, who later presented the celebrated BBC television series *The Ascent of Man* (1974). Whilst in Operations Research during the Second World War, Bronowski had been an official observer of the after-effects of the atomic bombing of Nagasaki, an experience of which he recalled in *Science and Human Values*: 'We looked up and saw the power of which we had been proud loom over us like the ruins of Nagasaki … the power of science for good and for evil.'[9]

The series of atmospheric tests of early hydrogen bombs by Britain in the central Pacific between 1956 and 1958, known as Operation Grapple, led to growing opposition to nuclear weapons in Britain. Galvanized by an article in the *New Statesman* magazine by J.B. Priestley on 'Britain and the Nuclear Bombs' on 2 November 1957, the Campaign for Nuclear Disarmament (CND) was formed on 17 February 1958. The supporters of CND included scientists, academics, journalists, writers, religious leaders and artists, in particular Barbara Hepworth, Patrick Heron, Henry Moore and Ben Nicholson. At Easter 1958 the Direct Action Committee organized a 52-mile march from Trafalgar Square in London to the Atomic Weapons Research Establishment at Aldermaston, which thereafter was held as an annual Easter march starting at Aldermaston and ending in London, drawing 60,000 participants in 1959, and 150,000 in 1961 and 1962. As Eduardo Paolozzi noted: 'In a world dusted with Strontium 90, the note of impotent fury can again be heard.'[10] The abstract artist William Crozier (1930–2011) took part in the Aldermaston march, and between 1958 and 1959 painted a series of desolate landscapes of Essex which express a sense of unease and anxiety. Whilst there are no explicit narrative elements in the dark forms, the splashed and dripped technique of paintings such as *Essex Landscape* (Figure 9.2) were to reflect Crozier's concerns about nuclear war at this time and his engagement with Art Informel in continental Europe.

The philosopher Bertrand Russell addressed the central question of man's survival in his 1961 book *Has Man a Future?* in which he observed: 'I am writing at a dark moment and it is impossible to know whether the human race will last long enough for what I write to be published.' In 1960 Russell had resigned from CND to form the direct action Committee of 100, which included the artists Augustus John and the German-born Gustav Metzger. Metzger had formulated a type of performance-based public art that he called 'Auto-Destructive Art' in 1959, which he later described as 'an attack on capitalist values and the drive to nuclear annihilation'.[11] Metzger's theories drew on Nietzsche's philosophies, reflecting Gilles Deleuze's commentary on Nietzsche: 'In auto-destruction the reactive forces themselves are denied and lead to nothingness.'[12] Metzger's manifesto *Auto-Destructive Art*, issued on 10 March 1960, declared:

> Man in Regent Street is auto-destructive.
>
> Rockets, nuclear weapons are auto-destructive.
>
> Auto-destructive art.
>
> The drop drop dropping of HH bombs.
>
> Not interested in ruins, (the picturesque).
>
> Auto-destructive art re-enacts the obsession with destruction, the pummelling to which individuals and masses are subjected.
>
> Auto-destructive art demonstrates man's power to accelerate disintegrative processes of nature and to order them.
>
> Auto-destructive art mirrors the compulsive perfectionism of arms manufacture – polishing to destruction point.
>
> Auto-destructive art is the transformation of technology into public art. The immense productive capacity, the chaos of capitalism and of Soviet communism, the co-existence of surplus and starvation; the increasing stock-piling of nuclear weapons – more than enough to destroy technological societies: the disintegrative effect of machinery and of life in vast built-up areas on the person …

Despite attempts to stop him, Metzger performed a demonstration of Auto-Destructive Art on 3 July 1961 at the International Union of Architects Conference at London's South Bank. His *Acid Action Painting* involved him painting, flinging and spraying hydrochloric acid onto three nylon canvases that were arranged behind each other. The instant disintegration of the canvases presented a powerful metaphor for the action of a nuclear attack on the human body. Metzger's approach shared affinities with other international artists at this time, such as Jean Tinguely who performed his self-destructing sculpture *Study for the End of the World 2* in March 1962 near a nuclear test site in Nevada, and the European Arte Nucleare movement which included Arman, Enrico Baj, Yves Klein, Asger Jorn, Piero Manzoni and Antonio Saura. In 1957 they had issued the Second Manifesto of Nuclear Art, which asserted: 'We Nuclears today denounce the final of conventions – style.'[13] Yves Klein, who saw materialism as one of the principal problems of the age and a consequence of the dematerializing possibilities of the atom bomb, declared:

> I am happy to be dealing with a problem that is so much of our time. One must – and this is not an exaggeration – keep in mind that we are living in the atomic age, where everything material and physical could disappear from one day to another, to be replaced by nothing but the ultimate abstraction imaginable.[14]

The work of the British Pop artists in the early 1960s reflected the uneasy, ambivalent relationship between consumerism and the nuclear threat.

David Hockney, then a student at Bradford School of Art, had taken part in the 1958 Aldermaston march and designed posters for CND, while Richard Hamilton had marched whilst carrying a cardboard cut-out of the American film star Marilyn Monroe. Much of Hamilton's work at this time explored the rhetoric of consumer persuasion in American film and advertising which, although broadly at odds with the politics of the Left associated with CND, were brought together in a work called *Hugh Gaitskell as a Famous Monster of Filmland* (1964, oil and photo-montage on panel, Arts Council Collection), which referenced American films to make a satirical comment on the leader of the Labour Party who had refused to commit to a non-nuclear policy, despite a vote at the 1960 Labour Party Conference in favour of unilateral nuclear disarmament. Hamilton painted over a newspaper photograph of Hugh Gaitskell, so that he resembles a photograph of the actor Claude Rains as he appeared in the 1943 film *The Phantom of the Opera*, which Hamilton had found on the cover of the American magazine *Famous Monsters of Filmland*.

Whilst artists such as Derek Boshier and Gerald Laing made incidental references to the tensions between the USA and the USSR, such as Laing's *Souvenir of the Cuba Missile Crisis* (1963, oil on wood, private collection) and Boshier's *Situation in Cuba* (1961, private collection), in reality very few British figurative artists made a direct response to the question of the bomb during the 1960s. The situation was not much different to the USA, where the American pop artists largely chose to appropriate images from mass media rather than creating direct political statements, presenting the bomb with kitsch immediacy as yet another aspect of modern American culture. Andy Warhol and Roy Lichtenstein had both chosen to present the mushroom cloud as the ubiquitous and depersonalized image of the atomic bomb in the West: Warhol's *Atomic Bombs* (1965, Daros Collection, Switzerland) featured 28 screen-printed mushroom clouds, while Lichtenstein's *Atom Burst* (1966, Collection of the Modern Art Museum of Fort Worth, USA) depicted in Ben-Day spots an image that had become so banal through reproduction that it had seemingly lost its charge.

Conspicuously absent from the majority of the artistic responses to the atomic age were the individual victims of a nuclear attack.[15] The narrative format of fictional novels and films had provided a readier format for such concerns to be addressed, principally: Nevil Shute's 1957 novel *On the Beach* and its subsequent film adaptation in 1959; Walter M. Miller's novel *A Canticle for Leibowitz* (1959); Eugene Burdick and Harvey Wheeler's 1962 novel *Fail-Safe* and its subsequent film adaptation by Sidney Lumet; and Peter George's 1958 novel *Red Alert*, which formed the basis for Stanley Kubrick's 1963 film masterpiece *Dr Strangelove or: How I Learned to Stop Worrying and Love the Bomb*, which ends with mushroom clouds exploding around the globe to the sound of the British wartime song 'We'll Meet Again'. Significantly Alain Resnais' film *Hiroshima Mon Amour* featured harrowing images of victims in sequences that drew attention to the human impact at Hiroshima.

Whereas most artists dealt with the spectacle of explosion, the British pop artist Colin Self (b. 1941) created a direct link between the weapons and the victims of nuclear attack. Self was spurred into action by the Cuban missile crisis in October 1962, when the USA and the Soviet Union stood on the brink of nuclear war over the discovery of medium-range ballistic missiles on new military bases in Cuba, 140 kilometres from the US coast. The US launched a naval blockade of Cuba and B-52 bombers, armed with nuclear weapons, were sent into the skies ready for action. Self was a student at the Slade School of Fine Art in London at the time of the crisis and had gone to the Eagle pub nearby with his fellow students, Michael Kenny and David Thompson, to mournfully await 'the end'. With the passing of the 2pm deadline for an ultimatum between President John F. Kennedy of the USA and Nikita Khrushchev, First Secretary of the Communist Party of the Soviet Union, like millions of others, Self was able to breathe a sigh of relief. His response was to paint a work entitled *Humanity Hanging by a Thread*, reflecting the comments of the Soviet General Anatoly Gribkov that 'Nuclear catastrophe was hanging by a thread'. Instead of the ubiquitous mushroom cloud, here we have a cancerous, decapitated head representing the situation that faced society at that time. It was painted in a raw 'unfinished' style to express the power of his own fear, and based on a news photograph of a sinister, grimacing man that Self had pasted into his 1962 Slade sketchbook.

Another painting in the same series, called *That's the Trouble with the Bastards*, relates to his feeling about the stand-off and features a hybrid of a modern F-4 nuclear bomber and a Pterodactyl, a flying reptile of the Mesozoic Era that hunted by swooping on fish. On one wing is the insignia of the US Air Force and the eponymous angry message like a graffiti slogan. Self believed that man would be the masters of their own extinction, and saw the Americans and Soviets as equal aggressors. At the same time Self created a related series of multiple-plate etchings such as *Bomber No. 2* (1963) and *4th Monument* (1964) printed from found metal parts to create menacing images from the detritus of modern culture, which are embossed with military insignia and form schematized missiles. These innovative etchings have the appearance of children's toys such as Meccano, but closer inspection of the transfers on the wings, such as the US Air Force insignia, makes it evident that these are concerned instead with the war games of adults.

In Self's *Waiting Women and Two Nuclear Bombers (Handley-Page Victors)*, which he painted in 1962–1963, a fashion parade of models taken from the pages of *Vogue* magazine appear in a landscape beneath two Handley-Page Victors – the jet bomber aircraft that provided Britain's nuclear deterrent, which was introduced in 1958 and designed for high-speed penetration of Soviet airspace to deliver a free-fall nuclear weapon. The suggestion is that women are the victims of warfare, and men by implication the perpetrators. Self observed how he portrayed men as:

aggressive, destructive, warmongering, 'God's law' violating, murderers of babies (an unborn baby is hidden on one bomber undercarriage). Womanhood is passively waiting, suffering, victims living quieter lives around the gaps between the brinkmanship and storms of the men (as World War II wives waited in my childhood.)[16]

The association between violence and predatory sexual behaviour was most strongly expressed in his iconic sculpture *Leopardskin Nuclear Bomber No. 1,* created in 1962–1963 from aluminium, steel, wood and fake leopard-skin. Self's use of fake animal fur, together with the phallic elongated pink fuselage of the plane, symbolizes male aggression. Like Meret Oppenheim's iconic fur-covered cup and saucer *Object (Le Déjeuner en fourrure)* (1936, Museum of Modern Art, New York), it is a subversive and surreal object, combining the Surrealist interest in 'Eros' and 'Thanatos' (sex and death) in a single form. Self explained that this hybrid sculpture was 'conceived to emphasize the deadliest aspects of a nuclear machine of destruction with the most fearsome associations of predatory animals'.[17] By using fake leopard-skin and other non-fine art materials, Self signified his belief that 'all art is sham', just like the situation it represented, whereby the population was subjugated through fear.[18] In the mid-1960s Self further explored the metaphorical association between the aggressive aspects of animals and human militarism in his fearsome drawing *Guard Dog on a Missile Base No. 1* (1965, Tate), which was created in response to the American deployment of nuclear missiles on airbases in England. Self had been staying with friends on a farm next to the American airbase at Brandon in West Norfolk, where he was shocked to see a huge Thor Intercontinental Ballistic Missile pointing skyward, 'like a nuclear Nelson's column'.[19] Later that night he was woken by the baying guard dogs. In the drawing, the ferocious animal's teeth mirror the sharpness of the bank of Bloodhound surface-to-air missiles, whose very name is a reference to hunting dogs. Self's drawings pose the same question as that of the great landscape historian W.G. Hoskins:

[What] has happened in the immemorial landscape of the English countryside? Airfields have flayed it bare wherever there are level, well-drained stretches of land above all in eastern England … Over them drones, day after day, the obscene shape of an atom-bomber, laying a trail like a filthy slug upon Constable and Gainsborough's sky. England of the Nissen hut, the 'pre-fab' and the electric fence, of the high barbed wire around some unmentionable devilment.[20]

Missiles formed the focus of other works by Self, such as his sculpture *Victor Valiant Nuclear Bomber and Nuclear Missiles,* which belonged to Terry Southern, the author of the screenplay for the film *Dr Strangelove.* Its subject matter and the materials were the opposite of what was considered 'high art' at that time, but Self intended his works to be anarchistic and unpalatable. His use of non-traditional art materials such as fake grass, wood and fibreglass reflects his belief that the situation in which society was oppressed by the threat of nuclear missiles was a fabrication.

Self had travelled to the USA in 1962 and 1965 and was fascinated to see fallout shelter signs on the city streets of New York. He produced a series of drawings of women in fallout shelters eating hamburgers and hot dogs or using the latest cosmetic machinery, such as *Fallout Shelter No. 2 (Public Shelter 110 – Woman on a Bar Stool with Double Cheeseburger)* (1965, private collection), to highlight how people used consumer products to avoid facing their fears. These drawings reflect Self's fascination with the subliminal way in which the manufactured products of a society often expressed the very things that were evident in that society, such as eroticism and fear, and his belief that: 'People appear to have forced fear of warfare from their everyday existence by developing a greed for these comforts.'[21]

Self did not intend the series to be misogynous, but rather to reflect his fascination with the odd behaviour of New York's startling, lonely women, the eroticism of advertising and machine design, the current political tensions and the implications of the fallout shelters. In his 'Notes November 1965–November 1967', Self listed his interests as including:

Upheavals, displacement and transference of emotion. Spiritual osmosis of sensation, thought, love or aggressive drives into new behaviour. The play on the spiritual osmosis in industrialized society or in wartime. Suppression of love, sexual force or whatever and its new (sometimes unconscious) emergence into new forms. Be it furniture or car design, new art or love. People starved of their natural environments. Adjusting to new environments and situations ... How people have adjusted to the raw truth of the thought of atomic warfare. The new culture incorporates crisis into its new behaviour patterns.[22]

He came to see the women collectively as 'Modern descendants of Liberty'[23] and the drawings as depicting 'the plight of the Cold War on Liberty herself'.[24] When Self showed his Fallout Shelter drawings that he had begun in New York to some of David Hockney's students at the University of Colorado, one of them gave him a fallout shelter sign as a memento, which he kept in his studio in the subsequent decades. The critic Christopher Finch suggested that Self had presented:

one side of the balance of terror and pleasure which, he suggests, defines the character of our age. The balance (or paradox) is crystallized in his series of fall-out shelter drawings ... as a background to all of these leisure activities appears the leitmotif of the fall-out shelter sign. The paradox is established. No overt moral is drawn but the fact is stated.[25]

Following his second trip to the USA in 1965, Self began a series of drawings and sculptures of hot dogs. Unlike the Pop images by Wayne Thiebaud and Andy Warhol of fast food, Self's meticulously rendered drawings of hot dogs were not straightforward celebrations of a contemporary icon, but rather a wider comment on society. In his 'Notes November 1965–November 1967' he wrote: 'To me a hot dog is as important a twentieth-century development as (say) a rocket. Both reflect a stage reached, are products of our condition.'[26]

Their form became a visual sexual innuendo but their charred appearance referenced the threat of nuclear holocaust.

Self adapted other aspects of American popular culture, such as imagery from children's cartoons, with black humour to convey an underlying political message. In *Pluto and the H-Bomb* (c. 1969, private collection), Self's carefully shaded pencil drawing of the Disney character eagerly fetches a cartoon hydrogen bomb, which is about to explode. The wry suggestion is that the idiotic dog, wagging his tale with innocent expectation of reward, has 'returned The Bomb back to humanity after the nuclear bomb was finally banned for good'.[27] There was a similar darkly humorous message in *Exploded Mickey* (1964–1965, private collection), a unique print created by sending an inflatable Mickey Mouse covered with paint through a press so that it looked like the popular character had been blown up. In a later drawing, Self superimposed *That's all Folks!* – the Looney Tunes slogan that appeared at the end of each cartoon – onto an atomic mushroom cloud to comment on the fact that, just as people do not take seriously the violence in cartoons, so too they do not take seriously the very real threat of nuclear war.

Between 1964 and 1974 Self worked on a hugely ambitious series of works entitled *1000 Temporary Objects of Our Time*, in which he spray-painted around objects to create a hypothetic record of existence in the event of its destruction. Self wanted to express the state of what he called 'nuclear non-being', recording the negative space that an object left behind, rather than the object itself. He decided to produce a suite of prints in which he placed an object on an etching plate and sprayed etching resist around it. The objects and figures depicted do not have hard edges as the result of Self's technique of spraying fine particles of atomized liquid, which enabled him to penetrate beneath the object or body to create a sense of movement that he described as 'Post-atomic Cubism'. Like Yves Klein's *Anthropométries* created in the late 1950s and early 1960s, which featured the imprint of a paint-smeared female nude on a bare canvas, Self's *Figure No. 2 (Nude Triptych)* (1971, Pallant House Gallery) suggested a fleeing figure caught in a blast of light or the thermal shadows of victims in Hiroshima or Nagasaki that remained on the pavements and walls as the only trace of their existence.

Self was one of the only artists in the West to create work directly inspired by the victims of Hiroshima. In 1966 he produced a series of ink drawings based on photographs of Hiroshima obtained from the Keystone press agency in London. These were not smooth-skinned, beautiful models with elegantly coiffed hair, but women who were horrifically scarred and burnt. He used some of the press photographs directly in a series of collages called *Fact and Fiction* in which he sought to show the impossible task of the artist in depicting such a reality. *Fact and Fiction 3* features a photograph of a female Hiroshima victim juxtaposed with a science-fiction monster that Self had created with newsprint. The reality of what man had inflicted on man was of course far more disturbing than anything the artist could create. The same female victim appears in another collage of the series, entitled *Fact and Fiction 4*, together with other photographs of bomber planes, nuclear missiles, radar dishes and atomic mushroom clouds. These aircraft and missiles are juxtaposed with photographs of birds of prey, expressing Self's ideas of the connection between predatory birds and animals and modern weaponry. At the bottom of the collage New York skyscrapers are placed near to the bombed city of Hiroshima, while a reproduction of a distorted figure by Francis Bacon is presented amidst a group of photographs of scarred Hiroshima victims, an inclusion that leaves the viewer wondering whether Self intended Bacon's uncompromising portrait to be the 'fiction' that exposes the horror of the surrounding 'fact'.

Self's paranoia concerning nuclear weapons culminated in an extraordinary sculpture entitled *Beach Girl Nuclear Victim* (1966, Imperial War Museum). This horrific life-size sculpture of a charred and mutilated victim of an atomic bomb attack was based on a real girl that Self saw sunbathing on a beach in California when he was visiting David Hockney in 1965. Constructed from fibreglass, hair, cinders and polyurethane paint on a plaster window dresser's dummy, Self described it as representing the culmination of his 'seven year journey ... through a Nuclear Winter of Non-Being'.[28] His ideas for the sculpture were first recorded in the sketchbook of his trip to the USA in summer 1965: 'California girl on beach dead from fallout, with bikini on or part on. With rockets over the top or near figure ... Revolving head (like hairdresser windows). Head half normal and half like Hiroshima with expensive capitalist wig on.'[29] It was a controversial exhibit at the Robert Fraser gallery in 1966 and in the words of the critic Christopher Finch:

> This corpse has a visual fascination which prevents it from being taken simply as a gesture of protest, a moral icon. It becomes a surreal object implying a wholly surreal climate; a plastic equivalent for the Theatre of Cruelty; a cataclysmic heir to Egyptian funerary art.[30]

In contrast to the 1950s 'Geometry of Fear' sculptures by Reg Butler, Geoffrey Clarke, Bernard Meadows and Eduardo Paolozzi, which were closely associated with the cultural climate of the Cold War, Colin Self's nuclear victim has an uncompromising directness. But Self's nuclear victims were in many ways too direct, too horrific, to be easily palatable by the wider art establishment, or even used as propaganda for the anti-bomb cause. Significantly, it was Henry Moore's almost classical *Falling Warrior* (1956–1957, bronze, Tate) that was proposed by John Berger as an emblem for CND, albeit unused. Moore's work encapsulated the tension between abstract and figurative responses to the Bomb, and the ambivalence of artists and the wider public. Moore had been an early supporter of CND, but in the 1960s he was commissioned to create a sculpture to commemorate the 25th anniversary of the first nuclear reaction in the exact location where the Manhattan Project team had split the atom in December 1942. Moore's sculpture *Nuclear Energy* (1964–1966, bronze, University of Chicago; see Chapter 2, Figure 2.5) sought to bring together both the positive and destructive forces of the atom. In doing so, Moore drew on the imagery that had been used by other artists to conflate the mushroom cloud, notions of shelter and protection, and the human skull. Moore later observed:

> The top of it is like some large mushroom, or a kink of mushroom cloud. Also it has a kind of head shape like the top to the skull but down below is more an architectural cathedral. One might think of the lower part of it being protective form and constructed for human beings and the top being more like the idea of the destructive side to the atom. So between the two, it might express to people, in a symbolic way, the whole event.[31]

In its symbolism Moore's humanist sculpture fused together the distinct artistic languages of abstraction and figuration which, during the immediate post-nuclear years, had seemed so mutually exclusive (like the seemingly state-sponsored opposition between Soviet Realism and American Abstract Expressionism). Artistic creation in the face of destruction is an inherently contradictory act that can seem simultaneously both defiant and futile. It recalls the words of J. Robert Oppenheimer (1904–1967), the so-called 'father of the atomic bomb', who linked creation with destruction in his observation after the Trinity test on 16 July 1945 that it had brought to mind words from the Bhagavad Gita: 'Now I am become Death, the Destroyer of Worlds.' In the following years, the rhetorical question posed by *Time* magazine in 1952 – 'How should a modern artist react to the atomic age?' – received numerous powerful responses. Most of the artworks created in response to the atomic bomb are rooted in the specific atmosphere of the Cold War, but they are far more than just historic documents of an age of fear. Despite the depressing subject matter, these works can nonetheless be seen as acts of defiant courage and optimism. In many ways they are timeless meditations about the human condition: how man faces death, and fear of death, but continues to create art.

Notes

1 'Outside is Everything', review of the 1952 Venice Biennale, *Time*, 26 May 1952.

2 Jackson Pollock, quoted in Francis Frascina, *Pollock and After: The Critical Debate* (London: Routledge, 2001), 122.

3 John Minton, Last 'Sketch Club' Speech given at the Royal College of Art, 1956 [from a transcript made on the occasion by Anne Martin], in David Mellor, *The Sixties Art Scene in London* (London: Barbican Art Gallery, 1993), 28.

4 See Paul S. Boyer, *By the Bomb's Early Light: American Thought and Culture at the Dawn of the Atomic Age* (New York: Pantheon, 1985), 5 and 8–9.

5 John Minton, letter to Judith Hollman, a student of Minton's at Camberwell (Cornwall, August 1945), Tate Gallery Archive 918.1.3.

6 Salvador Dalí, *The Unspeakable Confessions of Salvador Dalí as Told to Andre Parinaud*, trans. Harold J. Salemson (New York: William Morrow, 1976), 216 (see also 202).

7 Edward Burra, letter to Capt. Chappell 258507 from Springfield, 18 August 1945, in *Well, Dearie! The Letters of Edward Burra*, ed. William Chappell (London: Gordon Fraser, 1985), 127.

8 In the same fortnight, the first American intercontinental ballistic missile (ICBM) was launched from Cape Canaveral on 24 February and China launched its first liquid-propelled rocket on 19 February.

9 Jacob Bronowski, *Science and Human Values* (1956; London: Hutchinson & Co., 1961), 3. Jacob Bronowski (1908–1974) is sometimes also called 'Joseph'.

10 Michael Middleton, *Eduardo Paolozzi*, exhibition catalogue, 1962, unpaginated.

11 Gustav Metzger, *Auto-Destructive Art Machine Art Auto-Creative Art*, 23 June 1961, unpaginated.

12 Gilles Deleuze, 'Active and Reactive', in *The New Nietzsche*, ed. D.B. Allison, first published 1977 (Cambridge, MA: The MIT Press, 1985), 101.

13 *Second Manifesto of Nuclear Art*, 1956, unpaginated.

14 Yves Klein, 1958, in Otto Piene and Lawrence Alloway, *Zero* (Cambridge, MA: The MIT Press, 1973), 10.

15 Bertrand Russell, *Has Man a Future?* (Harmondsworth: Penguin Books, 1961), 127.

16 Colin Self, 'Final Voice', *Pallant House Gallery Magazine*, no. 4 (September 2004): 72.

17 Colin Self, letter to Marco Livingstone, 23 September 1987, quoted in Livingstone, *Pop Art: A Continuing History* (London: Thames and Hudson, 2002), 163.

18 Colin Self, email to Simon Martin, 6 May 2008.

19 Colin Self, interview with Simon Martin, 28 March 2008.

20 W.G. Hoskins, *The Making of the English Landscape* (Harmondsworth: Penguin, 1970), 299.

21 Marco Livingstone, *British Pop* (Bilbao: Museo de Bellas Artes, 2005), 447.

22 Colin Self, 'Notes November 1965–November 1967', in John Russell and Suzi Gablik, *Pop Art Redefined* (London: Hayward Gallery, 1969), 113–114.

23 Self, email to Simon Martin, 6 May 2008.

24 Ibid.

25 Christopher Finch, *Image as Language: Aspects of British Art 1950–1968* (Harmondsworth: Pelican, 1969), 128.

26 Self, 'Notes November 1965–November 1967'.

27 Colin Self, email to Simon Martin, 8 May 2008.

28 Self, letter to Marco Livingstone, 23 September 1987, quoted in Livingstone, *Pop Art*, 132.

29 Colin Self, Slade Sketchbook, 1961–1963, collection of the artist.

30 Finch, *Image as Language*, 132.

31 Henry Moore, quoted in David H. Katzive, 'Henry Moore's *Nuclear Energy*: The Genesis of a Monument', *Art Journal*, vol. 32, no. 3 (Spring 1973): 284–288, at 286.

Select Bibliography

Abbreviations

FAAG: Fondation Alberto et Annette Giacometti.

HMSO: Her Majesty's Stationery Office.

ICA: Institute of Contemporary Arts.

RAA: Royal Academy of Arts.

RCAHMS: Royal Commission on the Ancient and Historical Monuments of Scotland.

TGA: Tate Gallery Archive

TNA: The National Archives, Kew, Richmond, Surrey, UK.

Printed Primary Sources

Advising the Householder on Protection Against Nuclear Attack. London: V&A Publishing, 2008 (original pamphlet issued by HMSO in 1963).

'After the Atom Bomb: An Astonishing Rebirth'. *Picture Post*, 24 August 1946.

Alloway, Lawrence. *Nine Abstract Artists: Their Work and Theory.* London: Tiranti, 1954.

Arendt, Hannah. *The Human Condition.* Chicago: University of Chicago Press, 1958.

Arnold, James R. 'Effects of the Recent Bomb Tests on Human Beings'. *Bulletin of the Atomic Scientists*, vol. X, no. 9 (November 1954): 347–348.

Artaud, Antonin. *Van Gogh Le Suicidé de la Société* (February 1947). Paris: K éditeur, 1947. In *Artaud Anthology*, ed. Jack Hirschman. San Francisco: City Lights Books, 1965.

'Atom Parade: As Britain Explodes Her First Atomic Bomb, We Review Names that Make Atomic History'. *Picture Post*, 11 October 1952.

Barbara Hepworth: Carvings and Drawings. Introduction by Herbert Read. London: Lund Humphries, 1952.

Beebe, William. 'Evolution of a Lizard into a Bird'. *Proceedings of the Eighth American Scientific Congress*, vol. 3 (1942): 296.

Beebe, William. *Half Mile Down*. New York: Harcourt, Brace and Company, 1934.

Benjamin, Walter. *Charles Baudelaire: A Lyric Poet in the Era of High Capitalism*, trans. Harry Zohn. London: New Left Books, 1973.

Berger, John. 'John Bratby'. *New Statesman and Nation*, 25 September 1954.

Berger, John. 'Machine-Life Painting'. *The New Statesman and Nation*, 18 April 1953.

Bernal, J.D. 'Architecture and the Scientist'. *RIBA Journal*, 3rd series, vol. 44, no. 16 (June 1937): 805–812.

Bernal, J.D. 'Art and the Scientist'. In *Circle: International Survey of Constructive Art*, ed. J.L. Martin, Ben Nicholson and Naum Gabo, 119–123. London: Faber and Faber, 1937.

Bernal, J.D. 'My Time at the Royal Institution, 1923–27'. International Union of Crystallography. Accessed 12 December 2012. <http://www.iucr.org/people/crystallaphers/bernal2>.

Bernal, J.D., and Helen Megaw. 'The Function of Hydrogen in Intermolecular Forces'. *Proceedings of the Royal Society*, Series A, 151: 873 (2 September 1935): 384–420.

Borthwick, Alastair. 'The Conquest of Power'. *Scotland's Magazine*, 1951. RCAHMS, Bib. no: S7237 MS2329/SCT/25/1/14.

Bragg, William L. *Atomic Structure of Minerals*. Ithaca, NY: Cornell University Press; London: Humphrey Milford; Oxford: Oxford University Press, 1937.

Bratby, John. *Breakdown*. London: Hutchinson, 1960.

Bronowski, Jacob. *The Ascent of Man*. London: British Broadcasting Corporation, 1980.

Bronowski, Jacob. *The Common Sense of Science*. Harmondsworth: Penguin Books, 1951.

Bronowski, Jacob. 'The Creative Process'. *Leonardo*, vol. 18, no. 4 (1985): 245–248.

Bronowski, Jacob. *The Face of Violence: An Essay with a Play*. London: Turnstile Press, 1954.

Bronowski, Jacob. *The Origins of Knowledge and Imagination*. New Haven and London: Yale University Press, 1978.

Bronowski, Jacob. *The Poet's Defence*. Cambridge: Cambridge University Press, 1939.

Bronowski, Jacob. 'Responsibilities of Scientist and Public'. *Atomic Scientists Journal*, vol. 5, no. 1 (September 1955): 4–11.

Bronowski, Jacob. *Science and Human Values*. London: Hutchinson & Co., 1961.

Bronowski, Jacob. 'Science as a Humanistic Discipline'. *Leonardo*, vol. 18, no. 4 (1985): 261–264.

Bronowski, Jacob. 'Science in the New Humanism'. *Leonardo*, vol. 18, no. 4 (1985): 265–269.

Bronowski, Jacob. *A Sense of the Future: Essays in Natural Philosophy*, ed. Piero A. Ariotti. Cambridge, MA and London: The MIT Press, 1977.

Bronowski, Jacob. *Spain 1939: Four Poems*. Hull: Andrew Marvell Press, 1939.

Bronowski, Jacob. 'A Twentieth-Century Image of Man'. *Leonardo*, vol. 18, no. 4 (1985): 275–278.

Bronowski, Jacob. *William Blake: A Man Without a Mask*. New York, NY: Haskell House Publishers Ltd, 1967.

Bronowski, Jacob. *William Blake and the Age of Revolution*. London: Kegan & Paul Routledge, 1972.

Bronowski, Rita. 'Bruno: A Personal View'. *Leonardo*, vol. 18, no. 4 (1985): 223–225.

Brumwell, J.R.M., ed. *This Changing World: A Series of Contributions by Some of our Leading Thinkers, to Cast Light upon the Pattern of the Modern World*. London: Routledge, 1944.

Burchett, Wilfred. 'The Atomic Plague: "I Write This as a Warning to the World"'. *Daily Express*, 5 September 1945.

Camus, Albert. 'Editorial'. *Combat*, 8 August 1945.

Camus, Albert. *The Myth of Sysiphus*, trans. Justin O'Brian (1955). Harmondsworth: Penguin, 2000.

Camus, Albert. *Le Mythe de Sisyphe: Essai sur l'absurde*. Paris: Gallimard, 1942.

Cannetti, Elias. *Crowds and Power*, trans. Carol Stewart. New York, NY: The Nooneday Press, 1991. (First printed: *Masse und Macht*. Hamburg: Claassen Verlag, 1960; first English translation in 1962).

Carson, Rachel. *Silent Spring* (1962). London: Penguin, 2000.

Chisholm, Martin. 'The Navy in the Atom Age'. *Picture Post*, 25 October 1947.

Clark, Kathleen. *Positioning in Radiography*. London: Heinemann Medical Books, 1939.

Connolly, Cyril. 'Comment'. *Horizon*, vol. 12, no. 72 (December 1945): 365–367.

Connolly, Cyril. *The Missing Diplomats*. London: Queen Ann Press, 1952.

Dalí, Salvador. *The Unspeakable Confessions of Salvador Dalí as Told to Andre Parinaud*, trans. Harold J. Salemson. New York: William Morrow, 1976.

De Beauvoir, Simone. *La force des choses 1944–51* (1963). In *Force of Circumstance*, trans. Richard Howard. Harmondsworth: Penguin, 1968.

Delmer, Sefton. *Trail Sinister: An Autobiography Volume 1*. London: Secker & Warburg, 1961.

Del Renzio, Toni. 'Review of Graham Sutherland Exhibition at the Redfern Gallery'. *Art News and Review*, vol. 4, no. 23 (1952): 4.

Denvir, Bernard. 'The Geometry of Fear'. *Art*, no. 29 (June 1956): 3.

Duché, Jean. 'Andre Breton nous parle'. *Le Figaro literaire*, 5 October 1946.

Evening Telegraph, 8 August 1945.

Finch, Christopher. *Image as Language: Aspects of British Art 1950–1968*. Harmondsworth: Pelican, 1969.

Fitzgibbon, Theodora. *With Love: An Autobiography, 1938–46*. London: Century, 1982.

Fleming, Ian. *Casino Royale*. London: Jonathan Cape, 1952.

'Functional Sculpture'. *Architect's Journal*, 7 June 1951. RCAHMS, Bib. no: 67627, MS2329/SCT/25/2/62.

Gabo, Naum. 'On Constructive Realism' In Katherine S. Dreier, James Johnson Sweeney and N. Gabo, *Three Lectures on Modern Art*, 63–87. New York and London: Kennikat, 1971 (1949).

The Gallup International Public Opinion Poll: Great Britain 1937–1975, ed. George Gallup, vol. 1. New York: Random House, 1976.

Gibson, William. 'Introduction'. In *Barbara Hepworth: Sculptress*. London: Faber and Faber for Shenval, 1946.

Giedion, Siegfried. *Building in France: Building in Iron, Building in Ferro-concrete* (1928), trans. J. Duncan Berry. Santa Monica: The Getty Center for the Study of Art and the Humanities, 1995.

Giedion, Siegfried. *The Eternal Present: The Beginnings of Art.* New York: Pantheon Books, 1962.

Giedion, Siegfried. 'Space Conception in Prehistoric Art'. In *Explorations in Communication*, ed. Edward Carpenter and Marshall McLuhan. Boston: Beacon Press, 1960.

'Glasgow's Festival "Segment" a Piece of Wizardry: Kelvin Hall Gets Final Touches'. *Glasgow Herald*, 25 May 1951. RCAHMS, Bib. no: 67611, MS2329/SCT/25/2/36.

Goldberg, Alfred. 'The Atomic Origins of the British Nuclear Deterrent'. *International Affairs*, vol. 40, no. 3 (July 1964): 409–428.

Gottlieb, Carla. 'The Meaning of the Bull and Horse in *Guernica*'. *Art Journal*, vol. XXIV, no. 2 (Winter 1964–1965): 106–112.

Grigson, Geoffrey. 'Authentic and False in the New "Romanticism"'. *Horizon*, vol. xvii, no. 99 (March 1947): 203–213.

Haldane, J.B.S. 'H-Bomb Dust: Is It a Danger to the Living or Unborn?' *Picture Post*, 23 April 1955.

Hammacher, A.H. *Barbara Hepworth*. London: Thames and Hudson, 1968.

Hart, Liddell. 'Is Our Atom Strategy Suicidal?' *Picture Post*, 19 February 1955.

Hart, Liddell. 'Six Crucial Questions of the Day Including – Will There Be War?' *Picture Post*, 24 July 1948.

Hawkes, Jacquetta. *A Land.* London: Cresset Press, 1951. Reprint, Harmondsworth, Middlesex: Penguin Books Ltd, 1959.

Heilmann, Gerhard. *The Origin of Birds*. London: H.F. & G. Witherby, 1926.

Heron, Patrick. 'Fruit or Thorn?' *New Statesman and Nation*, 24 January 1953.

Hersey, John. *Hiroshima*. New York: Alfred A. Knopf, Inc.; London: Penguin, 1946.

Hersey, John. 'Hiroshima'. *The New Yorker*, 31 August 1946.

Hester, A.G. Editorial. *Atomic Scientists Journal*, vol. 5, no. 1 (September 1955): 1–2.

Hinton, Harold. 'Urey Fears Entry of Bombs in Ships'. *New York Times*, 9 February 1950.

Hodin, J.P. *The Painter Alan Reynolds*. London: The Redfern Artist Series, 1962.

Horrocks, Brian. 'The Lesson of Korea – Could This Happen in Western Germany?' *Picture Post*, 12 August 1950.

Hoskins, W.G. *The Making of the English Landscape*. London: Hodder & Stoughton, 1955; Harmondsworth: Penguin, 1970.

Hulton, Edward. 'The Last Picture Post'. *Picture Post*, 1 June 1957.

The Hulton Readership Survey 1948, ed. J.W. Hobson and H. Henry. London: Hulton Press, 1948.

The Hulton Readership Survey 1953, ed. J.W. Hobson and H. Henry. London: Hulton Press, 1954.

Hunter, Sam. 'Anatomy of Horror'. *Magazine of Art* (January 1952).

'A Hydrogen Atomic Bomb: Controversy in U.S.' *The Times*, 13 January 1950.

'The Hydrogen Bomb: A Drawing Illustrating Its Destructive Power, Forms of Protection, and Characteristics of Hydrogen Bomb Explosions'. *Illustrated London News*, 5 July 1958.

'An Imagined Cataclysm Becomes Fact'. *Picture Post*, 25 August 1945.

Ironside, Robin. *Painting Since 1939*. London: British Council, 1946.

Kirchwey, Mary (Freda). 'One World or None'. *The Nation*, 18 August 1945.

Lanyon, Peter. 'A Sense of Place'. *Painter and Sculptor*, vol. 5, no. 2 (Autumn 1962): 3–7.

Laurence, William. *Dawn over Zero: The Story of the Atomic Bomb*. New York: Knopf, 1946.

Laurence, William. 'Ending of All Life by Hydrogen Bomb Held a Possibility'. *New York Times*, 27 February 1950.

Leiris, M. 'Thoughts around Alberto Giacometti,' trans. D. Cooper. *Horizon*, vol. 114 (June 1949).

Leiris, Michel. *Journal: 1922–1989*, ed. Jean Jamin. Paris: Gallimard, 1992.

Leishman, J.B. 'The Poet's Defence'. Review of *The Poet's Defence* by Jacob Bronowski, *The Review of English Studies*, vol. 16, no. 63 (July 1940): 369–372.

Mainichi Shimbun, 21 August 1945.

'Man Enters the Atom Age'. *Picture Post*, 25 August 1945.

'The Man Who Saw It Coming'. *Picture Post*, 25 August 1945.

Mangeot, Sylvain. 'Ten Years after Hiroshima: Japan's Gentle Protest'. *Picture Post*, 19 March 1955.

Martin, J. Leslie. 'The State of Transition'. In *Circle: International Survey of Constructive Art*, ed. J.L. Martin, Ben Nicholson and Naum Gabo, 215–219. London: Faber and Faber, 1971 (1937).

Masters, Dexter, and Katharine Way, eds. *One World or None: A Report to the Public on the Full Meaning of the Atomic Bomb*. New York: McGraw-Hill Book Company, 1946.

Melville, Robert. 'Alan Reynolds'. *The Studio*, vol. 150, no. 751 (October 1955): 113.

Melville, Robert. 'Exhibitions'. *Architectural Review* (May 1958): 348.

Melville, Robert. 'Exhibitions: Paintings and Sculpture'. *Architectural Review*, vol. 116, no. 696 (December 1954): 400.

Metzger, Gustav. *Auto-Destructive Art Machine Art Auto-Creative Art*. 23 June 1961.

Michaelis, Anthony. 'Film Studies and Armament'. *Quarterly of Film, Radio and Television*, vol. 6, no. 3 (1952): 239.

Morley, Derek Wragge. 'Can Man Survive the Hydrogen Bomb?' *Picture Post*, 18 February 1950.

Nash, Paul. 'The Life of the Inanimate Object'. *Country Life*, 1 May 1937.

A National Survey of Readership of Newspapers and Magazines on a Continuing Basis, ed. Institute of Practitioners in Advertising. London: Institute of Practitioners in Advertising, 1956.

Newman, James R. *What is Science? Twelve Eminent Scientists and Philosophers Explain their Various Fields to the Layman*. London: Victor Gollancz Limited, 1956.

Nicolson, Harold. *Public Faces*. London: Constable, 1932.

O'Brien, C.C., and F. Topolski. *The United Nations Sacred Drama*. New York: Simon and Schuster, 1968.

O'Donovan, Patrick. 'Festival Journey – 6, Power and No Glory'. Review of the *Exhibition of Industrial Power*. RCAHMS, Bib no: 67661, MS2329/SCT/25/2/90.

Orwell, George. 'You and the Atomic Bomb'. *Tribune*, 19 October 1945.

Osborne, John. *Look Back in Anger*. London: Faber, 1975.

Ozenfant, Amédée. *Art*. Paris: Budry, 1928.

Ozenfant, Amédée. *The Foundations of Modern Art* (1928), trans. John Rodker. New York: Brewer, Warren and Putnam, 1931; New York: Dover, 1952.

Philpott, Trevor. 'Guided Robots Go to War: All Over the World Scientists Are Pushing Guided Missile Research to the Limit'. *Picture Post*, 22 November 1952.

Philpott, Trevor, and John Chillingworth. 'A "Picture Post" Team Reports from Hiroshima Ten Years After'. *Picture Post*, 6 August 1955.

Piper, David. 'The New Realism'. *Architectural Design* (April 1957).

Popper, Karl. *The Open Society and its Enemies*. Princeton, NJ: Princeton University Press, 1950.

'A Possible Destroyer of the World Illustrated Diagrammatically: How the Heavy Hydrogen Bomb Works'. *Illustrated London News*, 4 February 1950.

'La Premiere Bombe Atomique'. *Combat*, 7 August 1945.

Priestley, J.B. 'Britain and the Nuclear Bombs'. *New Statesman*, 2 November 1957.

'Princess Opens Glasgow Exhibition: Story of Man's Achievement "One Long Tale of Marvels"'. *Edinburgh Evening News*, 28 May 1951.

Raine, Kathleen. *Autobiographies*. London: Skoob, 1991.

Raine, Kathleen. *Stone and Flower: Poems, 1935–43*. London: Nicholson & Watson, 1943.

Raison, Tim. 'Who Is Right in the Row about Civil Defence?' *Picture Post*, 16 October 1954.

Read, Herbert. *A Concise History of Modern Sculpture*. London: Thames and Hudson, 1964.

Read, Herbert. 'Great Britain'. In Marcel Brion et al., *Art Since 1945*, 221–250. London: Thames and Hudson, 1958.

Read, Herbert. *The Green Child*. London: Heinemann, 1935; London: Robin Clark, 1989.

Read, Herbert. *Lynn Chadwick*. Amriswil: Bodensee-Verlag, 1958.

Read, Herbert. *The Philosophy of Modern Art*. London: Faber and Faber, 1982 (1951).

Read, Herbert, ed. *Unit One*. London: Cassell & Co., 1934.

Richards, J.M. 'The Exhibition Buildings'. *Architectural Review*, special issue on the South Bank Exhibition, vol. 110, no. 656 (August 1951): 123–135.

Ritchie-Calder, Peter (Lord). 'The Sociological Consequences of Nuclear Energy'. In *Economic and Social Consequences of Nuclear Energy*, ed. Roger Mellor Makins Sherfield, 76–86. London: Oxford University Press, 1972.

Roddam, J., and F. Topolski. *Topolski's Chronicle for Students of World Affairs*. London: George G. Harrap, 1958.

Russell, Bertrand. *Has Man a Future?* Harmondsworth: Penguin Books, 1961.

Russell, Bertrand. 'Science and Human Life'. In *What is Science? Twelve Eminent Scientists and Philosophers Explain their Various Fields to the Layman*, ed. James R. Newman, 6–23. London: Victor Gollancz, 1956.

Sartre, Jean-Paul. *L'existentialisme est un humanisme*, Club Maintenant, 29 October 1945. Published as *L'existentialisme est un humanisme*. Paris: Editions Nagel, 1946.

Sartre, Jean-Paul. 'La République du Silence'. *Les Lettres Français*, 9 September 1944.

'Science: Atomic X Ray'. *Time*, 24 September 1945.

'Science Must Warn Mankind'. *Picture Post*, 23 April 1955.

'Search for Smuggled A-Bombs'. *Bulletin of the Atomic Scientists*, vol. 6, nos 8–9 (1950): 288.

Shalett, Sidney, and Bertrand Zadie. 'Operation Crossroads'. *New York Times*, 17 February 1946.

'The Shape of Wars to Come'. *Picture Post*, 26 November 1955.

Shute, Nevil. *On the Beach*. London: Heinemann, 1957.

Snow, Charles Percy. *The Two Cultures*. Cambridge: Cambridge University Press, 1959.

Stimson, Henry. 'The Decision to Use the Atomic Bomb'. *Harper's Magazine*, vol. 194, no. 1161 (February 1947).

Stobbs, John. 'Civil Defence: Do We Need It? Because We Haven't Got It'. *Picture Post*, 4 October 1952.

Sutherland, Graham. 'Landscapes and Figures'. *The Listener*, 26 July 1962.

Sutherland, Graham. 'Thoughts on Painting'. *The Listener*, 6 September 1951.

Sutherland, Graham. 'A Welsh Sketchbook'. *Horizon*, vol. v, no. 28 (April 1942): 225–236.

Sylvester, David. 'The Auguries of Experience'. *Tiger's Eye*, vol. 1, no. 6 (15 December 1948): 48–51.

Sylvester, David. 'Exhibitions'. *New Statesman*, 1 October 1960.

Sylvester, David. *Interviews with Francis Bacon*. London: Thames and Hudson, 1993.

Sylvester, David. 'The Kitchen Sink'. *Encounter* (December 1954).

Sylvester, David. 'Round the London Galleries'. *The Listener*, 16 April 1953.

Tapié, Michel. *Un art autre où il s'agit de nouveaux dévidages du réel.* Paris: Gabriel-Giraud et fils, 1952.

Thirring, Hans. *Atomkrieg und Weltpolitik.* Vienna: Danubia, 1948.

Thirring, Hans. *Die Geschichte der Atombombe.* Vienna: Neues Österreich, 1946.

Thompson, D'Arcy Wentworth. *On Growth and Form.* 1917. Revised edition, 1942; reprint, Cambridge: Cambridge University Press, 1992.

Thompson, David. 'A Decade of British Sculpture'. *Cambridge Opinion*, no. 37 (January 1964): 24–32.

'To Make You Think: The Effect of an Atomic Bomb on a British City'. *Illustrated London News*, 13 July 1946.

Topolski, Feliks. *Britain in Peace and War.* London: Methuen, 1941.

Topolski, Feliks. 'Cavalcade of the Commonwealth.' *Topolski Chronicle*, vol. 10 (70), no. IV (1956).

Topolski, Feliks. 'The Citizen-Artist'. Paper presented at the Arts and Amenities Group of the Parliamentary Labour Party at the House of Commons, London, 23 February 1954. In *Topolski Chronicle*, vols 3–4, no. III (1955).

Topolski, Feliks. *Confessions of a Congress Delegate.* London: London Gallery Editions, 1949.

Topolski, Feliks. 'Contemporary Comment and Caricature'. *The Burlington Magazine for Connoisseurs*, vol. 83, no. 484 (1943): 165–168.

Topolski, Feliks. *Fourteen Letters.* London: Faber, 1988.

Topolski, Feliks. *The London Spectacle.* London: John Lane, 1935.

Topolski, Feliks. *Three Continents 1944–1945.* London: Methuen, 1946.

'Tragic Painters'. *The Times*, 18 June 1938, 12.

'War's Ending'. *Life*, 20 August 1945.

Watson, Francis. 'Recent Paintings by Graham Sutherland'. *The Listener*, 8 February 1951.

'What is De Gaulle Up To?' *Picture Post*, 26 April 1947.

'What the Atom Bomb Means If Mankind Should Now Fail to Control Its Use'. *Picture Post*, 25 August 1945.

'When Atom Bomb Struck – uncensored'. *Life*, 29 September 1952.

Winnicott, D.W. 'This Feminism'. In *Home Is Where We Start From: Essays by a Psychoanalyst*, comp. and ed. Claire Winnicott, Ray Shepherd and Madeleine Davis, 183–194. Middlesex: Penguin, 1986.

Winnicott, D.W. 'The Mother's Contribution to Society'. In *Home Is Where We Start From: Essays by a Psychoanalyst*, comp. and ed. Claire Winnicott, Ray Shepherd and Madeleine Davis, 123–127. Middlesex: Penguin, 1986.

Winocour, Jack. 'The Ten Minutes That Shook America'. *Picture Post*, 2 July 1955.

Wyatt, Woodrow. 'Guided Missile Race: What Britain Should Do'. *Picture Post*, 10 March 1956.

Wyndham, John. *The Day of the Triffids*. London: Penguin, 1951.

Exhibition Catalogues

Ades, Dawn. 'A Web of Images'. In *Francis Bacon*, exh. cat. London: Tate Gallery, Thames & Hudson, 1985.

Alan Reynolds, exh. cat. New York: Durlacher Gallery, 1954 and 1959.

Alan Reynolds, exh. cat. London: The Leicester Galleries, 1958.

Alan Reynolds, exh. cat. London: Redfern Gallery, 1952, 1953, 1954, 1956, 1960, 1962.

Alan Reynolds, Towards a Retrospective: From Early Landscapes to Abstraction, exh. cat. London: Agnews & Sons, 1982.

Alberto Giacometti: la collection du Centre Georges Pompidou, Musée national d'art moderne: Saint-Etienne, Musée d'art moderne, exh. cat. Paris: Centre Pompidou, 1999.

Alberto Giacometti, Exhibition of Sculptures, Paintings, Drawings, exh. cat. New York: Pierre Matisse Gallery, 1948.

Artaud, Antonin. 'Le Visage Humain'. In *Portraits et Dessins par Antonin Artaud*, exh. cat. Paris: Galerie Pierre, 1947.

Atomic Scientists' Association. *Atom Train: Guide to the Travelling Exhibition on Atomic Energy*, exh. cat. London: Atomic Scientists' Association, 1947.

Bacon: A Terrible Beauty, exh. cat. Dublin City Art Gallery, The Hugh Lane. Göttingen: Steidl, 2009.

Banks, Robin, Michael Harrison and Peter Adam. *Prunella Clough: The Late Paintings and Selected Earlier Works*, exh. cat. London: Annely Juda Fine Art, 2000.

Barbara Hepworth Centenary, ed. Chris Stephens, exh. cat. London: Tate, 2003.

Barker, Ian. *Prunella Clough: Unseen Reliefs, Drawings and Prints*, exh. cat. London: Annely Juda Fine Art, 2003.

Be-Bomb: The Transatlantic War of Images and All that Jazz, 1946–1956, exh. cat. Barcelona: Museu d'Art Contemporani de Barcelona; Madrid: Centro de Arte Reina Sofia, 2007.

Bernal, J.D. Foreword to *Barbara Hepworth*, exh. cat. London: Alex Reid and Lefevre Gallery, 1937.

Boyd, W. *Graham Sutherland*, exh. cat. Bernard Jacobson Gallery Ltd. London: Pale Green Press, 1993.

Causey, Andrew. *Paul Nash's Photographs: Document and Image*, exh. cat. London: Tate Gallery, 1973.

Compton, Michael. 'Gabo: Mathematics and Science'. In *Naum Gabo: Sixty Years of Constructivism*, exh. cat. London: Tate Gallery, 1987.

Cox, Ian. *The South Bank Exhibition: A Guide to the Story It Tells.* London: HMSO, 1951.

Craddock, Sacha, and Isabelle King. *Prunella Clough*, exh. cat. London: Camden Arts Centre, 1996.

Crowley, David, and June Pavitt, eds. *Cold War Modern: Design 1945–1970*, exh. cat. London: Victoria and Albert Museum, 2008.

Curtis, Barry, and Katharine Stout, eds. *Art and the Sixties: This Was Tomorrow*, exh. cat. London: Tate Britain, 2004.

Exhibition of Industrial Power, Kelvin Hall, Glasgow, Festival of Britain, exh. cat. London: HMSO, 1951. TNA, WORK 25/230/E1/A1/5.

Exhibition of Science South Kensington: A Guide to the Story It Tells, exh. cat. London: HMSO, 1951. TNA, WORK 25/230/E1/A1/3.

Garlake, Margaret. 'Identifying the Geometry of Fear'. In *Henry Moore and the Geometry of Fear*, exh. cat. London: James Hyman Fine Art, 2003.

Geometry of Fear: Works from the Arts Council Collection, exh. cat. London: South Bank Centre, 2007.

Germaine Richier, exh. cat. London: Hanover Gallery, 1955.

Germaine Richier, exh. cat. Venice: Peggy Guggenheim Collection, 2006.

Gooding, Mel. *The Fifties: Art from the British Council Collection*, exh. cat. London: British Council, 1998.

Gooding, Mel. *The Poetic Trace: Aspects of British Abstraction Since 1945*, exh. cat. New York: Adelson Galleries, 1992.

Gooding, Mel. *Prunella Clough: 50 Years of Making Art*, exh. cat. London: Annely Juda Fine Art, 2009.

Graham Sutherland, exh. cat. London: Tate Gallery, 1982.

Hammer, Martin. *Graham Sutherland: Landscapes, War Scenes, Portraits 1924–1950*, exh. cat. London: Dulwich Picture Gallery and Scala Publishers Ltd, 2005.

Hammer, Martin, and Christina Lodder. 'Naum Gabo and the Constructive Idea of Sculpture'. In *Naum Gabo: The Constructive Idea*, exh. cat. London: South Bank Centre, 1987.

Harrison, Martin. *Transition: The London Art Scene in the Fifties*, exh. cat. London: Barbican Arts Gallery, 2002.

Henderson, Nigel. *Nigel Henderson: Paintings, Collages and Photographs*, exh. cat. London: Anthony d'Offay Gallery, 1977.

Heron, Patrick. *Prunella Clough: Recent Paintings 1980–89*, exh. cat. London: Annely Juda Fine Art, 1989.

Hill, Emma, Michael Middleton and Bryan Robertson. *Prunella Clough*, exh. cat. Cambridge: Kettle's Yard Art Gallery, 1999.

Hyman, James. 'Henry Moore and the Geometry of Fear'. In *Henry Moore and the Geometry of Fear*, exh. cat. London: James Hyman Fine Art, 2003.

Jackson, Lesley. *From Atoms to Patterns: Crystal Structure Designs from the 1951 Festival of Britain*, exh. cat. Shepton Beauchamp and Yeovil: Richard Dennis in association with the Wellcome Collection, 2008.

James, Philip. 'Sculpture Today'. In *Sculpture in the Open Air*, exh. cat. London: London County Council, 1960.

Johnstone, Isobel. *Prunella Clough, Spotlight: Arts Council Collection*, exh. cat. Essex: Arts Council England, 1999.

Kenneth King, Eduardo Paolozzi, William Turnbull, exh. cat. London: Hanover Gallery, 1950.

Livingstone, Marco. *British Pop*, exh. cat. Bilbao: Museo de Bellas Artes, 2005.

Martin, Simon, ed. *Colin Self: Art in the Nuclear Age*, exh. cat. Chichester: Pallant House Gallery, 2008.

Martin, Simon, ed. *Edward Burra*, exh. cat. London: Lund Humphries, 2012.

Martin, Simon, ed. *John Tunnard: Inner Space to Outer Space*, exh. cat. Chichester: Pallant House Gallery, 2010.

Mellor, David. 'Existentialism and Post-War British Art'. In *Paris Post War, Art and Existentialism 1945–1955*, ed. Frances Morris, exh. cat. London: Tate Gallery, 1993.

Mellor, David. *A Paradise Lost: The Neo-Romantic Imagination in Britain 1935–55*, exh. cat. London: Barbican Art Galleries, Lund Humphries, 1987.

Mellor, David. *The Sixties Art Scene in London*, exh. cat. London: Barbican Arts Gallery, 1993.

Mellor, David, and Val Williams. *Too Short a Summer: The Photographs of Peter Rose Pulham, 1910–1956*, exh. cat. York: Impressions Gallery of Photography, 1979.

Middleton, Michael. *Eduardo Paolozzi*, exh. cat. 1962.

Middleton, Michael. *Prunella Clough: A Retrospective Exhibition*, exh. cat. London: Whitechapel Art Gallery, 1960.

The Mirror and the Square: An Exhibition of Art Ranging from Realism to Abstraction, exh. cat. London: Artists International Association, 1952.

Morris, Lynda, and Robert Radford. *AIA: The Story of the Artists International Association 1933–53*, exh. cat. Oxford: Museum of Modern Art Oxford, 1983.

Naum Gabo: The Constructive Process, exh. cat. London: Tate Gallery, 1976.

Official Book of the Festival of Britain, 1951. London: HMSO, 1951.

Oil Paintings and Gouaches by P Rose Pulham, exh. cat. London: London Gallery, 1948.

Osbert Lancaster, Richard Eurich, Alan Reynolds, Ethel Walker, Derwent Lees, exh. cat. London: Redfern Gallery, 1952.

Packer, William. *Contemporary British Landscape*, exh. cat. London: Flowers East, 1999.

Paul Klee 1879–1940, exh. cat. London: National Gallery, 1945.

Peter Lanyon, exh. cat. London: Arts Council, 1968.

Peter Lanyon, exh. cat. St Ives: Tate St Ives, 2010–2011.

Picasso and Modern British Art, exh. cat. London: Tate, 2012.

A Private World: Photographs by Paul Nash, exh. cat. London: Fisher Fine Art, 1978.

Prunella Clough: Recent Paintings, exh. cat. London: Arts Council of Great Britain, 1977.

Rapaport, Brooke Kamin, and Kevin Stayton, eds. *Vital Forms: American Art and Design in the Atomic Age, 1940–1960*, exh. cat. New York: Brooklyn Museum of Art in association with H.N. Abrams, 2001.

Read, Herbert. Introduction to *Barbara Hepworth: Carvings and Drawings*. London: Lund Humphries, 1952.

Read, Herbert. 'Introduction' to *An Exhibition of Paintings and Drawings of Sutherland*, exh. cat. London: Arts Council and Tate Gallery, 1953, unpaginated.

Read, Herbert. Introduction to *Sculpture in the Open Air*, exh. cat. London: London County Council, 1963.

Read, Herbert. 'New Aspects of British Sculpture'. In *The XXVIth Venice Biennale: The British Pavilion*, exh. cat. London: British Council, 1952.

Robbins, David, ed. *The Independent Group: Postwar Britain and the Aesthetics of Plenty*, exh. cat. Hanover, NH: Hood Museum of Art, 1991.

Robertson, Bryan. *Prunella Clough: New Paintings 1979–82*, exh. cat. London: Warwick Arts Trust, 1982.

Robertson, Bryan. *'45–'99: A Personal View of British Painting and Sculpture*, exh. cat. Cambridge: Kettle's Yard Art Gallery, 2000.

Spalding, Julian. *Forgotten Fifties*, exh. cat. Sheffield: Graves Art Gallery, 1984.

Spencer, Charles. 'Robert Adams'. In *Robert Adams: Retrospective Exhibition*, exh. cat. London: Arts Council, 1971.

Stewart, Angus. *Seeing the World Sideways: Prunella Clough*, exh. cat. London: The Fine Art and Antiques Fair, 2004.

Sutherland in Wales, exh. cat. Picton Castle, Haverfordwest, Dyfed, 1976.

Topolski, Feliks. 'Notes by the Artist'. In *Feliks Topolski's Memoir of the Century*, exh. cat. London: The Greater London Council, c. 1984.

Tufnell, Ben, ed. *Prunella Clough*, exh. cat. London: Tate Britain, 2007.

Zeugnisse der Angst in der modernen Kunst, exh. cat. Mathildenhöhe Darmstadt, 1963.

Archival

Advertising flyer for the *Exhibition of Industrial Power*, Kelvin Hall, Glasgow, 28 May–18 August 1951. TNA, WORK 25/230/E1/A3/3.

Archbishop of Canterbury, speech, 17 July 1950, quoted in the Festival of Britain press release, *Coming Events in Britain*, with an introduction by General Lord Ismay. TNA, WORK 25/250 E3/B32.

The Atomic Bombings of Hiroshima and Nagasaki by the Manhattan Engineer District, 29 June 1946, National Archive Washington DC, United States Air Force Strategy Bombing Survey. RG 243 Series H.

Atomic mural, unexecuted scheme, 1949, from the Basil Spence Archive. RCAHMS, Item DP012308.

Bataille, Georges. 'Apocalypse', *Documents* (1929), 2; *Apocalypse of Saint Sever*, C11th, folio 145v, Bibliotheque Nationale de France.

Blandy, William. 'Introduction'. In 'Operation Crossroads,' n.d., the Papers of Sir James Chadwick, 1914–1974. Churchill Archives Centre, Churchill College, University of Cambridge, Cambridge, United Kingdom. CHAD IV/3/18.

The British Mission to Japan. *The Effects of the Atomic Bombs at Hiroshima and Nagasaki*. London: HMSO, 1946.

Bronowski, Jacob. 'The Shape of Science in the Arts'. Institute of Contemporary Arts, London, lecture transcript (2 August 1951). TGA 955/1/7/21.

Chadwick, James. 1914–1974, Churchill Archives Centre, Churchill College, University of Cambridge, Cambridge, United Kingdom. CHAD IV/3/18, unpaginated.

Chadwick, James. 'Initial Report by Sir James Chadwick', 16 July 1945. TNA, CAB126/250.

Cockcroft, John. 'John Cockcroft – Banquet Speech'. In *Les Prix Nobel en 1951*, ed. Arne Holmberg. Stockholm: Nobel Foundation, 1952. <http://www.nobelprize.org/nobel_prizes/physics/laureates/1951/cockcroft-speech.html>.

De C. Nan Kivell, Rex. Letter to Humphrey Brooke, 21 March 1956. RAA/SEC/15.

Dimbleby, Richard. On Belsen, 19 April 1945.

Eames, Charles and Ray. *Powers of Ten*. 9 mins. USA: Office of Charles and Ray Eames, 1968/1977.

'Feliks Topolski: Chronicles 1930–1980'. 18 July–17 August 1980, Institute of ContemporaryArts, London, Tate Britain Archive. TGA 955/15/337.

Frisch, Otto. 'Eye-Witness Report of Nuclear Explosion, July 16, 1945', n.d. TNA, CAB126/250.

Giacometti, A. Letter to I. Rawsthorne, February 1949. TGA 9612.1.2.8.

Hepworth, Barbara. Letter to J.D. Bernal, undated [c. 1936/1937], J.D. Bernal papers, University Library (Add. 8287/J84 [Box 83]), University of Cambridge.

Home Office. *Civil Defence Manual of Basic Training, Volume II: Atomic Warfare, Pamphlet No. 6*. London: HMSO, 1950.

Home Office. *The Hydrogen Bomb*. London: HMSO, 1957.

Home Office. *Nuclear Weapons*. London: HMSO, 1956.

'The Hydrogen Bomb'. Civil Defence Pamphlet. London: HMSO, 1957.

'The Hydrogen Bomb'. Civil Defence Poster, 1957, INF 13/281. TNA.

Institute of Contemporary Arts, London. 'Proposals for an Exhibition: "Growth and Form"'. TGA 955.1.12.26, 3/36 [n/d].

Lambert, Patty. 'What I Remember about Isabel'. Manuscript. Private collection.

Lanyon, Peter, talking to Lionel Miskin about his early life, transcribed audio tapes. TGA TAV 215AB 1962.

Lanyon, Peter. 'Recorded Talk' for the British Council, 1963.

Paolozzi, Eduardo. Interview with Frank Whitford, National Life Story Collection: Artists' Lives (10 June 1993; 4 May 1994; 30 December 1994; and 5 January 1995). British Library, C466/17: 249–50.

Rawsthorne, I. Diary, 1947. TGA 9612.2.7.

Rawsthorne, Isabel. Autobiography, unpublished manuscript, private collection.

Rawsthorne, Isabel. Autobiographical notes. TGA 9612.2.4.1 and TGA MS 9612.2.4.33.

Rawsthorne, Isabel. Letter to A. Giacometti, 1948. FAAG, inv. 2003-1897.

Rawsthorne, Isabel. Letter to Alberto Giacometti, Autumn 1950. FAAG, inv. 2003-1903.

Rawsthorne, Isabel. Letter to Peter Rose Pulham, late 1948. TGA 9612.1.3.13.

Rawsthorne, Isabel. Letter to P. Rose Pulham, early 1949. TGA 9612.1.3.65; 37 Quai des Grands Augustins.

Rawsthorne, Isabel. Letter to P. Rose Pulham, 14 June 1949. TGA 9612.1.3.26.

Rawsthorne, Isabel. Letter to P. Rose Pulham, June 1949. TGA 9612.1.3.31.

Rawsthorne, Isabel. Letter to P. Rose Pulham, late 1952. TGA 9612.1.3.74.

Rawsthorne, Isabel. Letter to P. Rose Pulham, 21 October 1953. TGA 9612.1.3.97.

Rawsthorne, Isabel. Postcard to A. Giacometti, 8 May 1947. FAG, 20003-1890.

Report of the British Mission to Japan on the Effects of the Atomic Bombs at Hiroshima and Nagasaki. London: HMSO, 1946.

Resnais, Alain. *Hiroshima Mon Amour*. 90 mins. France, Japan, 1959.

Thompson, Lawrence. 'Portrait of the Artist as a Displaced Person'. Unpublished interview between the author and John Bratby, found in the artist's papers at Julian Hartnoll Gallery, London.

Topolski, Feliks. 'Draft letter to Ransom Collection, University of Texas at Austin regarding *Twenty Greats*', c. 1960. The Estate of Feliks Topolski Archive, uncatalogued.

Topolski, Feliks. 'Letter to Bertrand Russell', 10 October 1960. The Estate of Feliks Topolski Archive, uncatalogued.

Topolski, Feliks. Untitled draft regarding 'picturing' and the camera (c. 1950–1960). The Estate of Feliks Topolski Archive, uncatalogued.

Tuck, James, to James Chadwick, 17 July 1945. TNA, CAB126/250.

Unattributed. 'The Art of Feliks Topolski', c. 1951. The Estate of Feliks Topolski Archive, uncatalogued.

United States Air Force Strategy Bombing Survey, National Archive Washington DC. RG 243 Series H.

Waller, Professor Ivar. 'Nobel Prize in Physics 1951, Presentation Speech'. In *Nobel Lectures, Physics 1942–1962*. Amsterdam: Elsevier Publishing Company, 1964. <http://www.nobelprize.org/nobel_prizes/physics/laureates/1951/press.html>.

Ward, Edward. On Buchenwald, BBC Home Service, 1 April 1945.

Television and Radio Broadcasts

The Ascent of Man, directed by Adrian Malone, Dick Gilling, Mick Jackson, David John Kennard (written and narrated by Jacob Bronowski), 1973. London: BBC Worldwide Ltd, 2005.

Effects of the Atomic Bomb. Nichiei Kyo Productions, premiered Tokyo, July 1952.

Face to Face. 1959–1962. London: BBC Worldwide Ltd, 2009.

A Short Vision, directed by Peter and Joan Foldes. British Film Institute, 1956. <http://www.screenonline.org.uk/film/id/442000/>.

Watkins, Peter. *The War Game*. Britain: BBC, 1965.

Secondary Sources

Allison, D.B., ed. *The New Nietzsche*. Cambridge, MA: The MIT Press, 1985.

Andrews, Julian, ed. *Correspondences: Selected Writings on Art*. Picton and Geneva: The Graham and Kathleen Sutherland Foundation, 1982.

Appleyard, Bryan. *The Pleasures of Peace: Art and Imagination in Post-War Britain*. London and Boston: Faber and Faber, 1989.

Arnold, Lorna. *A Very Special Relationship: British Atomic Weapon Trials in Australia*. London: HMSO, 1987.

Arnold, Lorna, and Katherine Pyne. *Britain and the H-Bomb*. Basingstoke: Palgrave, 2001.

Banham, Mary, and Bevis Hillier, eds. *A Tonic to the Nation: The Festival of Britain, 1951*. London: Thames and Hudson, 1976.

Barlow, Anne J. 'Barbara Hepworth and Science'. In *Barbara Hepworth Reconsidered*, ed. David Thistlewood, 95–107. Liverpool: Tate Gallery, 1996.

Bartter, Martha. 'Nuclear Holocaust as Urban Renewal'. *Science-Fiction Studies*, vol. 13, no. 2 (1986): 148–158.

Beevor, Antony, and Artemis Cooper. *Paris After the Liberation, 1944–1949*. London: Doubleday, 1994.

Bernstein, Barton. 'Compelling Japan's Surrender Without the A-Bomb, Soviet Entry, or Invasion: Reconsidering the US Bombing Survey's Early Surrender Conclusions'. *Journal of Strategic Studies*, vol. 18, no. 2 (1995): 101–148.

Berryman, Larry. 'Prunella Clough'. *Arts Review*, 5 May 1989.

Berthoud, Roger. *Graham Sutherland: A Biography*. London: Faber & Faber, 1982.

'Bio-Bibliography of Jacob Bronowski'. *Leonardo*, vol. 18, no. 4 (1985): 282–287.

Boal, Iain A. 'Ground Zero: Henry Moore's *Atom Piece* at the University of Chicago'. In *Henry Moore: Critical Essays*, ed. Jane Beckett and Fiona Russell, 221–241. Aldershot: Ashgate, 2003.

Bowness, Alan. 'The Drawings of Barbara Hepworth'. In *Barbara Hepworth: Drawings from a Sculptor's Landscape*, 15–25. London: Cory Adams and Mackay, 1966.

Boyer, Paul. *By the Bomb's Early Light: American Thought and Culture at the Dawn of the Atomic Age*. New York: Pantheon, 1985. 2nd edition, Chapel Hill, NC: University of North Carolina Press, 1994.

Boyer, Paul. *When Time Shall Be No More: Prophecy Belief in Modern American Culture*. Cambridge, MA: Belknap Press of Harvard University Press, 1992.

Brennan, Marcia. *Modernism's Masculine Subjects: Matisse, the New York School and Post-Painterly Abstraction*. London and Cambridge, MA: The MIT Press, 2004.

Brians, Paul. *Nuclear Holocausts: Atomic War in Fiction, 1895–1984*. Kent, OH: Kent State University Press, 1987.

Broderick, Mick, and Robert Jacobs. 'Nuke York, New York: Nuclear Holocaust in the American Imagination from Hiroshima to 9/11'. *The Asia-Pacific Journal*, vol. 10, issue 11, no. 6 (2012). <http://japanfocus.org/-Mick-Broderick/3726>.

Brown, Andrew. *J.D. Bernal: The Sage of Science*. Oxford: Oxford University Press, 2005.

Burchett, Wilfred. *Shadows of Hiroshima*. London: Verso Books, 1983.

Burstow, Robert. 'Aesthetics, 1950–75'. In *Sculpture in 20th Century Britain: Vol. 1: Identity, Infrastructure, Aesthetics, Display, Reception*, ed. Penelope Curtis, 167–184. Leeds: Henry Moore Institute, 2003.

Burstow, Robert. 'Bernard Meadows'. In *Sculpture in 20th Century Britain: Vol. 2: A Guide to Sculptors in the Leeds Collections*, ed. Penelope Curtis, 232–235. Leeds: Henry Moore Institute, 2003.

Burstow, Robert. 'Butler's Competition Project for a Monument to "The Unknown Political Prisoner": Abstraction and Cold War Politics'. *Art History*, vol. 12, no. 4 (December 1989): 472–496.

Burstow, Robert. '*The Geometry of Fear*: Herbert Read and British Modern Sculpture'. In *Herbert Read: A British Vision of World Art*, ed. Ben Read and David Thistlewood, 119–132. Leeds: Leeds City Art Galleries in association with The Henry Moore Foundation and Lund Humphries, 1993.

Burstow, Robert. 'The Limits of Modernist Art as a "Weapon of the Cold War": Reassessing the Unknown Patron of the Monument to the Unknown Political Prisoner'. *Oxford Art Journal*, vol. 20, no. 1 (Spring 1997): 68–80.

Button, Virginia. *The Aesthetic of Decline: English Neo-Romanticism c. 1935–1956*. PhD thesis, Courtauld Institute of Art, University of London, 1991.

Calvocoressi, Richard. 'Public Sculpture in the 1950s'. In *British Sculpture in the Twentieth Century*, ed. Sandy Nairne and Nicholas Serota, 134–153. London: Whitechapel Art Gallery, 1981.

Caputi, Jane. *Gossips, Gorgons and Crones: The Fates of the Earth*. Santa Fe, NM: Bear & Company Publishing, 1993.

Causey, Andrew. *Paul Nash*. Oxford: Oxford University Press, 1980.

Causey, Andrew. *Paul Nash Writings on Art*. Oxford: Oxford University Press, 2000.

Causey, Andrew. *Peter Lanyon*. Henley-on-Thames: Aiden Ellis Publishing, 1971.

Causey, Andrew. *Peter Lanyon: Modernism and the Land*. London: Reaktion Books, 2006.

Caute, David. *The Dancer Defects: The Struggle for Cultural Supremacy During the Cold War*. Oxford: Oxford University Press, 2003.

Cavanagh, Terry, and Alison Yarrington. *Public Sculpture of Leicestershire and Rutland*. Liverpool: Liverpool University Press, 2000.

Chambers, Emma. *An Indolent and Blundering Art: The Etching Revival and the Redefinition of Etching in England 1838–1892*. Aldershot: Ashgate, 1999.

Chapman, James. *British Comic Books: A Cultural History*. London: Reaktion Books, 2011.

Chapman, James. '*Quatermass* and the Origins of British Television sf'. In *British Science Fiction Television: A Hitchhiker's Guide*, ed. John Cook and Peter Wright, 21–51. London: I.B. Tauris, 2006.

Chappell, William, ed. *Well, Dearie! The Letters of Edward Burra*. London: Gordon Fraser, 1985.

Cheetham, Mark A. *The Rhetoric of Purity: Essentialist Theory and the Advent of Abstract Painting*. Cambridge: Cambridge University Press, 1991.

Clark, Adrian. 'Neglected British Painters of the 20th Century: Prunella Clough and Gerald Wilde'. *The British Art Journal*, vol. V, no. 2 (2004): 107–108.

Cockcroft, Eva. 'Abstract Expressionism: Weapon of the Cold War'. *Art Forum*, vol. 12, no. 10 (June 1974): 39–41.

Cocroft, Wayne D., and Roger J.C. Thomas. *Cold War: Building for Nuclear Confrontation 1946–1989*, ed. P.S. Barnwell. London: English Heritage, 2003; Swindon: English Heritage, 2004.

Collins, Judith. 'Spotlight on Prunella Clough'. *Make*, no. 87 (March–May 2000): 12–13.

Collins, Judith, ed. *Writing on the Wall: Women Writers on Women Artists*. London: Weidenfeld & Nicolson, 1993.

Colomina, Beatriz. *Domesticity at* War. Barcelona: Actar, 2006.

Conekin, Becky E. *'The Autobiography of a Nation': The 1951 Festival of Britain.* Manchester and New York: Manchester University Press, 2003.

Cooper, Douglas. *The Work of Graham Sutherland*. London: Lund Humphries, 1961.

Coupe, Geoff. 'My Father, the Bomb and Me'. Geoff Coupe's Blog, 10 December 2010. <http://gcoupe.wordpress.com/2010/12/10/my-father-the-bomb-and-me/>.

Cowling, Elizabeth. 'Elective Affinities: Giacometti in the Palace of Bones'. *Apollo* (October 2003): 34–42.

Curtis, Barry. 'One Continuous Interwoven Story (the Festival of Britain)'. In *The Block Reader in Visual Culture*, ed. Jon Bird et al., 209–220. London: Routledge, 2008.

Curtis, Barry. 'War Games: Cold War Britain in Film and Fiction'. In *Cold War Modern: Design 1945–1970*, ed. David Crowley and Jane Pavitt, 122–127. London: V&A Publishing, 2008.

Dallas, Gregor. *1945: The War that Never Ended*. London and New Haven: Yale University Press, 2006.

Davis, Tracy. *Stages of Emergency: Cold War Nuclear Civil Defense.* Durham, NC: Duke University Press, 2007.

Davisson, Darrell D. *Art After the Bomb: Iconographies of Trauma in Late Modern Art*. Bloomington, IN: Authorhouse, 2008.

Decamous, Gabrielle. 'Nuclear Activities and Modern Catastrophes: Art Faces the Radioactive Waves'. *Leonardo*, vol. 44, no. 2 (2011): 124–132.

Deriu, David. 'The Ascent of the Modern Planeur: Aerial Images and Urban Imagination in the 1920s'. In *Imagining the City*, ed. Christian Keen, Catherine Emdem and David Midgley, 189–211. Oxford: Peter Lang, 2006.

Divine, Robert A. *Blowing on the Wind: The Nuclear Test Ban Debate, 1954–1960.* New York: Oxford University Press, 1978.

Donzelot, Jacques. *The Policing of Families*, trans. Robert Hurley. London: The Johns Hopkins University Press, 1997.

Drabble, Margaret. 'Cloudy Courage'. *The Guardian Review*, 14 April 2007.

Eisenbud, Merril, and Thomas F. Gesell. *Environmental Radioactivity: From Natural, Industrial, and Military Sources*, fourth edition. San Diego, CA and London: Academic Press, 1997.

Elliott, Patrick. 'Drawings by Alberto Giacometti. Paris'. *The Burlington Magazine*, vol. 143, no. 1178 (May 2001): 297–318.

Farmer, Rod. '*Life*'s Atom Bomb: The Image of Nuclear War in *Life* Magazine, 1945–1955'. *New England Journal of History*, vol. 52, no. 3 (1995).

Farrell, James. 'The Crossroads of Bikini'. *Journal of American Culture*, vol. 10, no. 2 (1987): 55–66.

Fer, Briony. *On Abstract Art*. New Haven and London: Yale University Press, 1997.

Fer, Briony, David Batchelor and Paul Wood. *Realism, Rationalism, Surrealism: Art between the Wars*. New Haven and London: Yale University Press, 1993.

Festing, Sally. *Barbara Hepworth: A Life of Forms*. London: Viking, 1995.

Forgan, Sophie. 'Atoms in Wonderland'. *History and Technology*, vol. 19, no. 3 (2003): 177–196.

Forgan, Sophie. 'Festivals of Science'. *The British Journal for the History of Science*, vol. 31, no. 2 (June 1998): 117–240.

Foucault, Michel. *The History of Sexuality I: The Will to Knowledge*. London: Penguin, 1998.

Foucault, Michel. 'Las Meninas'. In *The Order of Things*, 3–18. London: Routledge, 2001.

Franklin, Bruce. *War Stars: The Superweapon and the American Imagination*. New York: Oxford University Press, 1988.

Frascina, Francis. *Pollock and After: The Critical Debate*. London: Routledge, 2001.

Gale, Matthew, and Chris Stephens. *Barbara Hepworth: Works in the Tate Gallery Collection and the Barbara Hepworth Museum St Ives*. London: Tate Gallery, 1999.

Gamwell, Lyn. *Exploring the Invisible: Art, Science, and the Spiritual*. Princeton: Princeton University Press, 2002.

Gardiner, Margaret. *Barbara Hepworth: A Memoir*. London: Lund Humphries, 1994; Edinburgh: Salamander, 1982.

Garlake, Margaret. *The Drawings of Peter Lanyon*. Aldershot: Ashgate, 2003.

Garlake, Margaret. *New Art, New World: British Art in Postwar Society*. New Haven and London: Yale University Press, 1998.

Garlake, Margaret. *Peter Lanyon*. St Ives Artists series. London: Tate Gallery Publishing, 1998.

Garlake, Margaret. 'Prunella Clough'. *Art Monthly*, July–August 1996.

Garlake, Margaret. 'Prunella Clough 1919–1999'. *Art Monthly*, 2000.

Garlake, Margaret. *The Relationship between Institutional Patronage and Abstract Art in Britain c. 1945*–1956. PhD thesis, Courtauld Institute of Art, University of London, 1987.

Garlake, Margaret. 'Review: Prunella Clough, Cambridge'. *The Burlington Magazine*, vol. 141, no. 1159 (October 1999): 639–640.

Garlake, Margaret. 'Revisiting the Geometry of Fear'. In *Rereading Read: New Views on Herbert Read*, ed. Michael Paraskos, 134–143. London: Freedom Press, 2007.

Garlake, Margaret. *The Sculpture of Reg Butler*. Much Hadham and Aldershot: Henry Moore Foundation in association with Lund Humphries, 2006.

Gaskin, Fiona. 'British Tachisme in the Post-War Period, 1946–1957'. In *Artists and Patrons in Post-War Britain*, ed. Margaret Garlake, 17–54. Aldershot: Ashgate, 2001.

Gilroy, Paul. *Postcolonial Melancholia*. New York: Columbia University Press, 2005.

Goin, Peter. *Nuclear Landscapes*. Baltimore: The Johns Hopkins University Press, 1991.

Goldsmith, Maurice. *Sage: A Life of J.D. Bernal*. London: Hutchinson, 1980.

Gooding, Mel, ed. *Patrick Heron: Painter as Critic*. London: Tate Publishing, 1998.

Goodman, Michael. *Spying on the Nuclear Bear: Anglo-American Intelligence and the Soviet Bomb*. Stanford: Stanford University Press, 2007.

Gowing, Margaret, and Lorna Arnold. *The Atomic Bomb*. London: Butterworth, 1979.

Grant, Matthew. *After the Bomb: Civil Defence and Nuclear War in Britain, 1945–68*. Basingstoke: Palgrave Macmillan, 2010.

Gregory, Jane, and Steve Miller. *Science in Public: Communication, Culture and Credibility*. Cambridge, MA: Basic Books, 1998.

Grobstein, Clifford. 'Bronowski, Progression and Human Progress'. *Leonardo*, vol. 18, no. 4 (1985): 226–228.

Guilbaut, Serge. *How New York Stole the Idea of Modern Art: Abstract Expressionism, Freedom, and the Cold War*. Chicago: University of Chicago Press, 1983.

Gusterson, Hugh. *Nuclear Rites: A Weapons Laboratory at the End of the Cold War*. Berkeley, CA: University of California Press, 1996.

Hacker, Barton C. *Elements of Controversy: The Atomic Energy Commission and Radiation Safety in Nuclear Weapons Testing, 1947–1974*. Berkeley, CA and London: University of California Press, c. 1994.

Hadler, Mona. 'The Bomb in the Postwar Era: From the Sublime to Red Hot Candy'. *Source: Notes in the History of Art*, vol. 21, no. 1 (Fall 2001): 38–43.

Hales, Peter Bacon. 'The Atomic Sublime'. *American Studies*, vol. 32, no. 1 (1991): 5–31.

Hammer, Martin. *Bacon and Sutherland*. New Haven and London: Yale University Press, 2005.

Hammer, Martin, and Christina Lodder. *Constructing Modernity: The Art and Career of Naum Gabo*. New Haven and London: Yale University Press, 2000.

Hammer, Martin, and Christina Lodder. 'Hepworth and Gabo: A Constructive Dialogue'. In *Barbara Hepworth Reconsidered*, ed. David Thistlewood, 109–133. Liverpool: Tate Gallery, 1996.

Hardwig, John. 'Epistemic Dependence'. *Journal of Philosophy*, vol. 82, no. 7 (1985): 335–349.

Hardwig, John. 'Evidence, Testimony, and the Problem of Individualism'. *Social Epistemology*, vol. 2, no. 4 (1988): 309–321.

Harrison, Martin. *In Camera: Francis Bacon: Photography, Film and the Practice of Painting*. London: Thames and Hudson, 2005.

Harrison, Martin. *Transition: The London Art Scene in the Fifties*. London: Merrell, 2002.

Harrison, Michael. *Alan Reynolds: The Making of a Concretist Artist.* London: Lund Humphries, 2011.

Harwood, Elain, and Alan Powers, eds. 'Festival of Britain'. Special issue, *Twentieth Century Society Journal*, no. 5, 2001.

Hayes, J. *Graham Sutherland*. Oxford: Phaidon, 1980.

Hearn, Jeff. *The Violences of Men: How Men Talk About and How Agencies Respond to Men's Violence to Women*. London: Sage, 1998.

Hennessy, Peter. *The Secret State: Whitehall and the Cold War*. London: Allen Lane, 2002.

Hewison, Robert. *Culture and Consensus: England, Art and Politics Since 1940.* London: Methuen, 1995.

Hewison, Robert. *In Anger: Culture in the Cold War 1945–60*. London: Weidenfeld and Nicolson, 1981.

Hobsbawm, Eric. *Age of Extremes: The Short Twentieth Century 1914–1991*. London: Michael Joseph, 1994.

Hogg, Jonathan. '"The Family that Feared Tomorrow": British Nuclear Culture and Individual Experience in the Late 1950s'. *British Journal for the History of Science*, vol. 45, special issue (December 2012): 535–549.

Hohl, Reinhold. *Alberto Giacometti: Sculpture, Painting, Drawing*. London: Thames and Hudson, 1972.

Holton, Gerald. 'On the Integrity of Science: The Issues Since Bronowski'. *Leonardo*, vol. 18, no. 4 (1985): 229–232.

Hopkinson, Tom, ed. *Picture Post 1938–50*. London: Chatto & Windus: Hogarth Press, 1984.

Hubbard, Sue. 'Prunella Clough: An Artist's Artist'. *Contemporary Visual Arts* (2000).

Hughes, Jeff. 'The Strath Report: Britain Confronts the H-Bomb, 1954–1955'. *History and Technology*, vol. 19, no. 3 (2003): 257–275.

Hughes, Jeff. 'What is British Nuclear Culture?' *British Journal for the History of Science*, vol. 45, special issue (December 2012): 495–518.

Hulks, David. 'Despair, or Defiance: The Double Inflection in Herbert Read's Geometry of Fear'. In *Rereading Read: New Views on Herbert Read*, ed. Michael Paraskos, 144–151. London: Freedom Press, 2007.

Hutchings, Peter. *Hammer and Beyond: The British Horror Film.* Manchester: Manchester University Press, 1993.

Hyman, James. *The Battle for Realism: Figurative Art in Britain During the Cold War, 1945–1960.* New Haven and London: Yale University Press, 2001.

Jacobi, Carol. 'Allies in Art: Pablo Picasso and British Painting 1940–1944'. *Tate Papers* (Summer 2013).

Jacobi, Carol. 'Cat's Cradle: Francis Bacon and the Art of Isabel Rawsthorne'. *Visual Culture in Britain*, vol. 10, no. 3(December 2009): 294–296.

Jacobi, Carol. *Out of the Cage: The Art of Isabel Rawsthorne*. London and New Haven: Yale University Press, forthcoming, 2014.

Johnson, Paige. 'Safeguarding the Atom: The Nuclear Enthusiasm of Muriel Howorth'. *British Journal for the History of Science*, vol. 45, special issue (December 2012): 551–571.

Jolivette, Catherine. *Landscape, Art and Identity in 1950s Britain*. Aldershot and Burlington, VT: Ashgate, 2009.

Jolivette, Catherine. 'Science, Art and Landscape in the Nuclear Age'. *Art History*, vol. 35, no. 2 (April 2012): 253–269.

Jordanova, Ludmilla. *The Look of the Past: Visual and Material Evidence in Historical Practice*. Cambridge: Cambridge University Press, 2012.

Kahn, Herman. *On Thermonuclear War*. New Brunswick, NJ: Transaction Publishers, 2007.

Kaldor, Mary. *The Imaginary War: Understanding the East–West Conflict*. Oxford: Basil Blackwell, 1990.

Katzive, David H. 'Henry Moore's *Nuclear Energy*: The Genesis of a Monument'. *Art Journal*, vol. 32, no. 3 (Spring 1973): 284–288.

Key, Joan. 'The Metamorphosis of Attrition'. *Women's Art Magazine* (June–July 1996): 5–6.

Kirby, M., and R. Capey. 'The Area Bombing of Germany in World War II: An Operational Research Perspective'. *The Journal of Operational Research Society*, vol. 48, no. 7 (July 1997): 661–677.

Kirby, M., and R. Capey. 'The Air Defence of Great Britain, 1920–1940: An Operational Research Perspective'. *The Journal of Operational Research Society*, vol. 48, no. 6 (June 1997): 555–568.

Kirkpatrick, Diane. *Eduardo Paolozzi*. London: Studio Vista, 1970.

Kirsch, Scott. 'Watching the Bombs Go Off: Photography, Nuclear Landscapes and Spectator Democracy'. *Antipode*, vol. 29, no. 3 (1997): 227–255.

Klein, Jacky. 'Naum Gabo: Construction in Space (Crystal)'. Tate Gallery. September 2002, accessed 19 December 2012. <http://www.tate.org/art/artworks/gabo-construction-in-space-crystal-t06978>.

Klein, Jacky. 'Naum Gabo: Construction in Space with Crystalline Centre'. Tate Gallery. September 2002, accessed 19 December 2012. <http://www.tate.org/art/artworks/gabo-construction-in-space-with-crystalline-centre-t06977>.

Klein, Melanie. 'Early Stages of the Oedipus Conflict'. In *Love, Guilt and Reparation and Other Works, 1921–45*, 186–198. London: Hogarth Press and the Institute of Psychoanalysis, 1981.

Klein, Melanie. 'The Importance of Symbol Formation in the Development of the Ego'. In *Love, Guilt and Reparation and Other Works, 1921–45*, 219–228. London: Hogarth Press and the Institute of Psychoanalysis, 1981.

Klein, Melanie. 'On the Theory of Anxiety and Guilt'. In *Envy and Gratitude and Other Works, 1946–1963*, 25–42. London: Hogarth Press and the Institute Of Psychoanalysis, 1987.

Klein, Melanie, and Joan Riviere. *Love, Hate and Reparation*. New York: W.W. Norton and Company Inc./The Norton Library, 1964.

Krenn, Michael L. *Fall-out Shelters for the Human Spirit: American Art and the Cold War*. Chapel Hill, NC and London: University of North Carolina Press, 2005.

Kuhn, Thomas. *The Structure of Scientific Revolutions*. Chicago: University of Chicago Press, 1970.

Kynaston, David. *Austerity Britain, 1945–1951*. London: Bloomsbury, 2007.

Kynaston, David. *Family Britain, 1951–1957*. London: Bloomsbury, 2010.

Laing, Stuart. *Representations of Working-Class Life 1957–1964*. Basingstoke: Macmillan, 1986.

Lászlό, Helga. 'The Realistic Sculpture of Laszlo Peter Peri'. MA dissertation, trans. John Lloyd, University of Budapest, 1990.

Laucht, Christoph. 'Atoms for the People: The Atomic Scientists' Association, the British State and Nuclear Education in the Atom Train Exhibition, 1947–1948'. In *British Nuclear Culture*, ed. Jonathan Hogg and Christoph Laucht, special issue of the *British Journal for the History of Science*, vol. 45, no. 4 (December 2012): 591–608.

Lee, David. 'Prunella Clough in Profile'. *Art Review* (May 1996).

Leja, Michael. *Reframing Abstract Expressionism: Subjectivity and Painting in the 1940s*. New Haven and London: Yale University Press, 1993.

Leo, Vincent. 'The Mushroom Cloud Photograph: From Fact to Symbol'. *Afterimage*, vol. 13, nos 1–2 (1985): 9.

Light, Michael. *100 Suns*. New York: Knopf, 2003.

Lindey, Christine. *Art in the Cold War: From Vladivostok to Kalamazoo 1945–1962*. London: New Amsterdam Books, 1990.

Littler, Jo. '"Festering Britain": The 1951 Festival of Britain, Decolonisation and the Representation of the Commonwealth'. In *Visual Culture and Decolonisation in Britain*, ed. Simon Faulkner and Anandi Ramamurthy, 21–42. Aldershot: Ashgate, 2006.

Livingstone, Marco. *Pop Art: A Continuing History*. London: Thames and Hudson, 2002.

Lloyd, John. 'Correspondence: Peter Peri's Politics'. *Art Monthly*, no. 104 (March 1987): 25.

MacLear, Kyo. *Beclouded Visions: Hiroshima-Nagasaki and the Art of Witness*. Albany, NY: SUNY Press, 1999.

Martin, Simon. *John Tunnard: Inner Space to Outer Space*. Chichester: Pallant House Gallery, 2010.

Martin, Simon, and Marco Livingtone. *Colin Self: Art in the Nuclear Age*. Chichester: Pallant House and AVA Publishing, 2008; Lausanne and Worthing: AVA, 2008.

Massey, Anne. *The Independent Group: Modernism and Mass Culture in Britain, 1945–1959*. Manchester: Manchester University Press, 1995.

Matthews, J.H. *The Surrealist Mind*. Selinsgrove: Susquehanna University Press; London and Toronto: Associated University Presses, 1991.

Mazlish, Bruce. 'The Three and a Half Cultures'. *Leonardo*, vol. 18, no. 4 (1985): 233–236.

McCloskey, Joseph, F. 'British Operational Research in World War II'. *Operations Research*, vol. 35, no. 3 (May–June 1987): 453–470.

McEnaney, Laura. *Civil Defense Begins at Home: Militarization Meets Everyday Life in the Fifties*. Princeton, NJ: Princeton University Press, 2000.

Melly, George. *Don't Tell Sybil: An Intimate Memoir of ELT Mesens*. London: Heinemann, 1997.

Mercer, Kobena, ed. *Discrepant Abstraction*. London and Cambridge, MA: The MIT Press, 2006.

Mertins, Detlef. 'The Enticing and Threatening Face of Prehistory: Walter Benjamin and the Utopia of Glass'. *Assemblage*, no. 29 (April 1996): 6–23.

'MI5 "said Bronowski was a risk"'. *BBC News* online, 4 April 2011, accessed 30 April 2012. <http://news.bbc.co.uk/today/hi/today/newsid_9444000/9444270.stm>.

Miller, Tyrus. '"Glass before its time, premature iron": Architecture, Temporality, and Dream in Benjamin's Passagenwerk'. In *Walter Benjamin and the Arcades Project*, ed. Beatrice Hanssen, 240–263. London: Continuum International Publishing Group, 2006.

Mitchell, Juliet. *Psychoanalysis and Feminism*. London: Allen Lane, 1974.

Moffat, Isabelle. '"A Horror of Abstract Thought": Postwar Britain and Hamilton's 1951 "Growth and Form" Exhibition'. *October*, no. 94, The Independent Group (Autumn 2000): 89–112.

Monteyne, David. *Fallout Shelter: Designing for Civil Defense in the Cold War*. Minneapolis, MN: University of Minnesota Press, 2011.

Neve, Christopher. *Unquiet Landscape: Places and Ideas in Twentieth-Century British Painting*. London: Faber and Faber, 1990.

Nicholas, Siân. 'Being British: Creeds and Cultures'. In *The British Isles: 1901–1951*, ed. Keith Robbins, 102–135. Oxford: Oxford University Press, 2002.

Nuttall, Jeff. *Bomb Culture*. London: MacGibbon & Kee, 1968.

Oakes, Guy. *The Imaginary War: Civil Defense and American Cold War Culture*. New York: Oxford University Press, 1994.

Ortolano, Guy. *The Two Cultures Controversy: Science, Literature and Cultural Politics in Postwar Britain*. Cambridge: Cambridge University Press, 2009.

Pace, David. 'Old Wine – New Bottles: Atomic Energy and the Ideology of Science in Post-War France'. *French Historical Studies*, vol. 17, no. 1 (Spring 1991): 49.

Parkinson, Gavin. *Surrealism, Art and Modern Science*. London and New Haven: Yale University Press, 2008.

Paul, Gerhard. '"Mushroom Clouds": Entstehung, Struktur und Funktion einer Medienikone des 20. Jahrhunderts im interkulturellen Vergleich'. In *Visual History: Ein Studienbuch*, ed. Gerhard Paul, 243–264. Göttingen: Vandenhoeck & Ruprecht, 2006.

Peat, Alan, and Brian A. Whitton. *John Tunnard: His Life and Work*. Aldershot: Scolar Press, 1997.

Pellegrino, Charles. *The Last Train from Hiroshima: The Survivors Look Back*. New York: Henry Holt, 2010.

Petersen, Stephen. 'Explosive Propositions: Artists React to the Atomic Age'. *Science in Context*, vol. 17, no. 4 (2004): 579–609.

Petersen, Stephen. *Space-Age Aesthetics: Lucio Fontana, Yves Klein, and the Postwar European Avant-Garde*. Philadelphia, PA: Pennsylvania University Press and The Getty Research Institute, 2009.

Piene, Otto, and Lawrence Alloway. *Zero*. Cambridge, MA: The MIT Press, 1973.

Reed, David. *The Popular Magazine in Britain and the United States 1880–1960*. London: British Library, 1997.

Richards, Lee. *The Black Art: British Clandestine Psychological Warfare against the Third Reich*. Peacehaven: Psywar, 2010.

Riley, Denise. '"The Free Mothers": Pronatalism and Working Women in Industry at the End of the Last War in Britain'. *History Workshop Journal*, vol. 11, no. 1 (1981): 59–118.

Robertson, Bryan. 'Happiness Is the Light'. *Modern Painters*, vol. 9, no. 2 (Summer 1996): 16–20.

Robertson, Bryan, John Russell and Lord Snowdon. *Private View: The Lively World of British Art*. London: Thomas Nelson, 1965.

Rogers, Everett M., and Nancy R. Bartlit. *Silent Voices of World War II: When Sons of the Land of Enchantment Met Sons of the Land of the Rising Sun*. Santa Fe, NM: Sunstone Press, 2005.

Rosenthal, Peggy. 'The Nuclear Mushroom Cloud as Cultural Image'. *American Literary History*, vol. 3, no. 1 (1991): 63–92.

Russell, Christine. 'The Man Behind "The Ascent of Man": Jacob Bronowski'. *Bioscience: American Institute of Biological Sciences*, vol. 25, no. 1 (January 1975): 9–12.

Russell, John, and Suzi Gablik. *Pop Art Redefined*. London: Hayward Gallery, 1969.

Saint-Amour, Paul. 'Bombing and the Symptom: Traumatic Earliness and the Nuclear Uncanny'. *diacritics*, vol. 30, no. 4 (2000): 61.

Saint-Amour, Paul K. 'Modernist Reconnaissance'. *Modernism/Modernity*, vol. 10, no. 2 (April 2003): 349–380.

Saltman, Paul. 'The Softest Hard Sell: Bronowski's Approach to Communicating Science'. *Leonardo*, vol. 18, no. 4 (1985): 243–244.

Sandbrook, Dominic. *Never Had It So Good: A History of Britain from Suez to the Beatles*. London: Abacus, 2006.

Sandbrook, Dominic. *White Heat: A History of Britain in the Swinging Sixties*. London: Little, Brown, 2006.

Schmitt, Frederick. 'On the Road to Social Epistemic Interdependence'. *Social Epistemology*, vol. 2 (1988): 297–307.

Shapiro, Charles S. 'Jacob Bronowski: A Retrospective'. *Leonardo*, vol. 18, no. 4 (1985): 217–222.

Shepard, J. 'A Game of Chance: The Media and Techniques of Francis Bacon'. In *Francis Bacon: A Terrible Beauty*, 88–121. Göttingen: Steidl, 2009.

Singerman, Howard. *Art Subjects: Making Artists in the American University*. Berkeley, CA and London: University of California Press, c. 1999.

Smith, Melissa. 'Architects of Armageddon: The Home Office Scientific Advisers' Branch and Civil Defence in Britain, 1945–68'. *British Journal for the History of Science*, vol. 43, no. 2 (June 2010): 149–180.

Smith, Melissa. '"What to Do If It Happens": Planners, Pamphlets and Propaganda in the Age of the H-Bomb'. *Endeavour*, vol. 33, no. 2 (June 2009): 60–64.

Smith, P.D. *Doomsday Men: The Real Dr Strangelove and the Dream of the Superweapon.* London: Lane, 2007.

Smith, Sean Kenneth. '*Picture Post*, 1938–1945: Social Reform and Images of Britain at War'. PhD dissertation, Stanford University, 1992.

Solomon, Frederic, and Robert Q. Marston. *The Medical Implications of Nuclear War.* Washington, DC: National Academy Press, 1986.

Spalding, Frances. 'Prunella Clough, London and Newtown'. *The Burlington Magazine*, vol. 138, no. 1121 (1996): 557–558.

Spalding, Frances. *Prunella Clough: Regions Unmapped.* London: Lund Humphries, 2012.

Spencer, Robin, ed. *Eduardo Paolozzi: Writings and Interviews.* Oxford and New York: Oxford University Press, 2000.

Spencer, Stephanie. *Gender, Work and Education in Britain in the 1950s.* Basingstoke: Palgrave Macmillan, 2005.

Stephens, Chris. 'Henry Moore's *Atom Piece*: The 1930s Generation Comes of Age'. In *Henry Moore: Critical Essays*, ed. Jane Beckett and Fiona Russell, 243–256. Aldershot: Ashgate, 2003.

Stephens, Chris. '"The morrow we left behind": Landscape and the Rethinking of Modernism, 1939–53'. *Art History*, vol. 35, no. 2 (April 2012): 217–233.

Stephens, Chris. *Peter Lanyon: At the Edge of Landscape.* London: 21 Publishing, 2000.

Stephens, Chris, and Martin Hammer. '"Seeing the Story of One's Time": Appropriations from Nazi Photography in the Work of Francis Bacon'. *Visual Culture in Britain*, vol. 10, issue 3 (2009): 315–351.

Swann, Brenda, and Francis Aprahamian, eds. *J.D. Bernal: A Life in Science and Politics.* London: Verso, 1999.

Sylvester, David. *About Modern Art: Critical Essays 1948–96.* London: Chatto & Windus, 1996.

Szasz, Ferenc. *Atomic Comics: Cartoonists Confront the Nuclear World.* Reno, NV: University of Nevada Press, 2012.

Szasz, Ferenc. *Larger than Life: New Mexico in the Twentieth Century.* Albuquerque, NM: University of New Mexico Press, 2006.

Tassi, Roberto. *Graham Sutherland Complete Graphic Works.* London: Thames and Hudson, 1978.

Tate Gallery 1976–78: Illustrated Catalogue of Acquisitions. London: Tate, 1979.

Taylor, Bryan. 'Nuclear Pictures and Metapictures'. *American Literary History*, vol. 9, no. 3 (1997): 567–597.

Taylor, Bryan. '"Our Bruised Arms Hung Up as Monuments": Nuclear Iconography in Post-Cold War Culture'. *Critical Studies in Media Communication*, vol. 20, no. 1 (2003): 1–34.

Taylor, Richard. *Against the Bomb: The British Peace Movement, 1958–65*. Oxford: Clarendon, 1988.

Thomas-Symonds, Niklaus. *Attlee: A Life in Politics*. London: I.B. Tauris, 2009.

Thompson, Frank. *Dan Dare: Pilot of the Future. The Red Moon Mystery*. London: Titan Books, 2004.

Thornton, Stephen T., and Andrew Rex. *Modern Physics for Scientists and Engineers*, fourth edition. Boston: Brooks/Cole, Carnegie Learning, 2006.

Titus, Constandina. 'The Mushroom Cloud as Kitsch'. In *Atomic Culture: How We Learned to Stop Worrying and Love the Bomb*, ed. Scott Zeman and Michael Amundson, 101–123. Boulder, CO: University of Colorado Press, 2004.

Treves, Toby. 'Gisha Koenig'. In *Sculpture in 20th-Century Britain: A Guide to the Sculptors in the Leeds Collections*, ed. Penelope Curtis. Leeds: Henry Moore Institute, 2003.

Turner, Victor. *The Forest of Symbols: Aspects of Ndembu Ritual*. Ithaca, NY: Cornell University Press, 1967.

Walker, John A. *Cultural Offensive: America's Impact on British Art Since 1945*. London: Pluto Press, 1998.

Walker, Samuel. *Prompt and Utter Destruction: Truman and the Use of Atomic Bombs against Japan*. Chapel Hill, NC: University of North Carolina Press, 2004.

Weart, Spencer. *Nuclear Fear: A History of Images*. Cambridge, MA: Harvard University Press, 1988; 2009.

Weart, Spencer. *The Rise of Nuclear Fear*. Cambridge, MA: Harvard University Press, 2012.

Welbourne, Michael. 'The Community of Knowledge'. *Philosophical Quarterly*, vol. 31, no. 125 (1981): 302–314.

Wiesinger, Véronique. *Alberto Giacometti, Isabel Nicholas, Correspondences*. Paris: Fondation Alberto et Annette Giacometti, 2007.

Williams, Raymond. *The Country and the City*. London: Chatto and Windus, 1973.

Williams, Raymond. *Resources of Hope: Culture, Democracy, Socialism*. London: Verso, 1989.

Wilson, Elizabeth. *Only Halfway to Paradise: Women in Post-War Britain 1945–1968*. London: Tavistock, 1980.

Wilson, Sarah. 'Germaine Richier, Disquieting Matriarch'. *Sculpture Journal* (December 2005), *Correspondences*: 51–70.

Wilson, Sarah. 'Paris Post War: In Search of the Absolute'. In *Paris Post War: Art and Existentialism*, ed. Frances Morris. London: Tate Gallery, 1993.

Winnicott, D.W. *Deprivation and Delinquency*, ed. Claire Winnicott, Ray Shepherd and Madeleine Davis. London: Tavistock, 1984.

Wood, Denis, and John Fels. *The Natures of Maps: Cartographic Constructions of the Natural World*. Chicago and London: University of Chicago Press, 2008.

Yacowar, Maurice. *The Great Bratby: A Portrait of John Bratby RA*. London: Middlesex University Press, 2008.

Yavenditti, Michael J. 'John Hersey and the American Conscience: The Reception of 'Hiroshima'. *Pacific Historical Review*, vol. 43, no. 1 (1974): 24–49.

Yorke, Malcolme. *The Spirit of the Place: Nine Neo-Romantic Artists and Their Times*. London: Tauris Parke Paperbacks, 2001.

Zelizer, Barbie. *Remembering to Forget: Holocaust Memory Through the Camera's Eye*. Chicago: University of Chicago Press, 2000.

Zeman, Scott. '"Taking Hell's Measurements": *Popular Science* and *Popular Mechanics* Magazines and the Atomic Bomb from Hiroshima to Bikini'. *Journal of Popular Culture*, vol. 41, no. 4 (2008): 695–711.

Index

Note: **bold** page numbers indicate figures; **bold** numbers preceded by **Pl.** refer to colour plates; numbers in brackets preceded by *n* are chapter endnote numbers.

For Product Safety Concerns and Information please contact our
EU representative GPSR@taylorandfrancis.com Taylor & Francis
Verlag GmbH, Kaufingerstraße 24, 80331 München, Germany